Metabolism of nucleotides, nucleosides and nucleobases in microorganisms

Metabolism of nucleotides nucleosides and nucleobases in microorganisms

Edited by

AGNETE MUNCH-PETERSEN

University Institute of Biological Chemistry B, Copenhagen, Denmark

1983

ACADEMIC PRESS

A Subsidiary of Harcourt Brace Jovanovich, Publishers

London · New York

Paris · San Diego · San Francisco · São Paulo

Sydney · Tokyo · Toronto

ACADEMIC PRESS INC. (LONDON) LTD.
24/28 Oval Road
London NW1

United States Edition published by
ACADEMIC PRESS INC.
111 Fifth Avenue
New York, New York 10003

British Library Cataloguing in Publication Data

Metabolism of nucleotides by microorganisms.
1. Nucleotides 2. Microbiological synthesis
I. Munch-Petersen, Agnete
661′.815 TP248.N8/

ISBN: 0-12-510580-0

Filmset by Latimer Trend & Company Ltd, Plymouth
Printed in Great Britain by Whitstable Litho

Contributors

KARIN HAMMER-JESPERSEN
Institute of Biological Chemistry B, University of Copenhagen, Denmark

KAJ FRANK JENSEN
Institute of Biological Chemistry B, University of Copenhagen, Denmark

HENRIK MØLLGAARD
NOVO Industri A/S, Enzyme Division, Copenhagen, Denmark

AGNETE MUNCH-PETERSEN
Institute of Biological Chemistry B, University of Copenhagen, Denmark

BENTE MYGIND
Institute of Biological Chemistry B, University of Copenhagen, Denmark

JAN NEUHARD
Institute of Biological Chemistry B, University of Copenhagen, Denmark

PER NYGAARD
Institute of Biological Chemistry B, University of Copenhagen, Denmark

Preface

The last decade has witnessed an immense expansion in the field of nucleic acid research. In parallel to this there has been an accumulation of data concerning the metabolism of compounds involved in, or related to, the cellular synthesis of nucleic acids. Procaryotic as well as eucaryotic cells have been shown to command numerous enzymatic reactions, through which they can scavenge nucleic acid precursors from the surroundings and metabolize them according to their immediate needs, whether they be for nucleic acid synthesis or as carbon or nitrogen sources.

Besides being involved in nucleic acid synthesis, the role of certain nucleotides as important regulatory compounds in cell metabolism has been well documented. In addition there has been an increasing pharmacological use of nucleoside and nucleotide analogs as antimetabolites in the treatment of disease. This in turn emphasizes the practical importance of knowledge about the enzymatic reactions which may be the targets for the therapeutic treatment, since the analogs are metabolized mainly by the same enzymes as are the natural intermediates in nucleic acid metabolism.

The main objective of the present book has been to present a detailed description of the anabolic and catabolic pathways which may function when nucleic acid precursors (or their analogs) are made available to growing cells. The book has developed as a joint project at our Institute. The different contributors are all past or present members of the staff and each author has been a researcher within the particular field which is treated in the chapter.

The first chapter deals with the metabolism of 5-phosphoribosyl 1-pyrophosphate, a key compound in nucleotide metabolism. While the enzyme responsible for the synthesis of this compound has been characterized in detail, its regulation on the genetic level is still quite obscure. Mutants with defects in the synthesis of 5-phosphoribosyl 1-pyrophosphate have now been isolated as mentioned in Chapter 1, but at present they have only been partially characterized. Chapters 2, 3, and 4 describe in detail the multitude of enzymatic reactions and pathways which are brought into operation when exogenous purine or pyrimidine compounds are utilized by the cells for nucleic acid synthesis. Each compound is individually treated and a practical aspect of this information is that it should enable the reader, who wishes to do experiments with labelled nucleic acids, to work out labelling procedures, optimally tailored to the enzymatic patterns found in the individual bacterial strains. The two last chapters deal specifically with the transport of the nucleosides into the cells and the subsequent intracellular catabolism. While the enzymatic background for the latter has been known in detail for some years, the transport systems are only beginning to be unravelled. It is known, however, that transport and catabolism are regulated by the same cellular control systems and thus closely linked in the metabolism of the cells. The control systems have proved to be of considerable complexity and they are described in some detail in Chapter 5.

Throughout the chapters the importance of well characterized mutants with defects in the different enzymes has been emphasized, and selection procedures have been described or referred to whenever possible.

An appendix concludes the book. It enlists the genes from *Salmonella typhimurium* and *Escherichia coli*, which are known to be involved in the metabolism of nucleic acid precursors and thus are relevant to the subjects treated in the book. In addition the appendix contains a linkage map, indicating the location of these genes on the circular chromosome of *S. typhimurium* and *E. coli*.

The dramatic developments within molecular genetics during the last few years perhaps tend to overshadow the enzymological and metabolic aspects of nucleic acid precursors in the cells, although they form the basis for nucleic acid synthesis and for many cellular control systems. It has been our aim to produce a useful book on these subjects, which are too specialized for textbooks and yet of importance for those who carry on experimental research in the field.

May 1983 *Agnete Munch-Petersen*

Contents

List of contributors v
Preface vii

1 Metabolism of 5-phosphoribosyl 1-pyrophosphate (PRPP) in *Escherichia coli* and *Salmonella typhimurium* 1
I. Introduction 1
II. Regulation of PRPP synthesis 3
III. Utilization of PRPP 10
IV. Intracellular concentration of PRPP (PRPP pools) . . 15
V. Mutants of *Escherichia coli* and *Salmonella typhimurium* defective in PRPP synthetase 16
VI. Conclusion 20

2 Utilization of preformed purine bases and nucleosides . . . 27
I. Introduction 28
II. *De novo* biosynthesis of inosine 5′-phosphate . . . 29
III. Purine nucleotide interconversions 31
IV. Salvage pathways in *E. coli* and *S. typhimurium* . . . 41
V. Purine salvage pathways in other microorganisms . . 54
VI. Individual enzymes of purine salvage 64
VII. Intracellular formation of purine nucleosides and nucleobases 69
VIII. Interplay between the purine *de novo* and salvage pathways . 73

3 Utilization of preformed pyrimidine bases and nucleosides . . 95
I. Introduction 96
II. Pathways in *Salmonella typhimurium* and *Escherichia coli* . 97
III. Pathways in other microorganisms 106
IV. Enzymes involved in pyrimidine salvage 118
V. Physiological aspects of pyrimidine salvage 133

4 Biosynthesis of deoxythymidine triphosphate 149
I. Introduction 149
II. Endogenous synthesis of dTTP 150
III. Synthesis of dTTP from exogenous thymine and thymidine . 179

5 Nucleoside catabolism 203
I. Introduction 203
II. Nucleoside catabolism in *Escherichia coli* and *Salmonella typhimurium* 204
III. Other microorganisms 238

6 Transport of nucleic acid precursors 259
I. Introduction 259
II. Transport of nucleosides in enteric bacteria 260
III. Transport of nucleosides in other bacteria 278
IV. Transport of nucleobases in bacteria 279
V. Transport of nucleosides and nucleobases in fungi . . 287

Appendix *Linkage map of* Escherichia coli *K12 and* Salmonella typhimurium *LT2* 307

Index 311

1

Metabolism of 5-phosphoribosyl 1-pyrophosphate (PRPP) in *Escherichia coli* and *Salmonella typhimurium*

KAJ FRANK JENSEN

I.	Introduction	1
II.	Regulation of PRPP synthesis	3
	A. Properties of PRPP synthetase (EC 2.7.6.1)	3
	B. Physiological evidence for regulatory properties of PRPP synthetase	7
	C. Control of enzyme synthesis	8
III.	Utilization of PRPP	10
	A. General	10
	B. PRPP consumption in the biosynthesis of nucleotides	11
VI.	Intracellular concentration of PRPP (PRPP pools)	15
V.	Mutants of *Escherichia coli* and *Salmonella typhimurium* Defective in PRPP synthetase	16
VI.	Conclusion	20

I. INTRODUCTION

The discovery of 5-phosphoribosyl-α-1-pyrophosphate (PRPP) was intimately connected to the metabolic function of the compound in nucleotide biosynthesis (33, 55). In searching for the origin of the phosphoribosyl moiety which is being added to preformed purine and pyrimidine bases during their conversion to nucleotides, Kornberg *et al.* (33) discovered and partially purified an enzymatic activity from pigeon liver extracts which synthetized PRPP from ribose 5-phosphate and ATP:

$$\text{ribose 5-phosphate} + \text{ATP} \rightarrow \text{PRPP} + \text{AMP} \qquad [I]$$

METABOLISM OF NUCLEOTIDES, NUCLEOSIDES
AND NUCLEOBASES IN MICROORGANISMS
ISBN: 0–12–510580–0

The following two reactions were used to monitor PRPP formation:

$$\text{orotate} + \text{PRPP} \rightleftharpoons \text{OMP} + \text{PP}_i$$
$$\text{OMP} \rightarrow \text{UMP} + \text{CO}_2$$

From the stoichiometry of reaction [I] and from studies of the decomposition of the product in acid solution the structure of PRPP was deduced (Fig. 1.1) and it was suggested that the pyrophosphoryl group of ATP was transferred intact to ribose 5-phosphate in the reaction. Evidence supporting this was provided shortly after by Khorana *et al.* (31) who showed that the phosphate group which is being esterified to the anomeric hydroxyl group of ribose 5-phosphate in the reaction derives from the β-position of ATP. The α-configuration at carbon number one of PRPP was deduced from the observation that PRPP is broken down to 5-phosphoribosyl-1.2-cyclic phosphate and orthophosphate in alkaline solution (31, 55, 67).

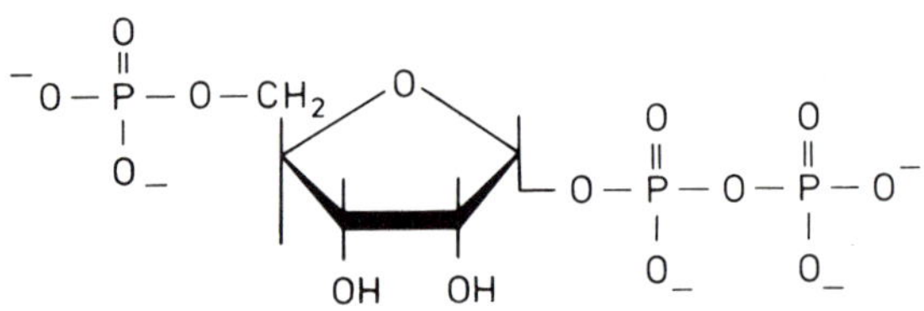

FIG. 1.1. Structure of 5-phosphoribosyl-α-1-pyrophosphate (PRPP); Haworth projection.

After the discovery of PRPP synthetase in pigeon liver Kornberg *et al.* (33) showed the enzyme to be present in various other tissues and organisms including *Escherichia coli.*

PRPP was shown to function in the *de novo* synthesis of pyrimidine and purine nucleotides (18, 19, 33), in the biosynthesis of histidine, tryptophan, and pyridine nucleotide coenzymes (45, 52, 53, 70), and was also found to participate in the utilization of preformed purine and pyrimidine bases for nucleotide synthesis (9, 34, 38, 55). Thus the PRPP synthetase reaction may be regarded as the first step in a highly branched biosynthetic pathway leading to the formation of these compounds (Fig. 1.2) and may therefore be expected to be subject to rigorous metabolic control.

The literature concerning PRPP synthesis in mammalian cells has been reviewed by Becker and Seegmiller (6). In this chapter we shall deal only with the synthesis and utilization of PRPP in the two enteric bacteria *Escherichia coli* and *Salmonella typhimurium*. For other microorganisms the knowledge of PRPP metabolism, notably PRPP synthesis, is scarce and sporadic.

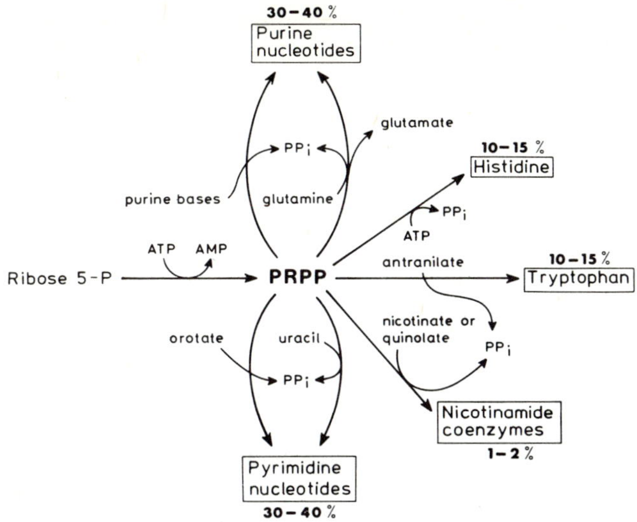

FIG. 1.2. Consumption of PRPP. The endproducts of the various biosynthetic pathways that utilize PRPP are shown in boxes. The numbers next to the boxes indicate the percentage of total PRPP which is consumed by the different pathways when *E. coli* or *S. typhimurium* is grown in a glucose salt minimal medium.

II. REGULATION OF PRPP SYNTHESIS

A. Properties of PRPP synthetase (EC 2.7.6.1)

1. Enzyme structure

PRPP synthetase has been purified to homogeneity and in high yield from an extract of *Salmonella typhimurium* (61). A molecular weight of the subunit of 31 000 was determined by a variety of methods (59). Proline was found to be the only aminoterminal residue indicating that the enzyme contains only one kind of polypeptide chains. This is also consistent with sulfhydryl titration data (56, 59).

The aggregated state of the native enzyme is not well defined. In the presence of inorganic phosphate the majority of the enzyme is found in two species with molecular weights of about 160 000 and 320 000 suggesting that the predominant forms are pentamers and decamers or possibly hexamers and dodecamers. The relative distribution of enzyme in the two major forms depends on the experimental conditions. In the electron microscope an asymmetric pentameric structure has been observed (59).

When inorganic phosphate is removed PRPP synthetase completely loses a defined quaternary structure and all aggregational states from the size of the subunit to very large nonspecific aggregates are observed (59). This situation is accompanied by a total and irreversible loss of activity (61).

2. Catalytic properties

The PRPP synthetase reaction [I] is reversible with an equilibrium constant in the direction of PRPP synthesis of about 29 (61). The enzyme requires orthophosphate for stability and activity (61). Nuclear magnetic resonance studies have revealed the presence of a bound orthophosphate ion in close proximity to the substrate binding site (17, 37). Also these studies showed that orthophosphate ions occupy the site for the 5-phosphate group of ribose 5-phosphate and the site for the γ-phosphate group of ATP when these substrates are absent from the enzyme (17).

Magnesium ions or other divalent metal ions are also needed for enzymatic activity. This ion has two functions in catalysis. First, the true substrate for the enzyme is the β,γ-bidentate complex MgATP, and the true product is MgPRPP. Second, an additional metal ions binds to each subunit of PRPP synthetase and is needed for enzymatic activity (Fig. 1.3). This was concluded from the observation that the exchange stable β,γ-bidentate complex

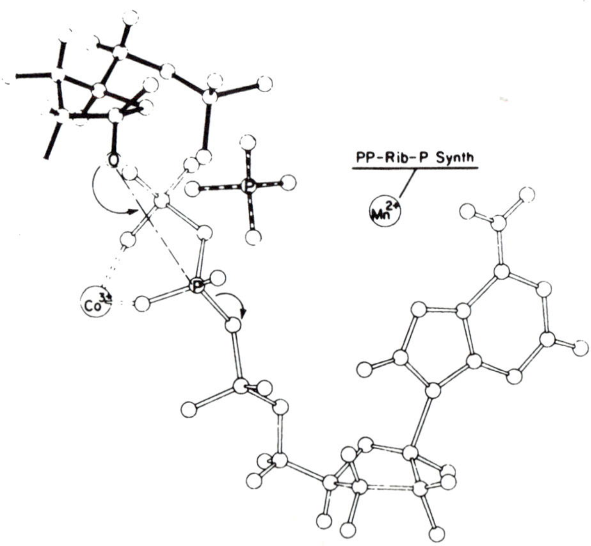

FIG. 1.3. Arrangement and conformations of substrates and activators at the active site of PRPP synthetase of *Salmonella typhimurium* consistent with the determined distances from enzyme-bound Mn^{2+} (17) and from the Cr^{3+} atom of exchange-stable, substitution inert α,β,γ-tridentate Cr-ATP (37). Ribose 5-phosphate is appropriately positioned for nucleophilic attack with inversion on the β-phosphorous atom of ATP. The broken line from the 1-α-D-ribose oxygen to the β-phosphorous atom of ATP shows the reaction coordinate. Reproduced (with permission) from Granot *et al.* (17).

$Co(III)(NH_3)_4$-ATP is a substrate for PRPP synthetase only if a second cation such as Mg^{2+} or Mn^{2+} is present (36). The conclusion was confirmed by nuclear magnetic resonance spectroscopy using paramagnetic ions (17, 37). These studies also revealed that the enzyme-bound metal ion does not directly interact with the bound substrates (17).

Kinetic studies indicate a sequential reaction mechanism for PRPP synthetase (62, 64). This implies that the reaction takes place as a direct substitution while both substrates are bound to the enzyme. The combined data agree best with an obligatory reaction order where binding of MgATP precedes binding of ribose 5-phosphate (14, 17, 61, 62).

When the reaction is carried out in $H_2{}^{18}O$, oxygen-18 is recovered in PRPP, but not in AMP. This indicates that a β P-O bond is broken and hence that the reaction proceeds via a direct nucleophilic attack (S_N2) by the α-1-OH group of ribose 5-phosphate on the β-phosphorous atom of the polyphosphate chain of ATP, displacing AMP and forming PRPP (43). Nuclear magnetic resonance studies using paramagnetic metal ions (17, 37) showed that the spatial arrangement of the substrates in the active centre of the enzyme is favourable for such a nucleophilic attack to occur; the distance between the α-1-OH group of ribose 5-phosphate and the β-phosphorous atom of ATP being about 3.8 Å (37). Figure 1.3 shows a model of the arrangement of substrates and activators in the active site of PRPP synthetase.

The enzyme is inactivated by oxidation of a single specific cystein residue (56) and by chemical modification of a single specific lysine residue (57). Since the reactivity of either of the two residues to chemical modification is reduced by the presence of substrates they are probably located near the active site, but their roles in catalysis are unknown (56, 57).

3. Regulatory properties

The effect of purine and pyrimidine nucleotides, tryptophan, histidine, and nicotinamide coenzymes on the reaction kinetics of purified PRPP synthetase has been investigated since these compounds may be regarded as endproducts of a branched biosynthetic pathway with the PRPP synthetase reaction being the first step (Fig. 1.2).

All pyrimidine and purine nucleotides except ADP and dADP inhibit the enzyme competitively with ATP. In general, purine nucleotides are stronger inhibitors ($K_i \sim 2$ mM) than pyrimidine nucleotides ($K_i \sim 5$ mM). Very weak competitive inhibition is observed with NAD^+, $NADP^+$, and ppGpp, while histidine and tryptophan have no effect on PRPP synthetase activity (66).

ADP is by far the most powerful inhibitor of PRPP synthetase activity. The K_i is 10 to 100 times smaller than for other purine nucleotides, dependent on the concentration of ribose 5-phosphate. The mode of action of this inhibitor is complex. At low ribose 5-phosphate concentrations ADP produces a linear competitive inhibition towards ATP, but at high concentrations of ribose 5-phosphate this inhibition becomes a parabolic function of the concentration of ADP, indicating that this compound interacts with more than one enzyme form in the reaction. The presence of ADP also induces a pronounced substrate inhibition exerted by ribose 5-phosphate. The higher the concentration of ADP is the lower is the concentration of ribose 5-phosphate needed to produce an evident substrate inhibition (Fig. 1.4).

Equilibrium dialysis experiments have shown that ADP can bind to the active site of PRPP synthetase in the absence of ribose 5-phosphate. This binding is sensitive to competition by ATP and by the reaction inert analog, α,β-methylene ATP. However, in the presence of ribose 5-phosphate an additional site becomes available for ADP, and the binding to this site is not subject to competition with α,β-methylene ATP (14). Thus ADP seems to be a

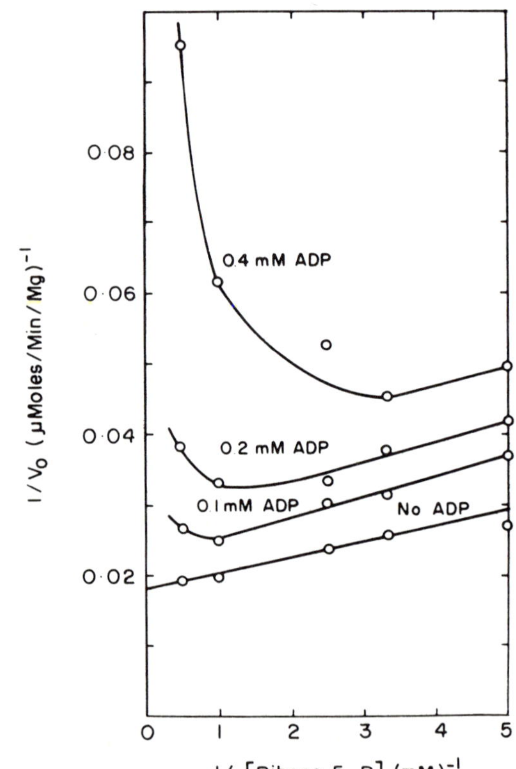

FIG. 1.4. Effect of ADP on the saturation kinetics of PRPP synthetase with ribose 5-phosphate. Reproduced with permission from Switzer and Sogin (66).

true allosteric inhibitor of PRPP synthetase (14, 66). DeoxyADP produces an inhibition pattern which is qualitatively similar to that of ADP, but it is a somewhat weaker inhibitor (66).

Since ATP is a substrate for the enzyme, AMP is a product, and ADP is a strong inhibitor, it has been suggested that PRPP synthetase activity is regulated by "energy charge" (2, 35). As the inhibition by ADP depends on the concentration of ribose 5-phosphate, the effect of "energy charge" on PRPP synthetase activity varies with the concentration of ribose 5-phosphate (66).

In summary, the *in vitro* studies indicate that PRPP synthetase activity is regulated by the entire pool of purine and, to a lesser extent, pyrimidine nucleotides. In particular ADP is a potent inhibitor. The supply of ribose 5-phosphate in combination with "energy charge" may also be of physiological significance.

B. Physiological evidence for regulatory properties of PRPP synthetase

Measurements of the intracellular concentration of PRPP in growing cells have given some information concerning the physiological significance of the regulatory properties found *in vitro* for PRPP synthetase.

An inverse relationship has been observed between the size of the PRPP pool and the pools of ribonucleoside triphosphates in *E. coli* K-12 under conditions where RNA synthesis is suddenly stimulated (3). The stimulation was accomplished by addition of leucine or chloramphenicol to a leucine-starved culture or by addition of methionine to an exponentially growing culture. At the onset of the enhanced RNA synthesis the nucleoside triphosphate pools drop transiently to about one third of their initial size, whereas the ADP pool remains unchanged. Concomitantly with this the PRPP pool increases substantially and returns to normal simultaneously with the purine nucleoside triphosphate pools. This rise in PRPP concentration may be interpreted as the result of a release of feed-back inhibition of PRPP synthetase activity caused by the fall in the purine ribonucleoside triphosphate pools (3, 66).

When a purine requiring mutant of *S. typhimurium (purF)* is starved for purine both the purine and the pyrimidine nucleoside triphosphate pools fall to low levels. Under these conditions PRPP accumulates rapidly and the concentration increases more than one hundred fold. This accumulation of PRPP continues for hours indicating that no factor which inhibits PRPP synthesis is operative in this condition (58).

When an adenine requiring mutant of *S. typhimurium* is starved for adenine, the adenine nucleotide pools fall to low levels while GTP and the pyrimidine nucleotide pools remain relatively normal. Also in this case PRPP accumulates (58).

During guanine starvation of *S. typhimurium (guaB)* mutants the guanine nucleotide pools are depleted while the pools of adenine nucleotides increase and the pyrimidine nucleotide pools remain relatively constant. This condition of starvation is accompanied by a reduction of the PRPP pool. Such a reduction of the PRPP pools is also seen during starvation of pyrimidine auxotrophic mutants, where the pyrimidine nucleotide pools decrease and the purine nucleotide pools increase (58).

Guanine requiring mutants of *S. typhimurium* that harbour the mutation in *guaA* respond differently to guanine starvation with regard to the behaviour of ATP and PRPP than the corresponding *guaB* mutants (Fig. 1.5). During starvation of a *guaA* strain the ATP pool initially falls and then increases again. In contrast, the PRPP pool first increases, then reaches a maximum and decreases again. The maximal PRPP pool is observed simultaneously with the minimal ATP pool (Fig. 1.5).

Thus it appears that starvation conditions which lead to *high* adenine nucleotide pools result in a *reduction* of the PRPP pool, while conditions which lead to reduced pools of adenine nucleotides result in an *accumulation* of PRPP. Hence the *in vivo* experiments demonstrate the presence of a physiologically significant inhibitor amongst the adenine nucleotides; this could well be ADP (58).

At the onset of a nitrogen starvation of *S. typhimurium* the PRPP pool is rapidly and totally depleted even though the specific activity of PRPP synthetase remains unchanged, the ribonucleoside triphosphate pools decrease, and the energy charge ratio remains constant. This indicates that the enzyme activity becomes immediately inhibited. This phenomenon is of interest as all compounds that require PRPP for their synthesis contain nitrogen. Since there is no known *in vitro* property of purified PRPP synthetase which might explain this phenomenon it was suggested that the synthesis of PRPP terminates due to decreased availability of ribose 5-phosphate (58).

At present there is no strong evidence which supports the hypothesis that energy charge *per se* might be a relevant physiological regulator of PRPP synthetase activity. However, experiments relating to this problem has been discussed by others (3, 58).

C. Control of enzyme synthesis

The specific activity of PRPP synthetase in extracts of *S. typhimurium* depends on the carbon source used for growth but appears not to be correlated to the growth rate (69).

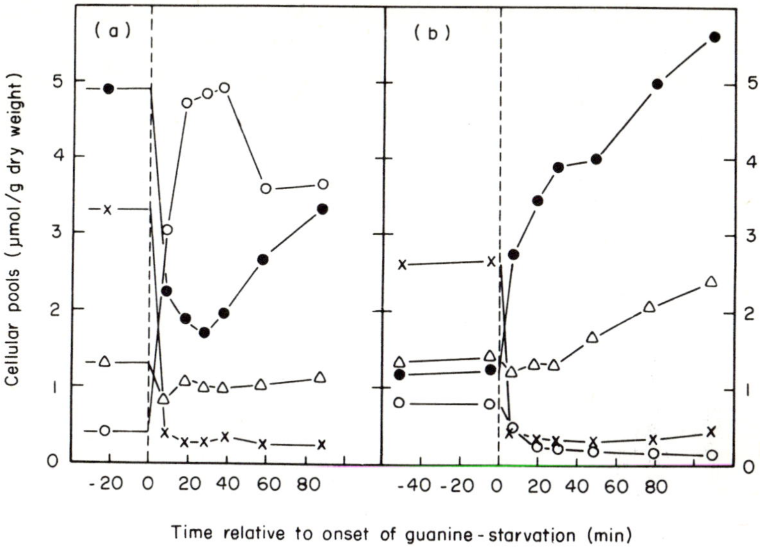

FIG. 1.5. Reciprocal relation between the pool of ATP and the pool of PRPP during starvation of guanine requiring mutants of *Salmonella typhimurium*. The bacteria were grown in a glucose-salt minimal medium supplemented with adenine (15 μg/ml) and guanine (6 μg/ml), labelled with [^{32}P]-orthophosphate, and allowed to exhaust their guanine supply (time equal zero min). At various times samples were withdrawn for determination of pools of nucleotides and PRPP (24). (a) Starvation of KP-1489 (*guaA*); ATP (●—●), GTP (× — ×), UTP (△—△), and 0.5 × PRPP (○—○). (b) Starvation of KP-1476 (*guaB*); 0.25 × ATP (●—●), GTP (× — ×), UTP (△—△), and PRPP (○—○). The pool of CTP responded essentially as the UTP pool and has not been shown for clarity reasons. (K. F. Jensen, unpublished results.)

The possibility that enzyme synthesis might be controlled by repression by some of the biosynthetic end products has been investigated in *S. typhimurium* (69). The addition of purines, pyrimidines, histidine, or tryptophan to the medium has only marginal repressive effects on the specific activity of PRPP synthetase (69).

Pyrimidine starvation was found to be the only starvation condition which leads to derepression of the synthesis of PRPP synthetase (69). By total starvation the level of PRPP synthetase increases two to three fold. Slowfeeding of pyrimidine auxotrophic mutants by the use of orotate as pyrimidine source may lead to a ten fold elevated enzyme level. Thus it appears that the synthesis of the enzyme is under repressive control by pyrimidine nucleotides (69). The repressing pyrimidine compound seems to be a uridine nucleotide rather than a cytidine nucleotide (49) since PRPP synthetase activity only increases by starvation for uridine nucleotides and not by starvation for cytidine nucleotides in mutants where the interconversion of uridine and cytidine compounds is prevented by mutation (46, 49). Furthermore, the function of UMP as the repressing metabolite seems to be ruled out since mutants which harbour leaky mutations in the gene encoding UMP kinase (*pyrH*) contain PRPP synthetase activity in the range of 100–150 per cent of normal. In these mutants the UMP pool is about 20 fold elevated while the UTP pool is reduced to about 25 per cent, and the CTP pool is normal (K. F. Jensen and P. Nygaard, unpublished observations).

Recently it was found that the level of PRPP synthetase is 2.3 fold enhanced in an *rpoBC* mutant of *S. typhimurium* which contains a four fold increased pool of UTP (27). This mutant also shows high constitutive expression of some *pyr* genes (27). This observation may indicate that the syntheses of PRPP synthetase and of the enzymes of the pyrimidine nucleotide *de novo* biosynthetic pathway share regulatory elements in common and that the control involves RNA polymerase.

III. UTILIZATION OF PRPP

A. General

The PRPP consuming reactions are all nucleophilic substitutions where the pyrophosphoryl group of PRPP is displaced by a nitrogen compound resulting in an inversion of the configuration at carbon number one of the ribosyl group and leading to the formation of a β-glycosylic bond (Fig. 1.2 and Table 1.1).

From the composition of enteric bacteria in steady state growth (39) it can be calculated that a culture of *S. typhimurium* or *E. coli* in a glucose-salt minimal medium growing with a generation time of 50 min will utilize about 7–8 μmol of PRPP per minute per gram dry weight. Out of this PRPP 70–80% is used for the synthesis of purine and pyrimidine nucleotides (about half for each), about 10–15% for each of the amino acids histidine and tryptophan, and only little (probably 1–2%) is used for the synthesis of nicotinamide coenzymes (58).* The cellular pool of PRPP under these conditions is about one μmol per gram dry weight; thus the PRPP pool turns over in a fraction of a minute (3, 4, 58).

* One per cent accounts for the synthesis of an amount of NAD^+ equivalent to the free NAD^+ pool. This pool is 4–5 μmol per gram dry weight in *E. coli* (K. F. Jensen and P. Nygaard, unpublished observations). However, NAD^+ is also consumed for DNA ligation (32). The ligation of Okazaki fragments during replication consumes less NAD^+ per generation than the free pool of NAD^+. However, it is difficult to estimate the consumption of NAD^+ for DNA ligations in DNA repair synthesis. The use of PRPP for NAD^+ synthesis also depends on the metabolic route by which the breakdown product, NMN, from DNA ligations is reutilized for NAD^+ synthesis (10, 11).

TABLE 1.1
PRPP utilizing reactions in *Salmonella typhimurium*.

Biosynthetic pathway	Enzyme	Reaction product(s)[a]	Gene
Purine salvage	Adenine phosphoribosyltransferase (EC 2.4.2.7)	AMP	*apt*
Purine salvage	Hypoxanthine phosphoribosyl-transferase (EC 2.4.2.8)	IMP	*hpt*
Purine salvage	Guanine phosphoribosyl-transferase (EC 2.4.2.–)	GMP	*gpt*
Purine *de novo* synthesis	Glutamine phosphoribosylamido-transferase (EC 2.4.2.14)	phosphoribosylamine (+glutamate)	*purF*
Pyrimidine salvage	Uracil phosphoribosyltransferase (EC 2.4.2.9)	UMP	*upp*
Pyrimidine *de novo* synthesis	Orotate phosphoribosyltransferase (EC 2.4.2.10)	OMP	*pyrE*
Pyridine salvage	Nicotinate phosphoribosyl-transferase (EC 2.4.2.11)	FMN	*pncB*
Pyridine *de novo* synthesis	Quinolate phosphoribosyl-transferase (EC 2.4.2.a)	FAMN (+CO_2)	*nadC*
Histidine synthesis	ATP phosphoribosyltransferase (EC 2.4.2.c)	N-1-phosphoribosyl-ATP	*hisG*
Tryptophan synthesis	Antranilate phosphoribosyl-transferase (EC 2.4.2.18)	phosphoribosyl-antranilate	*trpD*

[a] In addition to the products mentioned all enzymes produce pyrophosphate.

B. PRPP consumption in the biosynthesis of nucleotides

1. Pyrimidine nucleotide synthesis de novo

The pathway for the pyrimidine nucleotide synthesis in *S. typhimurium* is shown in Fig. 1.6. PRPP enters the pathway in the conversion of orotate to OMP, catalyzed by orotate phosphoribosyltransferase (*pyrE*). The first enzyme of the pathway, carbamoylphosphate synthetase (*pyrA*), is needed for the synthesis of arginine as well as for the synthesis of pyrimidine nucleotides. This enzyme is feedback-inhibited by UMP and activated by ornithine, by IMP, and by PRPP. The activation *in vitro* with the latter two compounds is about a factor two (1). The concentration of IMP needed to produce activation of the enzyme *in vitro* appears to be far above the physiological concentration range. In contrast to this the activation seen with PRPP occurs in the range of concentrations where PRPP is found intracellularly and may

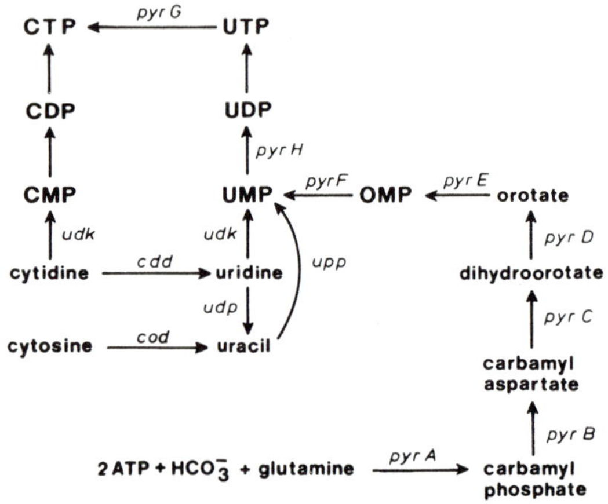

FIG. 1.6. Pathways for the biosynthesis of pyrimidine nucleoside triphosphates in *S. typhimurium*. The enzymes are indicated by their gene corresponding designations as follows: *cdd*, cytidine deaminase (EC 3.5.4.5); *cod*, cytosine deaminase (EC 3.5.4.1); *pyrA*, carbamylphosphate synthetase (EC 2.7.2.5); *pyrB*, aspartate transcarbamylase (EC 2.1.3.2); *pyrC*, dihydroorotase (EC 3.5.2.3); *pyrD*, dihydroorotate dehydrogenase (EC 1.3.3.1); *pyrE*, orotate phosphoribosyltransferase (EC 2.4.2.10); *pyrF*, OMP decarboxylase (EC 4.1.1.23); *pyrG*, CTP synthetase (EC 6.3.4.2); *pyrH*, UMP kinase (EC 2.7.4.–); *udk*, uridine kinase (EC 2.7.1.48); *udp*, uridine phosphorylase (EC 2.4.2.3); *upp*, uracil phosphoribosyltransferase (EC 2.4.2.9).

therefore be of a physiological significance (1, 26, 58). The second enzyme of the pathway, the aspartate transcarbamylase (*pyrB*), is feedback-inhibited by CTP and allosterically activated by ATP (48).

The flow through the pathway for the *de novo* synthesis of pyrimidine nucleotides is also regulated at the level of enzyme synthesis by pyrimidine and guanine nucleotides (25, 60). This regulation involves RNA polymerase (27). Thus it seems that the consumption of PRPP by this biosynthetic pathway is regulated by the supply of endogenously synthetized intermediate orotate, and that PRPP itself might participate in this control *via* its effect on carbamoylphosphate synthetase.

2. *Purine nucleotide synthesis* de novo

The pathway for the synthesis of purine nucleotides is shown in part in Fig. 1.7. PRPP is a substrate for the first enzyme of this pathway, the glutamine phosphoribosylpyrophosphate amidotransferase (*purF*). This enzyme is

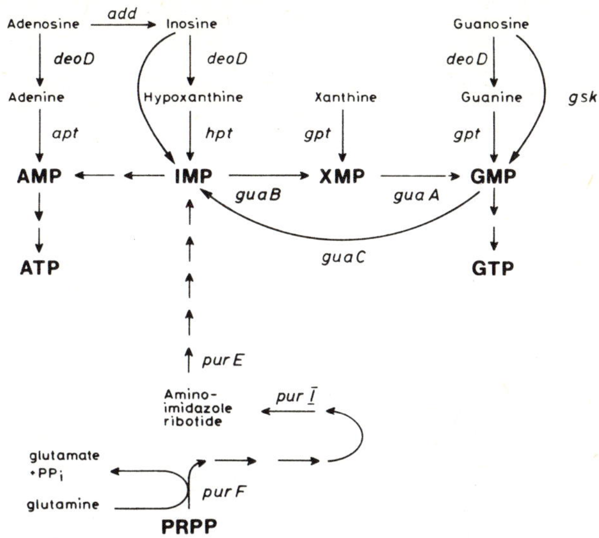

FIG. 1.7. Relevant steps in the biosynthesis of purine nucleotides in *S. typhimurium*. The enzymes are indicated by their corresponding gene designations as follows: *apt*, adenine phosphoribosyltransferase (EC 2.4.2.7); *deoD*, purine nucleoside phosphorylase (EC 2.4.2.1); *gpt*, guanine phosphoribosyltransferase (EC 2.4.2.–); *gsk*, guanosine kinase (EC 2.7.1.–); *guaA*, GMP synthetase (EC 6.3.4.1); *guaB*, IMP dehydrogenase (EC 1.2.2.24); *guaC*, GMP reductase (EC 1.6.6.8); *purE*, phosphoribosylaminomidazole carboxylase (EC 4.1.1.21); *purF*, amidophosphoribosyltransferase (EC 2.4.2.14); *purI*, phosphoribosylaminoimidazole synthetase (EC 6.3.3.1); and *hpt*, hypoxanthine phosphoribosyltransferase (EC 2.4.2.8).

feedback-inhibited in a synergistic and highly cooperative manner by AMP and GMP, in fact all purine nucleotides inhibit (41, 42). In the absence of feedback-inhibitors the enzyme of *E. coli* shows hyperbolic saturation kinetics for the substrate PRPP ($K_m = 67\ \mu M$). GMP is the most powerful inhibitor of this enzyme, and in the presence of this compound the enzyme shows positive cooperativity for PRPP, which means that the apparent K_m for PRPP increases with decreasing concentrations of PRPP (41). Thus variations of the intracellular concentrations of PRPP may influence the rate whereby this first enzyme of the purine nucleotide *de novo* synthesis operates and hence influence the flow through the entire pathway. In corollary to this it has been shown that addition of preformed purines to cultures of *E. coli* or *S. typhimurium* dramatically reduces the PRPP pool (Table 1.2) and also terminates flow through this biosynthetic pathway (15).

3. PRPP consumption in the presence of preformed "end products"

Addition of histidine and tryptophan to the growth medium will reduce the amount of PRPP needed for their biosynthesis to virtually nothing since the biosynthetic pathways stop operating due to feedback-regulation (7, 40).

TABLE 1.2
PRPP pools in *E. coli* K-12 and *S. typhimurium* LT-2 in a glucose-salt minimal medium in the absence or presence of purine supplement.

Purine base or purine nucleoside[a] (0.1 mM)	PRPP pool (μmol/g dry weight)	
	E. coli K-12[b]	*S. typhimurium* LT-2[c]
unsupplemented	1.3	1.24
adenine (adenosine)	0.04 (0.08)[a]	0.12
hypoxanthine (inosine)	0.08 (0.12)	0.32
guanine (guanosine)	0.17 (0.60)	0.69

[a] The numbers in parentheses give the PRPP pool in the presence of purine nucleoside.
[b] The *E. coli* K-12 strain ABLA was used. Data are taken from Bagnara and Finch (4), and pool sizes have been converted from μmol/g protein to μmol/g dry weight by dividing by two.
[c] K. F. Jensen, unpublished data. The pools were determined chromatographically as described (26).

In contrast to this the purine and pyrimidine nucleotides and the nicotinamide coenzymes cannot be fed intact to the bacteria due to permeability barriers (Chapters 2 and 3).

Although it is possible to feed high concentrations of nicotinamide mononucleotide as a source for the synthesis of NAD^+ in mutants lacking nicotinamide deamidase or nicotinate phosphoribosyltransferase and thus escape the consumption of PRPP for this purpose, the predominant rute for the synthesis of NAD^+ in wild type cells from exogenous precursors appears to involve the nicotinate phosphoribosyltransferase (11).

Exogenous purine and pyrimidine bases are converted to the nucleotides nearly exclusively *via* the phosphoribosyltransferases and thus use PRPP during their conversion to nucleotides (31, 48). Despite the presence of nucleoside kinases (5, 22, 29, 68, 71) exogenous nucleosides will not spare PRPP to any great extent. This is because the nucleosides are degraded by nucleoside phosphorylases to the corresponding nucleobase and ribose 1-phosphate. Subsequently the nucleobases are converted to nucleotides by phosphoribosyltransferases (21, 54). In exponentially growing *E. coli* cultures this latter pathway catalysed by nucleoside phosphorylases and phosphoribosyltransferases seems to predominate over the direct pathway of phosphorylation even while intact nucleoside is still available in the medium (K. F. Jensen, unpublished observations). Only in mutants lacking the proper nucleoside phosphorylases are nucleosides incorporated without cleavage of the glycosidic bond and nucleoside addition thus expected to have PRPP sparing effects (Chapters 2 and 3).

IV. INTRACELLULAR CONCENTRATION OF PRPP (PRPP POOLS)

In *E. coli* K-12 the PRPP pool is proportional to the growth rate over a range from 0.3 doublings per hour to 1.0 doublings per hour (Ref. 3, Fig. 1.8). The nucleotides show a similar dependency on growth rate (Fig. 1.8). However, the variations of the PRPP pool is greater than for the nucleotides. This has been interpreted to indicate that the supply of PRPP might be a rate-limiting factor for nucleotide biosynthesis (3).

In *S. typhimurium* the PRPP pool size also depends on the carbon source, but no strict correlation to growth rate has been found (69).

When purine bases are added to exponentially growing cultures of *E. coli* the PRPP pool immediately falls to low levels. The pool remains low until the purine base has been consumed and then returns to normal (4). Table 1.2 gives the steady-state pool in the presence of various purines. The interpretation of these results was that the depletion of the PRPP pool after addition of

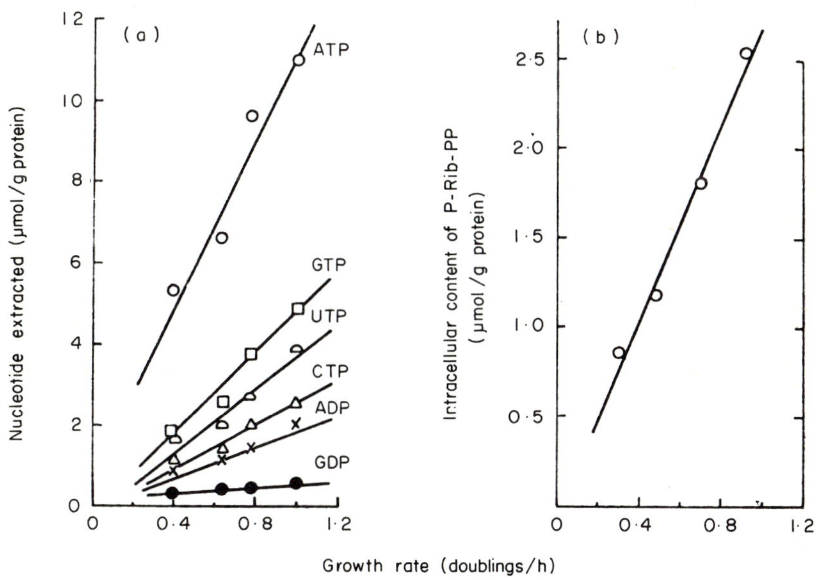

FIG. 1.8. Effect of growth rate on the intracellular pools of nucleotides (a) and PRPP (b) in *E. coli* K12. The following carbon sources were used to generate the different growth rates: glucose + casaminoacids, glucose, glycerol, and succinate. Reproduced from Bagnara and Finch (3) with permission. If the numbers on the ordinate axis are divided by two the pools will roughly be converted into (μmol per gram dry weight) which is the unit usually used in this chapter.

preformed purines is a major control factor is turning off *de novo* purine nucleotide synthesis in *E. coli* when preformed purines are available (4).

Addition of nucleosides has a similar effect on the PRPP pool as the purine bases (Table 1.2). This is in agreement with the notion that purine nucleosides are used for purine nucleotide synthesis virtually only after cleavage of the glycosidic bond (4).

Addition of adenine to *E. coli* cultures causes a temporary fall in the pools of UTP and CTP lasting for about 30 min. This indicates that the pyrimidine nucleotide *de novo* synthesis is retarded by the very low PRPP pool present under these conditions (Table 1.2). Subsequently the pyrimidine nucleotide pools returns to normal due to a slight derepression of the synthesis of orotate phosphoribosyltransferase and aspartate transcarbamylase (4, 8). Uracil and uridine has only minor effects upon the PRPP pool of *E. coli* indicating that a competition between *de novo* and salvage pathways for PRPP is not a factor which controls the turning off of *de novo* synthesis of pyrimidine nucleotides after addition of preformed pyrimidine bases or nucleosides (4, 8).

The response of the PRPP pool of *S. typhimurium* to addition of purine bases is similar to the response of *E. coli* but appears to be less dramatic (Table 1.2).

V. MUTANTS OF *ESCHERICHIA COLI* AND *SALMONELLA TYPHIMURIUM* DEFECTIVE IN PRPP SYNTHETASE

A. Selection

Studies on the physiological importance of variations of the PRPP supply have been hampered by the lack of mutants defective in PRPP synthetase. Such mutants are also needed in order to evaluate the physiological importance of the various potentially regulatory properties of PRPP synthetase that have been indicated by the *in vitro* studies on the purified enzyme.

Several attempts have been performed to isolate leaky or conditional mutants in PRPP synthetase of enteric bacteria by means of selecting for resistance to purine or pyrimidine base analogs, but without success.

Mutants with defects in the PRPP synthetase (in the following called *prs* mutants) have however been isolated in a rather unexpected way in both *E. coli* and *S. typhimurium* (23, 28 and personal communications with B. Hove-Jensen, B. Jochimsen, and P. Nygaard). The background for this selection is described below:

Purine requiring mutants of *E. coli* and *S. typhimurium* which harbour a mutation in *purE* and in addition lack purine nucleoside phosphorylase (*deoD*) (*see* Fig. 1.7) are unable to satisfy their growth requirement with guanosine even though the enzymes needed for guanosine utilization (guanosine kinase and GMP reductase) are present in the cells (12, 16, 23, 28). The reason for this behaviour is unclear, but it seems to relate to a toxicity of aminoimidazole ribotide which accumulates in *purE* strains during purine starvation (12).

When secondary mutants of the *purE,deoD*-strains are selected for ability to grow on guanosine as sole purine source many of these have acquired additional blocks in the purine *de novo* pathway prior to the formation of aminoimidazole ribotide, i.e. additional mutations in for instance *purI* or *purF*. A common feature of these mutants able to grow on guanosine is that the new mutation has prevented formation of aminoimidazole ribotide (Fig. 1.7), and it was realized that reduced activity of PRPP synthetase might also give rise to reduced formation of aminoimidazole ribotide. Hence, mutants with defective PRPP synthetases might also be found amongst the guanosine growers, and indeed this turned out to be the case (23, 28).

A mutant with defective PRPP synthetase activity in extracts has also been isolated by Pandey and Switzer (50). This mutant was obtained following enzymatic screening of a large number of temperature sensitive mutants.

The mutants exhibit many of the predicted phenotypic consequences of a reduced supply of PRPP, but since considerable divergence exists between the mutant characters they will be described individually below.

B. *E. coli prs*-mutant

The *prs* mutant, Sϕ 1172 (*purE,deoD,prs*) is able to satisfy its purine requirement with guanosine alone, but grows somewhat slower than the parent strain on purine bases. The growth rate increases by addition of histidine and tryptophan. This is in agreement with the notion that the mutant grows slowly due to shortage of PRPP.

Partially purified mutant enzyme shows a 17-fold increased K_m for ATP and a 6-fold increased K_m for ribose 5-phosphate (Table 1.3). In addition the mutant enzyme is only weakly inhibited by ADP; rather at low concentrations of ribose 5-phosphate the enzyme appears to be stimulated by ADP (23). This behaviour supports the proposed role of ADP as a true allosteric inhibitor of PRPP synthetase activity (14, 66).

When assayed at saturating substrate concentrations the mutant contains two to three times more PRPP synthetase activity than the parent. It is not obvious whether this increase in enzyme activity is due to the altered kinetic properties of the mutated enzyme, whether the synthesis of PRPP synthetase is internally derepressed, or whether the mutant harbours an additional mutation in a regulatory gene for enzyme synthesis. However, this trait cotransduces with the enzyme defect.

TABLE 1.3
Apparent Michaelis constants of *E. coli* mutant PRPP synthetase.[a]

Substrate	Enzyme source	
	Parent	Mutant
	K_m	K_m
Ribose 5-phosphate	45 μM	240 μM
ATP	60 μM	1000 μM

[a] Partially purified PRPP synthetase preparations were used. Data from Hove-Jensen and Nygaard (23).

During purine starvation the PRPP pool of the mutant decreases, whereas PRPP accumulates in the parent under similar conditions. This is consistent with the altered kinetic properties of the mutant enzyme, since the purine nucleotide pools including the substrate ATP fall drastically during purine starvation.

In contrast to the parent strain the mutant seems not to derepress synthesis of the first enzyme of the purine *de novo* biosynthetic pathway, glutamine PRPP amiditransferase, during purine starvation even though the purine nucleotide pools are depleted. This may indicate that a proper supply of PRPP is important for the regulation of *E. coli pur* gene expression and the observation is consistent with a recent proposal that the purine bases, hypoxanthine and guanine, control the level of *pur* gene expression in *S. typhimurium* (22a).

The *prs*-mutation of strain Sϕ 1172 is located at 26 min on the *E. coli* chromosome between *hemA* and *purB*, cotransducible with either of the two markers. During the genetic manipulations it became clear that the *prs*-mutation alone is not responsible for the ability of the mutant to grow rapidly with guanosine as sole purine source. This phenotype appears to be the sum of the *prs*-mutation and an additional mutation located in or near the gene (*gsk*) encoding guanosine kinase (23, 23a). The *prs* gene near *hemA* has been cloned on pBR322. The multicopy plasmid containing the gene results in about 25-fold elevation of PRPP synthetase activity and, when introduced into "mini-cells", the plasmid leads to the synthesis of a polypeptide with a molecular weight similar to that of the PRPP synthetase subunit (B. Hove-Jensen, personal communication). This strongly indicates that the gene near *hemA* encodes the amino acid sequence of PRPP synthetase.

C. *Salmonella typhimurium prs* mutant

This mutant GP 122 (*proAB,purE,deoD,prs*) grows much better with purine nucleosides than with hypoxanthine as purine source. Growth on hypoxanthine is stimulated by addition of histidine, but not by tryptophan. The reason for this behaviour is not clear. The growth rate of the mutant decreases with decreasing phosphate concentrations in the medium, and in the presence of only 0.3 mM orthophosphate the mutant hardly shows any growth at all. This is in line with the observation that phosphate is required for stability and activity of PRPP synthetase *in vitro* (28, 61). Revertants capable of forming colonies on 0.3 mM phosphate have lost the ability to use guanosine as sole purine source indicating that the two properties are due to the same mutational event (28).

This *S. typhimurium prs*-mutant contains 15–20% of parental PRPP synthetase activity. The residual enzyme shows normal K_m for ribose 5-phosphate, but appears to be more heat-labile than the parental enzyme in crude extracts (28).

The PRPP pool of the mutant is about 10 per cent of the parental pool in steady state growth in a medium containing 1 mM phosphate. The mutant accumulates aminoimidazole ribotide only very slowly during purine starvation indicating that the operation of the first part of the *de novo* biosynthetic pathway for purine nucleotides (Fig. 1.7) is strongly reduced because of the low PRPP pool (28). Two of the enzymes of the pyrimidine *de novo* biosynthetic pathway (Fig. 1.6), aspartate transcarbamylase and dihydroorotase, are elevated six and three fold, respectively, indicating that the low PRPP pool causes a partial pyrimidine starvation. The levels of the enzymes of the histidine and tryptophan biosynthetic pathways have not been determined, but it was found that the mutant excretes substantial quantities of antranilate into the medium. This excretion terminates when tryptophan is supplied to the culture indicating that the biosynthesis of tryptophan in the mutant is limited by the supply of PRPP (28). This result is somewhat puzzling, since of all PRPP consuming reactions the antranilate phosphoribosyltransferase appears to have the lowest K_m for PRPP, 7 μM *in vitro* (20).

The mutation responsible for the ability of the mutant strain to utilize guanosine has been reported to be located at 7 units on the *Salmonella* chromosome cotransducible with *proAB* (28). However, recent experiments have shown that the *prs* mutation of GP122 is located near *hemA* at 35 units on the *S. typhimurium* chromosome (B. Jochimsen, B. Hove-Jensen, B. Garber, and J. S. Gots, manuscript in preparation).

D. Temperature sensitive mutant of *S. typhimurium* with defective PRPP synthetase activity

This mutant isolated by brute force (50) is devoid of PRPP synthetase activity in extracts at all temperatures tested and seems to contain very little material (2–3%) detectable by antibodies raised against pure PRPP synthetase both at

30 and 42°C. However, since the mutant grows with near parental rate it seems most likely that it contains fairly normal amounts of PRPP synthetase, but that the enzyme is destroyed upon extraction.

The temperature sensitivity and the lack of PRPP synthetase activity in extracts is probably not due to the same mutational event since temperature resistant revertants have not regained PRPP synthetase activity.

The mutant contains normal PRPP pools when it grows at 30°C and this pool remains unchanged following a shift to 42°C. Thus it seems that growth does not terminate due to lack of PRPP.

The mutation(s) responsible for the lack of PRPP synthetase in extracts and for the temperature sensitivity of this mutant has been located near *hisW* on the Salmonella chromosome. The wild type is dominant over the mutant character since both traits (temperature sensitivity and lack of PRPP synthetase) are suppressed by introduction of the *E. coli* episome F′32. This observation further indicates that the corresponding gene has a similar location in *E. coli* (50).

VI. CONCLUSION

The enzymology of the PRPP synthetase reaction *in vitro* is well known mainly due to the work of Switzer and his collaborators. Their studies have revealed the fundamental kinetic and chemical features of the reaction and have also indicated a potential feedback mechanism to regulate enzyme activity via inhibition by purine nucleotides, in particular by ADP. Physiological evidence obtained by measuring pools of nucleotides and PRPP at different conditions of growth have demonstrated that this regulatory property of PRPP synthetase may well be of significance *in vivo*. On the other hand the factors which control the synthesis of the enzyme are still rather obscure. A uridine nucleotide has been proposed as a candidate for a repressing metabolite, and recent evidence may indicate that the synthesis of PRPP synthetase is co-regulated with the synthesis of the pyrimidine *de novo* biosynthetic enzymes and that RNA polymerase is involved in sensing the regulatory signals.

The analysis of the structure of PRPP synthetase indicates that all subunits of the native enzyme are identical, and thus that PRPP synthetase is encoded by a single gene. However, the possibility that the enzyme does contain more than one type of polypeptide chain cannot strictly be ruled out. A structural gene for PRPP synthetase is located near *hemA*. A mutation at this position (26 min) on the *E. coli* chromosome results in the synthesis of a defective enzyme with altered kinetic properties, and the presence of this gene on a multicopy plasmid leads to overproduction of PRPP synthetase. In addition a mutation at the corresponding position near *hemA* (35 min) in *S. typhimurium* has been shown to give rise to a PRPP synthetase defect. Still another mutation affecting the activity of PRPP synthetase in extracts of *Salmonella* have been located at 47 min. This gene is also a candidate for a structural *prs* gene since PRPP synthetase seems to function in the mutant cells, but is absent from extracts of the cells. However, other explanations of this peculiar phenotype are possible.

Hopefully the characterization of the properties of the three existing *prs* mutants will pave the way for the isolation of other mutants with more clearcut phenotypes, for instance temperature sensitive ones, to allow for an unambiguous localization of the structural gene(s) for the enzyme. In such *prs* mutants it should be possible to vary the intracellular supply of PRPP as a function of temperature. This will permit a study of the effect of PRPP pool on the operation of the various PRPP consuming biosynthetic pathways. The properties of the existing mutants indicate that such an investigation may be valuable. Likewise the characterization of the existing *prs* mutants may indicate conditions whereby it will be possible to isolate regulatory mutants: (a) some, where the amount of synthesis of the enzyme is abnormal, which will be of help in the determination of the factors involved in the control of enzyme synthesis, and (b) some, where the kinetic properties of PRPP synthetase are altered, which will be of help to evaluate the physiological importance of the regulatory properties of the enzyme that have been indicated by the *in vitro* characterization of the enzyme.

ACKNOWLEDGEMENTS

The author wants to thank Drs R. L. Switzer, B. Hove-Jensen, B. Jochimsen, and P. Nygaard for discussions and for communicating results prior to publication.

REFERENCES

1. Abd-El-Al, A. T. H., and Ingraham, J. L. (1975). Carbamylphosphate synthetase from *Salmonella typhimurium*. *J. Biol. Chem.* **250**, 4410–4417.
2. Atkinson, D. E. (1968). The energy charge of the adenylate pool as a regulatory parameter. Interaction with feedback modifiers. *Biochemistry* **7**, 4030–4034.
3. Bagnara, A. S., and Finch, L. R. (1973). Relationships between intracellular contents of nucleotides and 5-phosphoribosyl 1-pyrophosphate in *Escherichia coli*. *Eur. J. Biochem.* **36**, 422–427.
4. Bagnara, A. S., and Finch, L. R. (1974). The effect of bases and nucleotides and 5-phosphoribosyl 1-pyrophosphate in *Escherichia coli*. *Eur. J. Biochem.* **41**, 421–430.
5. Beck, C. F., Ingraham, J. L., Neuhard, J., and Thomassen, E. (1972). Metabolism of pyrimidines and pyrimidine nucleosides by *Salmonella typhimurium*. *J. Bacteriol.* **110**, 219–228.

6. Becker, M. A., and Seegmiller, J. E. (1979). Synthesis of phosphoribosylpyrophosphate in mammalian cells. *In* "Adv. in Enzymology and Related Areas of Molecular Biology" (A. Meister, *Ed.*), Vol. 49, pp. 281. J. Whiley and Sons, New York.
7. Brenner, M., and Ames, B. N. (1971). The histidine operon and its regulation. *In* "Metabolic Pathways" (H. J. Vogel, *Ed.*), 3rd Edition, Vol. 5, pp. 349–387.
8. Christopherson, R. I., and Finch, L. R. (1978). RCponse of the pyrimidine pathway of *Escherichia coli* K 12 to exogenous adenine and uracil. *Eur. J. Biochem.* **90**, 347–358.
9. Crawford, I., Kornberg, A., and Simms, E. S., (1957). Conversion of uracil and orotate to uridine 5′-phosphate by enzymes in Lactobacilli. *J. Biol. Chem.* **226**, 1093–1101.
10. Foster, J. W., Kinney, D. M., and Moat, A. G. (1979). Pyridine nucleotide cycle of *Salmonella typhimurium*: Isolation and characterization of *pncA*,*pncB*, and *pncC* mutants and utilization of exogenous nicotinamide adenine dinucleotide. *J. Bacteriol.* **137**, 1165–1175.
11. Foster, J. W., and Baskowsky-Foster (1980). Pyridine nucleotide cycle of *Salmonella typhimurium*: *In vivo* recycling of nicotinamide adenine dinucleotide. *J. Bacteriol.* **142**, 1032–1035.
12. Garber, B., Jochimsen, B., and Gots, J. S. (1980). Glutamine and related analogs regulate guanosine monophosphate reductase in *Salmonella typhimurium. J. Bacteriol.* **143**, 105–111.
13. Gibson, K. J., and Switzer, R. L. (1980). Structure of the divalent cation-nucleotide complex at the active site of phosphoribosylpyrophosphate synthetase. *J. Biol. Chem.* **255**, 694–696.
14. Gibson, K. J., Schubert, K. R., and Switzer, R. L. (1982). Binding of the Substrates and the Allosteric Inhibitor Adenosine 5′-Diphosphate to Phosphoribosylpyrophosphate Synthetase from *Salmonella typhimurium. J. Biol. Chem.* **257**, 2391–2396.
15. Gots, J. S. (1971). Regulation of purine and pyrimidine metabolism. *In* "Metabolic Pathway" (H. J. Vogel, *Ed.*), Vol. 5, pp. 225–255. Academic Press, London.
16. Gots, J. S., Benzon, C. F., Jochimsen, B., and Koduri, K. R. (1977). Microbial models and regulatory elements in the control of purine metabolism. *In* "Purine and Pyrimidine Metabolism". Ciba Foundation Symposium 48 (New York), pp. 23–41.
17. Granot, J., Gibson, K. J., Switzer, R. L., and Mildvan, A. S. (1980). NMR Studies of the nucleotide conformation and arrangement of substrates and activators on phosphoribosylpyrophosphate synthetase. *J. Biol. Chem.* **255**, 10931–10937.
18. Hartman, S. C., Levenberg, B., and Buchanan, J. M. (1955). Involvement of ATP, 5-phosphoribosylpyrophosphate and L-asaserine in the enzymatic formation of glycinamide ribotide intermediates in inosinic acid biosynthesis. *J. Am. Chem. Soc.* **77**, 501–503.
19. Hartman, S. C., Levenberg, B., and Buchanan, J. M. (1956). Biosynthesis of the purines. XI. Structure, enzymatic synthesis and metabolism of glycinamide ribotide and (α-N-formyl) glycinamide ribotide. *J. Biol. Chem.* **221**, 1057–1070.
20. Henderson, E. J., Zalkin, H., and Hwang, L. H. (1970). The antranilate synthetase-antranilate 5-phosphoribosylpyrophosphate phosphoribosyltransferase aggregate. Catalytic and regulatory properties of aggregated and unaggregated forms of antranilate 5-phosphoribosylpyrophosphate phosphoribosyltransferase. *J. Biol. Chem.* **245**, 1425–1431.

21. Hochstadt-Ozer, J. (1972). The regulation of purine utilization in bacteria. IV. Roles of membrane-localized and pericyptoplasmic enzymes in the mechanism of purine nucleoside transport across isolated *Escherichia coli* membranes. *J. Biol. Chem.* **247**, 2419–2426.
22. Hoffmeyer, J., and Neuhard, J. (1971). Metabolism of exogenous purine bases and nucleosides by *Salmonella typhimurium*. *J. Bacteriol.* **106**, 14–24.
22a. Honberg, U., and Jensen, K. F. (1983). Role of hypoxanthine and guanine in regulation of *Salmonella Hyphimurium pur* gene expression. *J. Bacteriol.* **153**, 837–845.
23. Hove-Jensen, B., and Nygaard, P. (1982). Phosphoribosylpyrophosphate Synthetase of *Escherichia coli*: Identification of a Mutant Enzyme. *Eur. J. Biochem.* **126**, 327–332.
23a. Hove-Jensen, B. (1983). Chromosomal location of the gene encoding phosphoribosylpyrophosphate synthetase in *Escherichia Coli*. *J. Bacteriol.* **154**.
24. Jackman, L. E., and Hochstadt, J. (1976). Regulation of purine utilization in bacteria. VI. Characterization of hypoxanthine and guanine uptake into isolated membrane vesicles from *Salmonella typhimurium*. *J. Bacteriol.* **126**, 312–326.
25. Jensen, K. F. (1979). Apparent involvement of purines in the control of expression of *Salmonella typhimurium pyr* genes: Analysis of a leaky *guaB* mutant resistant to pyrimidine analogs. *J. Bacteriol.* **138**, 731–738.
26. Jensen, K. F., Houlberg, U., and Nygaard, P. (1979). Thin-layer chromatographic methods to isolate ^{32}P-labeled phosphoribosyl-α-1-pyrophosphate (PRPP): Determination of cellular pools and assay of PRPP synthetase activity. *Analyt. Biochem.* **98**, 254–263.
27. Jensen, K. F., Neuhard, J., and Schack, L. (1982). RNA Polymerase Involvement in the Control of *Salmonella typhimurium pyr* Gene Expression. Isolation and Characterization of a Fluorouracil Resistant Mutant with High, Constitutive Expression of *pyrB* and *pyrE* due to a Mutation in *rpoBC*. *The EMBO Journal* **1**, 69–74.
28. Jochimsen, B., Garber, B., and Gots, J. S. (1980). Phosphoribosylpyrophosphate (PRPP) synthetase mutant in *Salmonella typhimurium*. *Adv. Exp. Med. Biol.* **122B**, 131–136.
29. Jochimsen, B., Nygaard, P., and Vestergaard, T. (1975). Location on the chromosome of *Escherichia coli* of genes governing purine metabolism. Adenosine deaminase (*add*), guanosine kinase (*gsk*) and hypoxanthine phosphoribosyltransferase (*hpt*). *Molec. Gen. Genet.* **143**, 85–91.
30. Karlström, O. (1968). Mutants of *Escherichia coli* defective in ribonucleoside and deoxyribonucleoside catabolism. *J. Bacteriol.* **95**, 1069–1077.
31. Khorana, H. G., Ferdandes, J. F., and Kornberg, A. (1958). Pyrophosphorylation of ribose 5-phosphate in the enzymnatic synthesis of 5-phosphoribosyl 1-pyrophosphate. *J. Biol. Chem.* **230**, 941–948.
32. Kornberg, A. (1980). "DNA Replication". W. H. Freeman and Co., San Francisco.
33. Kornberg, A., Liebermann, I., and Simms, E. (1955). Enzymatic synthesis and properties of 5-phosphoribosylpyrophosphate. *J. Biol. Chem.* **215**, 389–402.
34. Kornberg, A., Lieberman, I., and Simms, E. S. (1955). Enzymatic synthesis of purine nucleotides. *J. Biol. Chem.* **215**, 417–427.
35. Kungsløyr, L., Hageman, J. H., Fall, L., and Atkinson, D. E. (1968). Interaction between energy charge and product feedback in the regulation of biosynthetic enzymes. Aspartokinase, phosphoribosyladenosine triphosphate synthetase, and phosphoribosylpyrophosphate synthetase. *Biochemistry* **7**, 4035–4040.

36. Li, T. M., Mildvan, A. S., and Switzer, R. L. (1978). Studies of the stereochemistry and of the role of metal ions in the mechanism of phosphoribosylpyrophosphate synthetase from *Salmonella typhimurium*. *J. Biol. Chem.* **253**, 3918–3923.
37. Li, T. M., Switzer, R. L., and Mildvan, A. S. (1979). Kinetic and magnetic resonance studies of the interaction of the Cr-ATP complex with phosphoribosylpyrophosphate synthetase from *Salmonella typhimurium*. *Arch. Biochem. Biophys.* **193**, 1–13.
38. Lieberman, I., Kornberg, A., and Simms, E. (1955). Enzymatic synthesis of pyrimidine nucleotides. Orotidine-5′-phosphate and uridine-5′-phosphate. *J. Biol. Chem.* **215**, 403–415.
39. Maaløe, O., and Kjeldgaard, N. O. (1966). Control of macromolecular synthesis.
40. Margolin, P. (1971). Regulation of tryptophan synthesis. *Metabolic Pathways* **5**, 389–488.
41. Messenger, L. J., and Zalkin, H. (1979). Glutamine phosphoribosylpyrophosphate amidotransferase from *Escherichia coli*. Purification and properties. *J. Biol. Chem.* **254**, 3382–3392.
42. Meyer, Eckhart, and Switzer, R. L. (1979). Regulation of *Bacillus subtilis* glutamine phosphoribosylpyrophosphate amidotransferase activity by end products. *J. Biol. Chem.* **254**, 5397–5402.
43. Miller, G. A., Rosenzweig, S., and Switzer, R. L. (1975). Oxygen-18 studies of the mechanism of pyrophosphoryl group transfer catalyzed by phosphoribosylpyrophosphate synthetase. *Arch. Biochem. Biophys.* **171**, 732–736.
44. Molloy, A., and Finch, L. R. (1969). Uridine-5′-monophosphate pyrophosphorylase activity from *Escherichia coli*. *FEBS Letts.* **5**, 211–213.
45. Moyed, H. S., and Magasanik, B. (1957). The role of purines in histidine biosynthesis. *J. Am. Chem. Soc.* **79**, 4812–4813.
46. Neuhard, J., and Ingraham, J. (1968). Mutants of *Salmonella typhimurium* requiring cytidine for growth. *J. Bacteriol.* **95**, 2431–2433.
47. Nygaard, P., and Jensen, K. F. (1980). Chromatographic determination of PRPP-synthetase activity in human blood cells. *Adv. Exp. Med. Biol.* **122B**, 401–407.
48. O'Donovan, G. A., and Neuhard, J. (1970). Pyrimidine metabolism in microorganisms. *Bacteriol. Rev.* **34**, 278–343.
49. Olszowy, J., and Switzer, R. L. (1972). Specific repression of phosphoribosylpyrophosphate synthetase by uridine compounds in *Salmonella typhimurium*. *J. Bacteriol.* **110**, 450–451.
50. Pandey, N. K., and Switzer, R. L. (1981). Isolation of a temperature-sensitive mutant of *Salmonella typhimurium* with a closely-linked mutation affecting phosphoribosylpyrophosphate synthetase activity. *J. Gen. Microbiol.* **128**, 1863–1871.
51. Paulus, T. J., and Switzer, R. L. (1979). Characterization of pyrimidine repressible and arginine repressible carbamyl phosphate synthetases from *Bacillus subtilis*. *J. Bacteriol.* **137**, 82–91.
52. Preiss, J., and Handler, P. (1958). Biosynthesis of diphosphopyridine nucleotide. I. Identification of intermediates. *J. Biol. Chem.* **233**, 488–492.
53. Preiss, J., and Handler, P. (1958). Biosynthesis of diphosphopyridine nucleotide. II. Enzymatic Aspects. *J. Biol. Chem.* **233**, 493–500.
54. Rader, R. L., and Hochstadt, J. L. (1976). Regulation of purine utilization in bacteria. VII. Involvement of membrane-associated nucleoside phosphorylases in the uptake and the base-mediated loss of the ribose moiety of nucleosides by *Salmonella typhimurium* membrane vesicles. *J. Bacteriol.* **128**, 290–301.
55. Remy, C. N., Remy, W. T., and Buchanan, J. M. (1955). Biosynthesis of the

purines. VII. Enzymatic synthesis and utilization of α-5-phosphoribosyl-pyrophosphate. *J. Biol. Chem.* **217**, 885–895.

56. Roberts, M. F., Switzer, R. L., and Schuberts, K. R. (1975). Inactivation of *Salmonella* phosphoribosylpyrophosphate synthetase by oxidation of a specific sulfhydryl group with potassium permanganate. *J. Biol. Chem.* **250**, 5364–5369.
57. Roberts, M. F., and Switzer, R. L. (1978). Inactivation of *Salmonella* phosphoribosylpyrophosphate synthetase by specific chemical modification of a lysine residue. *Arch. Biochem. Biophys.* **185**, 391–399.
58. Sadler, W. C., and Switzer, R. L. (1977). Regulation of *Salmonella* phosphoribosylpyrophosphate synthetase activity *in vivo*, deductions from pool measurements. *J. Biol. Chem.* **252**, 8504–8511.
59. Schubert, K. R., Switzer, R. L., and Shelton, E. (1975). Studies of the quaternary structure and the chemical properties of phosphoribosylpyrophosphate synthetase from *Salmonella typhimurium*. *J. Biol. Chem.* **250**, 7492–7500.
60. Schwartz, M., and Neuhard, J. (1975). Control of expression of the *pyr* genes in *Salmonella typhimurium*: Effects of variations in uridine and cytidine nucleotide pools. *J. Bacteriol.* **121**, 814–822.
61. Switzer, R. L. (1969). Regulation and mechanism of phosphoribosylpyrophosphate synthetase. I. Purification and properties of the enzyme from *Salmonella typhimurium*. *J. Biol. Chem.* **244**, 2854–2863.
62. Switzer, R. L. (1971). Regulation and mechanism of phosphoribosylpyrophosphate synthetase. III. Kinetic studies of the reaction mechanism. *J. Biol. Chem.* **246**, 2447–2458.
63. Switzer, R. L. (1974). Phosphoribosylpyrophosphate synthetase and related pyrophosphokinases. *In* "The Enzymes" (P. D. Boyer, *Ed.*), Vol. X, pp. 607–629. Academic Press, London and New York.
64. Switzer, R. L., and Gibson, K. J. (1978). Phosphoribosylpyrophosphate synthetase (ribose-5′-phosphate pyrophosphokinase) from *Salmonella typhimurium*. *In* "Methods of Enzymology", Vol. 51, pp. 3–17. Academic Press, London and New York.
65. Switzer, R. L., and Simcox, P. D. (1974). Evidence against formation of a pyrophosphoryl enzyme intermediate in the phosphoribosylpyrophosphate synthetase reaction. *J. Biol. Chem.* **249**, 5304–5307.
66. Switzer, R. L., and Sogin, D. C. (1973). Regulation and mechanism of phosphoribosylpyrophosphate synthetase. V. Inhibition by end products and regulation by adenosine diphosphate. *J. Biol. Chem.* **248**, 1063–1073.
67. Tener, G. M., and Khorana, H. G. (1958). Phosphorylated sugars. VI. Synthesis of α-D-ribofuranose 1,5-diphosphate and α-D-ribofuranose 1-pyrophosphate 5-phosphate. *J. Am. Chem. Soc.* **80**, 1988–2004.
68. Valentin-Hansen, P. (1978). Uridine cytidine kinase from *Escherichia coli*. *In* "Methods in Enzymology", Vol. LI, pp. 308–314. Academic Press, London and New York.
69. White, M. N., Olszowy, Z., and Switzer, R. L. (1971). Regulation and mechanism of phosphoribosylpyrophosphate synthetase: Repression by end products. *J. Bacteriol.* **108**, 122–131.
70. Yanofsky, C. (1956). The enzymatic conversion of antranilic acid to indole. *J. Biol. Chem.* **223**, 171–184.
71. Zimmerman, E. F., and Magasanik, B. (1964). Utilization and interconversion of purine bases and ribonucleosides by *Salmonella typhimurium*. *J. Biol. Chem.* **239**, 293–300.

2

Utilization of preformed purine bases and nucleosides

PER NYGAARD

I. Introduction 28
II. *De novo* biosynthesis of inosine 5′-phosphate 29
III. Purine nucleotide interconversions 31
A. Biosynthesis of AMP from IMP 33
B. Biosynthesis of GMP from IMP 34
C. Phosphorylation of AMP and GMP to ATP and GTP . 36
D. Conversion of adenine nucleotides to IMP 37
E. Conversion of guanine nucleotides to IMP 39
F. Purine deoxyribonucleotide synthesis *de novo* 40
IV. Salvage Pathways in *E. coli* and *S. typhimurium* 41
A. Adenine 43
B. Adenosine 45
C. Hypoxanthine 46
D. Inosine 48
E. Guanine 48
F. Guanosine 49
G. Xanthine and xanthosine 51
H. Purine deoxyribonucleosides 52
I. Purine nucleotides 52
K. Regulation of purine salvage 53
V. Purine salvage pathways in other microorganisms 54
A. *Bacillus* 54
B. *Lactobacillus* 55
C. *Neisseria* 56
D. *Fungi* 56
E. Purine auxotrophs 59
VI. Individual enzymes of purine salvage 64
A. Purine phosphoribosyltransferases 64
B. Purine nucleoside kinases 66
C. Purine nucleoside phosphorylases and hydrolases . . 67
D. Nucleoside deoxyribosyltransferases 68
E. Deaminases 68

METABOLISM OF NUCLEOTIDES, NUCLEOSIDES
AND NUCLEOBASES IN MICROORGANISMS
ISBN: 0–12–510580–0

VII. Intracellular formation of purine nucleosides and nucleobases . 69
A. Degradation of RNA 69
B. Turnover of S-adenosylmethionine 70
C. Naturally occurring purine nucleoside analogs . . 71
VIII. Interplay between the purine *de novo* and salvage pathways . 73
A. Regulation of the synthesis of the *de novo* enzymes . . 74
B. Regulation of the activity of the *de novo* enzymes . . 75
C. Sporulation and spore germination 77

I. INTRODUCTION

A sufficient and balanced synthesis of purine nucleotides is necessary for the growth and function of cells. The purine nucleoside 5′-phosphates play a vital role in metabolism; they are intermediates in the biosynthesis of nucleic acids and histidine, and of cofactors such as NAD, FAD, folic acid, thiamine, deoxyadenosylcobalamin and coenzyme A. They constitute the main energy transducing system in the cells and are involved in the regulation of a large number of enzyme reactions. Specialized functions of purine nucleotides having their phosphate group(s) in other positions, e.g. 3′,5′-cyclic AMP and guanosine 5′-diphosphate-3′-diphosphate (ppGpp), include regulatory effects on gene expression.

Purine nucleotides are synthesized either *de novo* from small precursors or via *salvage pathways* from preformed purines. The salvage pathways mediate a continuous influx into the nucleotide pools of purine nucleosides and bases arising from the intracellular breakdown of unstable RNA. Purine compounds utilized through the salvage pathways may also originate from exogenous sources as catabolic products of nucleic acids and nucleotides in decaying cells. Finally, it should be mentioned that both nucleosides and nucleobases can diffuse out of the cell and thus cause a loss of valuable energy and precursors. This loss is prevented by the salvage enzymes, which return the nucleosides and bases to the cellular nucleotide pools. The nucleotides *per se* do not normally escape the cells inner compartment.

Nucleotide synthesis from preformed purines may be one of the most primitive biological reactions, since adenine and guanine are readily formed by non-enzymatic reactions under conditions which are likely to have existed on the earth prior to the appearance of living organisms.

Most organisms possess both the *de novo* pathway and the salvage pathways. Whereas the *de novo* pathway appears to be the same in all organisms, the salvage pathways seem far more diverse in their nature and distribution, since the different organisms vary greatly in their ability to utilize and to interconvert exogenously supplied purine derivatives. Some microorganisms are incapable of synthesizing purines *de novo* and require therefore an exogenous supply of a purine compounds.

The main purpose of this chapter is to discuss the different ways in which microorganisms may utilize preformed purine compounds for the synthesis of purine nucleotides. The chapter starts with a brief account of the *de novo* synthesis of IMP and its regulation. This was considered necessary for the understanding of the regulatory interplay between the endogenous biosynthetic reactions and the salvage reactions. Furthermore the metabolic conversion of IMP to other purine nucleotides are related in some detail, since these interconversion reactions are required for the synthesis of the nucleoside triphosphate precursors of nucleic acids in both the *de novo* and the salvage pathways.

The purine salvage systems have been studied in most detail in *Escherichia coli* and *Salmonella typhimurium*, and the major part of the following will deal with results obtained from these two organisms, but information, when available, about the salvage reactions in other microorganisms will also be included. For previous reviews on purine salvage readers are referred to references 71, 159, 180.

An important factor, which must be considered in dealing with purine salvage, is that many microorganisms are able to degrade the purine ring, i.e. they can utilize the purine nucleobases as nitrogen and/or carbon source. The choice between salvage and catabolism is subject to metabolic control in such a way that catabolism is promoted under conditions of nitrogen limitation (272, 279). The reactions which lead to catabolism of the purine structure will not be treated in this chapter. For a comprehensive review, see Vogel and Drift (272).

II. *DE NOVO* BIOSYNTHESIS OF INOSINE 5′-PHOSPHATE

The individual steps in the *de novo* pathway for purine synthesis are shown in Fig. 2.1. The structural genes coding for the individual enzymes of the pathway were first identified in mutants of *S. typhimurium*, and the chromosomal locations of the genes were determined (281). Essentially the same map position is found in *E. coli* (*see* Appendix). In this context it should be mentioned, though, that no mutants in the gene, encoding the enzyme which catalyzes the formylation of phosphoribosylglycinamide, have been isolated in either of the two organisms. Different genetic symbols for the genes encoding the purine *de novo* biosynthetic enzymes are used in *Neurospora crassa* (216) and yeast (7, 156). In contrast to *E. coli* and *S. typhimurium*, mutants deficient in the enzyme catalyzing the conversion of GAR to FGAR have been found in *Neurospora* (216) and yeast (7).

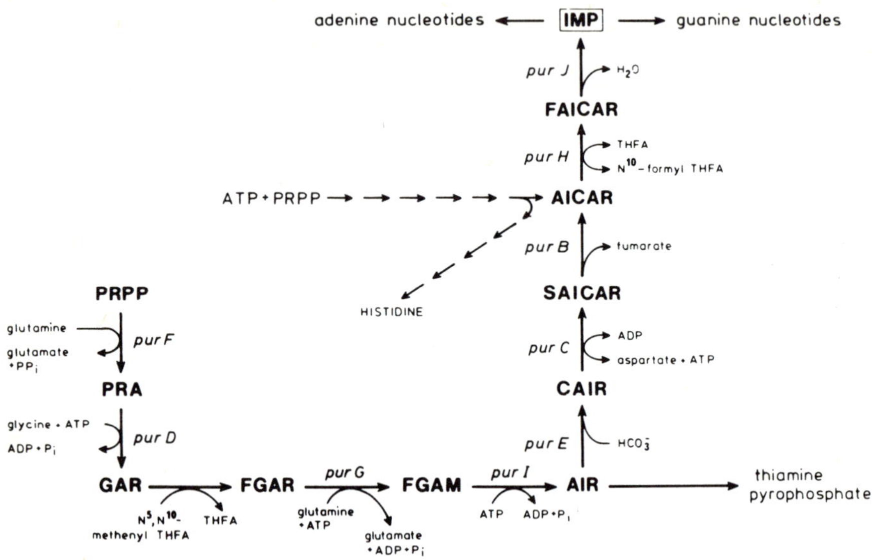

FIG. 2.1. *De novo* synthesis of IMP in *S. typhimurium* and *E. coli*. The individual enzymes of the pathways are identified by their gene symbols (*see* Appendix). The gene symbols in *E. coli* are identical except for *purI* and *purG* (*see* footnote to Appendix). The *purJ* gene has not been identified in *E. coli*. PRPP, phosphoribosylpyrophosphate; PRA, phosphoribosylamine; GAR, phosphoribosylglycinamide; FGAR, phosphoribosylformylglycinamide; FGAM, phosphoribosylformylglycinamidine; AIR, phosphoribosylaminoimidazole; CAIR, phosphoribosylcarboxyaminoimidazole; SAICAR, phosphoribosylaminoimidazolesuccinocarboxamide; AICAR, phosphoribosylaminoimidazolecarboxamide; FAICAR, phosphoribosylformamidoimidazolecarboxamide; THFA, tetrahydrofolic acid.

The regulation of purine nucleotide biosynthesis *de novo* is only partly understood. The control seems to be mediated by feedback inhibition of the initial enzymes by purine nucleotides, and by repression of the synthesis of the individual enzymes (71).

The first complete nucleotide formed in the *de novo* pathway is IMP. The major regulation of the pathway leading to IMP from PRPP (Fig. 2.1) is exerted on the enzyme PRPP aminotransferase (*purF*), which is subject to two modes of regulation: repression and feedback inhibition. With *E. coli*, *S. typhimurium* (107), *Aerobacter aerogenes* (196) and *Bacillus subtilis* (181), repression of the synthesis of this enzyme has been observed when purine compounds are added to the growth medium.

The feedback inhibition of PRPP amidotransferase by purine nucleotides is competitive with the substrate PRPP in all organisms studied. In the absence of the inhibitors, the PRPP saturation curve is hyperbolic, but it is converted to a sigmoidal pattern in the presence of the inhibitors. In enteric bacteria AMP and GMP are the strongest single inhibitors, in combination they show a synergistic inhibition (172, 195). PRPP amidotransferase has also been studied in yeasts (188, 234) and in *B. subtilis* (173). Evidence for feedback control in these organisms has been obtained, but the inhibition pattern differs from that of *E. coli* in that purine nucleotides other than AMP or GMP are strong inhibitors, and that there is no synergistic inhibition.

Control of purine nucleotide biosynthesis may also be exerted, at least in part, by the availability of PRPP. Thus addition of purines to growing cells of *E. coli* causes a depletion of the intracellular PRPP pool, and a simultaneous decrease in the *de novo* synthesis, which is compensated by salvage of the added purines (8).

The reactions leading to formation of AIR (Fig. 2.1) can also be viewed as a part of the *de novo* pathway for thiamine synthesis. Exogenous adenine may inhibit growth in wild type strains. This is due to repression of the *de novo* biosynthesis which results in starvation for thiamine. The growth inhibition may be relieved by addition of thiamine. Furthermore mutants which are blocked in one of the first five steps of the *de novo* pathway, and therefore require purines for growth, concurrently have a thiamine requirement (194).

Another feature which is of importance for purine biosynthesis is that one of the intermediates in the *de novo* pathway, namely AICAR (Fig. 2.1), is also formed as a byproduct from the *de novo* synthesis of histidine (19). This contribution to purine synthesis is regulated by histidine (*see* p. 37).

A comprehensive discussion of the *de novo* pathway for purine biosynthesis may be found in (84, 86).

III. PURINE NUCLEOTIDE INTERCONVERSIONS

Inosinic acid (IMP) is situated at a metabolic branchpoint: one branch leads to the formation of AMP and the other results in the synthesis of GMP. By the subsequent phosphorylations AMP and GMP are converted to ATP and GTP with an intermediate formation of ADP and GDP, respectively.

The purine deoxyribonucleotides are synthesized by reduction at carbon atom number 2 in the pentose moieties of the corresponding ribonucleotides. This reduction is catalyzed by one single enzyme, ribonucleotide reductase, which uses either ribonucleoside di- or triphosphates as substrates, dependent on the organism. Ribonucleotide reductases are non-specific with respect to the nucleobases. Fig. 2.2 summarizes the pathway involved in the formation of the nucleic acid precursors ATP, GTP, dATP, and dGTP from IMP.

In addition, most microorganisms contain enzymes which enable them to interconvert adenine and guanine nucleotides through IMP (Fig. 2.3). These reactions are important in adjusting the intracellular supplies of adenine and guanine nucleotides according to the immediate needs of the cells under different physiological conditions.

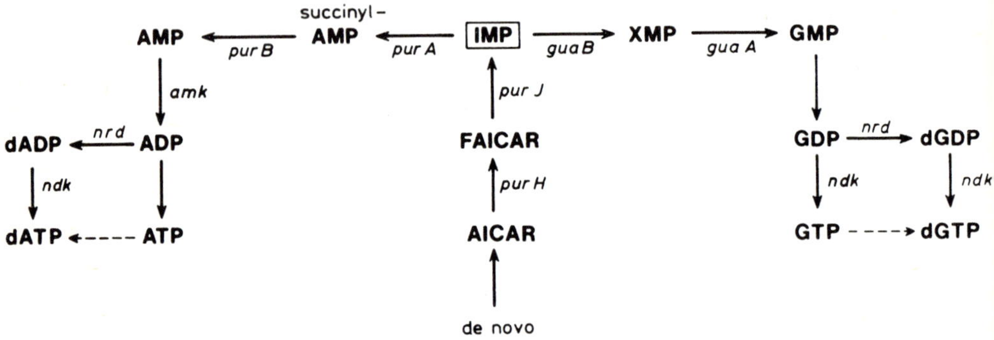

FIG. 2.2. Conversion of IMP to ATP, dATP, GTP and dGTP in *E. coli* and *S. typhimurium*. The individual enzymes are identified by their gene symbols (*see* Appendix). Broken arrows indicate reactions which do not occur in *E. coli* or *S. typhimurium*, but which replace the *nrd* reactions in some microorganisms.

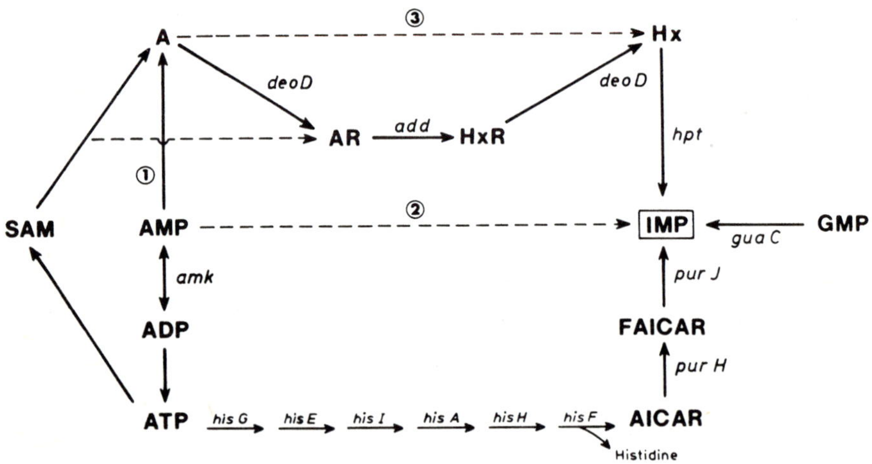

FIG. 2.3. Purine nucleotide interconversions. Full arrows, reactions occurring in *E. coli* and *S. typhimurium*; broken arrows, reactions absent from *E. coli* and *S. typhimurium*, but present in other microorganisms. Enzymes are identified by their gene symbols (*see* Appendix) or by numbers: (1), AMP nucleosidase; (2), AMP deaminase; (3), adenine deaminase. SAM, S-adenosylmethionine; A, adenine; Hx, hypoxanthine; AR, adenosine; HxR, inosine.

A. Biosynthesis of AMP from IMP

IMP is converted to AMP in two steps, with succinyl-AMP as an intermediate, catalyzed by adenylosuccinate synthetase and adenylosuccinate lyase, respectively (Fig. 2.2). The latter enzyme is bifunctional: it is also required in the *de novo* synthesis of IMP (p. 34). AMP synthesis may be regulated *in vivo* by feedback inhibition of adenylosuccinate synthetase, since this enzyme is inhibited *in vitro* by AMP and guanine nucleotides including ppGpp.

The structural genes which code for adenylosuccinate synthetase and for adenylosuccinate lyase, *purA* and *purB*, respectively, have been located on the chromosomes of *E. coli* and *S. typhimurium* (*see* Appendix). Phenotypically *purA* and *purB* mutants show an absolute and specific growth requirement for adenine. Little is known about the regulation of the expression of the *purA* and *purB* genes in *E. coli* and *S. typhimurium*. When wild type cells of *S. typhimurium* are grown in the presence of adenine, a two-fold repression of the synthesis of the *purB* encoded enzyme is observed (15).

In *B. subtilis* it has been shown that adenosine represses the synthesis of adenylosuccinate synthetase, and that adenosine, guanosine, and to a lesser extent inosine repress the synthesis of adenylosuccinate lyase (199). Studies with *Aerobacter aerogenes* revealed that adenine and guanine repress the synthesis of both adenylosuccinate synthetase and adenylosuccinate lyase (196). Purine starvation of a purine auxotroph *A. aerogenes* resulted in derepressed synthesis of both enzymes (196).

1. Adenylosuccinate synthetase (EC 6.3.4.4)

Adenylosuccinate synthetase catalyzes the following reaction:

$$\text{IMP} + \text{L-aspartate} + \text{GTP} \rightarrow \text{adenylosuccinate} + \text{GDP} + \text{phosphate}$$

It has been purified from *E. coli* (53, 229), *Azotobacter vinelandii* (165), *B. subtilis* (108), and *Schizosaccharomyces pombe* (190). For the *Azotobacter* enzyme the molecular weight was estimated to 110 000 by gel filtration and to 54 000 by SDS gel electrophoresis (165). The enzyme is inhibited by AMP and GDP. Inhibition by GDP is competitive with respect to GTP and noncompetitive with IMP and L-aspartate (53, 108, 190, 286).

Studies with the *E. coli* enzyme has shown that ppGpp is also a potent inhibitor (257). The inhibitor constants for AMP, GDP, ppGp and ppGpp were found to be 100 μM (286), 12 μM, 20 μM, and 50 μM respectively (211, 257). Since these values all are within the intracellular concentration ranges it may indicate that ppGpp, GDP, and AMP all contribute to the regulation of AMP synthesis *in vivo*.

2. Adenylosuccinate lyase (EC 4.3.2.2)

Adenylosuccinate lyase is a bifunctional enzyme that catalyzes two non-sequential reactions in the *de novo* synthesis of AMP (Figs. 2.1. and 2.2) i.e.

$$\text{SAICAR} \rightarrow \text{AICAR} + \text{Fumarate}$$
$$\text{Succinyl-AMP} \rightarrow \text{AMP} + \text{Fumarate}$$

The enzyme has been purified from yeast (28) and from *Neurospora crassa* (282). No cofactor requirement was observed. The *Neurospora* enzyme has a molecular weight of 200 000 and is composed of six to eight apparently identical subunits.

B. Biosynthesis of GMP from IMP

The biosynthesis of GMP from IMP, via xanthosine 5′-phosphate (XMP), is catalyzed by IMP dehydrogenase and XMP amidotransferase. GMP synthesis may be regulated by feedback inhibition of IMP dehydrogenase, since the enzyme is inhibited by GMP and ppGpp. In *E. coli* and *S. typhimurium* the structural genes coding for the two enzymes, *guaB* and *guaA*, respectively, constitute an operon (198, 235). In *E. coli* the gene order is promoter-*guaB*-*guaA* (142, 279). Phenotypically *guaA* and *guaB* mutants show an absolute growth requirement for guanine. A *guaB* mutant can satisfy its guanine requirement by xanthine, whereas a *guaA* mutant cannot.

The synthesis of the two enzymes is controlled by a guanine compound. When *guaA* or *guaB* strains are starved for guanine, derepression of the synthesis of IMP dehydrogenase or XMP amidotransferase, respectively, is observed (171, 197). When wildtype cultures of *E. coli* are supplemented with guanine, the synthesis of both enzymes is repressed (171, 197). Repression of the synthesis of IMP dehydrogenase by guanosine has been found in *B. subtilis* and *Neurospora* (76, 199).

Mehra and Drabble (171) have found that expression of the *gua*-operon is induced in *E. coli* following addition of adenine to the growth medium, even in the presence of repressing concentrations of guanine. This may indicate, that the *gua*-operon is under dual control: repression by a guanosine nucleotide and induction by an adenine nucleotide.

1. IMP dehydrogenase (EC 1.2.1.14)

$$\text{IMP} + \text{NAD}^+ + \text{H}_2\text{O} \rightarrow \text{XMP} + \text{NADH} + \text{H}^+$$

IMP dehydrogenase has been studied most extensively in *E. coli* (64, 139) and in *Aerobacter aerogenes* (90, 91, 92). The enzyme from both these sources consists of two subunits with molecular weights of 58 000 (64) and 38 000 (90), respectively. They require a monovalent cation such as K^+ for activity and are activated by sulfhydryl reagents.

The *E. coli* enzyme is subject to inhibition by GMP ($K_i = 80\ \mu M$) and ppGpp ($K_i = 30\ \mu M$) (59). The inhibition by GMP is competitive with respect to IMP and non competitive with respect to NAD^+ (141), this has also been observed with the enzyme from *A. aerogenes* (91). The enzyme reaction has been investigated kinetically and by isotope exchange experiments in *A. aerogenes*. The mechanism appears to be partially random in which IMP and K^+ can bind randomly to the free enzyme while NAD^+ does not react unless K^+ or both K^+ and IMP are present on the enzyme (92). GMP inhibition of IMP dehydrogenase from *S. typhimurium* (24), *B. subtilis* (287) and *Schizosaccharomyces pombe* has also been observed (221).

IMP dehydrogenase from *B. subtilis* appears to be membrane bound (285). Whether this is the case in other microorganisms has not been established.

2. *GMP Synthetase* (EC 6.3.4.1)

GMP synthetase catalyzes the ATP dependent amination of XMP with either glutamine or ammonia as the aminogroup donor (143, 212):

$$\text{XMP} + \text{ATP} + \text{glutamine} \xrightarrow{Mg^{2+}} \text{GMP} + \text{AMP} + \text{pyrophosphate} + \text{glutamate}$$
(XMP amidotransferase activity)

$$\text{XMP} + \text{ATP} + \text{NH}_3 \xrightarrow{Mg^{2+}} \text{GMP} + \text{AMP} + \text{pyrophosphate}$$
(XMP aminase activity)

The enzyme from *E. coli* is composed of two apparently identical subunits of molecular weight 63 000 (233). Each subunit can utilize glutamine and ammonia for GMP synthesis. Treatment of GMP synthetase with 6-diazo-5-oxo-L-norleucine (DON), which is known to bind covalently to the glutamine-binding site of amidotransferases (213, 292), inactivates both the glutamine and the ammonia dependent activity. In contrast iodoacetamide selectively inactivates the glutamine dependent activity suggesting distinct binding sites for ammonia and glutamine.

If ATP is replaced by adenosine and pyrophosphate the enzyme will act as a glutaminase, provided XMP and Mg^{2+} are also present. Under these conditions no GMP is formed (292). Since pyrophosphate is a competitive inhibitor of the complete reaction with respect to ATP, the pyrophosphate requirement of the glutaminase reaction may indicate, that the ATP-site of the enzyme must be occupied either by ATP or by adenosine plus pyrophosphate in order to catalyze cleavage of glutamine to glutamate.

The uncoupling effect of adenosine on the complete reaction is the basis for the strong inhibition by adenosine and by the naturally occurring adenosine analogs psicofuranine and decoyinine ($K_i < 1$ μM) (*see* Fig. 2.10) (255). GMP is a competitive inhibitor with respect to XMP, $K_i = 0.33$ mM (256).

C. Phosphorylation of AMP and GMP to ATP and GTP

AMP and GMP are converted to ADP and GDP by separate kinases. ATP is generated from ADP by the energy yielding reactions of oxidative phosphorylation and glycolysis, whereas GDP is converted to GTP by nucleoside diphosphokinase or by pyruvate kinase (p. 157).

Nucleoside monophosphate kinases are highly specific towards the nucleobase moiety of their substrates but they will catalyze the phosphorylation of both ribo- and deoxyribonucleoside monophosphates (96). In contrast, nucleoside diphosphate kinases have very broad substrate specificities, and will phosphorylate any nucleoside 5′-diphosphate using any nucleoside 5′-triphosphate as a phosphate donor (65, 224, 242).

The existence of kinases highly specific for AMP and for GMP implies that IMP and XMP cannot be converted to the corresponding diphosphates. This may be of great physiological importance, since any IDP and XDP formed would be phosphorylated to the corresponding triphosphates by nucleoside diphosphokinase, and hence be incorporated into nucleic acids with deleterious effects.

Temperature sensitive mutants with mutations in the structural genes for adenylate kinase (*amk* in *S. typhimurium* and *adk* in *E. coli*) and guanylate kinase (*gmk* in *S. typhimurium*) have been isolated and the mutations located on the chromosomes (B. Jochimsen, personal communication and 67) (*see* Appendix). At the nonpermissive temperatures such mutants cannot grow in any medium, presumably due to ATP shortage (67, 265).

Certain temperature sensitive mutations in the *adk* gene result in a temperature sensitive lipid synthesis and a thermolabile glycerol 3-phosphate acyltransferase (35). These results suggest that the adenylate kinase may be directly involved in phospholipid biosynthesis through formation of a complex with glycerol 3-phosphate acyltransferase (34, 35, 68).

Mutants with a defective nucleoside diphosphokinase have been isolated in *S. typhimurium* (66) and in *S. pombe* (45), but only few details are at present available concerning the properties of the *S. typhimurium* mutant. The mutant of *S. pombe* is temperature sensitive and produces a defective nucleoside diphosphokinase (45).

1. Adenylate kinase (EC 2.7.4.3)

$$\text{AMP} + \text{ATP} \rightleftharpoons 2\text{ADP}$$

Purified adenylate kinase from *E. coli* and yeast (107, 110) appears to be a single polypeptide chain with a molecular weight of 27 000. The equilibrium constant for the reaction in the direction of formation of ADP is 1.16 (104). The enzyme requires Mg^{2+} and is highly specific for AMP and dAMP, but shows somewhat less specificity for the base of the nucleoside triphosphate substrate. The highest activity is found with ATP (110).

2. *Guanylate kinase* (EC 2.7.4.8)

$$\mathrm{GMP + ATP \rightleftharpoons GDP + ADP}$$

Guanylate kinase has been purified from *E. coli*, and a molecular weight of 88 000 has been found (209). Mg^{2+} or Mn^{2+} ions are required for activity. The enzyme phosphorylates GMP and dGMP specifically. ATP and dATP are the preferential phosphate donors.

D. Conversion of adenine nucleotides to IMP

Adenine nucleotides may be converted to IMP in most, if not all microorganisms. This conversion is of physiological importance because it enables the cells to synthesize guanine nucleotides from adenine nucleotides under conditions where the supply of guanine nucleotides may be limiting. There is also evidence that removal of AMP from the intracellular pool under certain conditions is of importance in stabilizing the energy charge of the cells (29).

Several pathways for the synthesis of IMP from endogenous adenine nucleotides exist. However, their distribution in different microorganisms differs significantly.

1. The histidine pathway

The first enzyme of the histidine biosynthetic pathway, ATP phosphoribosyltransferase (EC 2.4.2.17), catalyzes the formation of phosphoribosyl-ATP from ATP and PRPP. Subsequently phosphoribosyl-ATP is converted in 10 steps to histidine. ATP contributes one carbon and one nitrogen atom to the imidazole ring of the histidine molecule (19, 160). As a byproduct of the pathway phosphoribosylaminoimidazolecarboxamide (AICAR) is formed. This compound is an intermediate in the purine *de novo* pathway (Fig. 2.3). Thus, under conditions where a microorganism must synthesize histidine *de novo*, a proportional amount of IMP is formed from ATP through this pathway.

In enteric bacteria the first enzyme of the histidine biosynthetic pathway, i.e. ATP phosphoribosyltransferase, is feedback inhibited by histidine (10, 167), and the synthesis of all the histidine biosynthetic enzymes, coded for by the *his*-operon, is repressed by histidine. The synthesis of IMP from ATP through this pathway will therefore be determined mainly by the intracellular concentration of histidine, rather than by purine nucleotide pools. As a consequence the presence of histidine in the growth medium prevents the formation of IMP from ATP through the histidine pathway (19, 160).

Studies on purified ATP phosphoribosyltransferase from *S. typhimurium* have indicated that in the presence of partially inhibitory concentrations of histidine, ppGpp is a potent inhibitor of the enzyme (182). The physiological significance of this observation is unclear, although it may suggest that the formation of IMP from ATP through this pathway is under stringent control (*see* p. 32).

2. Pathways involving adenine or adenosine

Considerable amounts of adenine nucleotides are continuously broken down to adenine and/or adenosine during normal growth. Through the action of the salvage pathway enzymes the adenine and adenosine thus formed may be reutilized for IMP synthesis. As shown in Fig. 2.3 this may occur differently in different microorganisms. In *E. coli* and *S. typhimurium* adenine may be converted to hypoxanthine through the intermediate formation of adenosine and inosine catalyzed by purine nucleoside phosphorylase and adenosine deaminase (*see* Fig. 2.5). In microorganisms containing adenine deaminase, adenine can be deaminated directly to hypoxanthine (*see* Fig. 2.3). In all organisms hypoxanthine is subsequently converted to IMP through the action of hypoxanthine phosphoribosyltransferase.

Hydrolysis of AMP to adenosine. A number of 5′nucleotidases and non-specific phosphatases which will cleave AMP to adenosine have been identified in microorganisms (47, 48). Since the majority of these enzymes are located either extracellularly or in the periplasmic space, they are probably not involved in the degradation of intracellular AMP, but rather in the utilization of exogenous nucleotides.

Hydrolysis of AMP to adenine. A highly specific cytoplasmic AMP nucleosidase (AMP glycosylase), (EC 3.2.2.4), which hydrolyzes AMP to adenine and ribose-5-phosphate has been found in a number of microorganisms. It has been studied most extensively in *Azotobacter vinelandii* (44, 146, 186) and in *E. coli* (147). The *A. vinelandii* enzyme requires MgATP as a positive effector, and phosphate inhibits the enzyme in competition with MgATP. In addition phosphate causes inactivation of the enzyme, and promotes a change in the molecular weight of the enzyme from 320 000 to 110 000 (238).

The *E. coli* enzyme is an oligomer with a molecular weight of 280 000 and a subunit molecular weight of 52 000. In the presence of the activator, MgATP, or the inhibitor, inorganic phosphate, the enzyme shows positive cooperativity with respect to the substrate AMP. Kinetic investigations of MgATP activation give a constant for half maximum activation varying from 120 μM in the presence of low AMP concentration to 2 μM when AMP is present at near saturating concentrations (147).

S-adenosylmethionine turnover. In all reactions where S-adenosylmethionine is a substrate, whether it functions as methyldonor in transmethylations or, following decarboxylation, it acts as aminopropyl donor in spermine and spermidine biosynthesis; the purine moiety is eventually liberated either as adenine or as adenosine (*see further* p. 71). In *E. coli* S-adenosylhomocysteine, the demethylated product of S-adenosylmethionine, is cleaved to yield adenine, whereas the cleavage in yeast and fungi results in the formation of adenosine (Fig. 2.3).

3. Deamination of AMP to IMP

Certain microorganisms contain an AMP deaminase (EC 3.5.4.6) which catalyzes the conversion of AMP to IMP. These microorganisms are devoid of AMP nucleosidase activity (146).

The AMP deaminase has been purified from yeast and shown to consist of four subunits with a molecular weight of 83 000 each (185). ATP is a positive effector and inorganic phosphate inhibits the enzyme allosterically. The enzyme requires monovalent cations for optimal activity. AMP is the preferred substrate, dAMP is hydrolyzed at 7% of the rate observed for AMP (290). Kinetic analysis has shown that the enzyme contains two binding sites for AMP and two for inorganic phosphate, suggesting that the tetrameric enzyme consists of two functional units (185).

AMP deaminase activity is absent from *E. coli* and *S. typhimurium*. Enzymes which catalyze the deamination of ATP, ADP, AMP, NAD and adenosine have been purified from *Desulfovibrio desulfuricans* (294) and *Microsporum audonini* (3). The metabolic significance of these enzymes is at present unknown.

Some microorganisms appear to contain only one pathway, i.e. the histidine pathway, for converting adenine nucleotides to IMP. Thus in the presence of exogenous histidine such microorganisms are unable to synthesize guanine nucleotides from adenine nucleotides. In *Neisseria* this appears to be due to lack of AMP deaminase and inability to deaminate adenine and adenosine (120).

E. Conversion of guanine nucleotides to IMP

Guanine nucleotides may be converted to IMP and AMP under conditions of unbalance between the guanine nucleotide and adenine nucleotide pools. The only known enzyme which catalyzes a conversion of guanine nucleotides (i.e. GMP) to IMP is GMP reductase. *E. coli*, *S. typhimurium* and *B. subtilis* mutants lacking GMP reductase activity, are unable to utilize guanine as the sole purine source (116, 169).

The levels of GMP reductase in *E. coli* and *S. typhimurium* increase with increasing concentrations of guanine or guanosine in the growth medium (12, 197), but the nature of the actual inducer has not been established. In *E. coli*, but not in *S. typhimurium*, cyclic AMP and the cyclic AMP binding protein are also required for the induction of GMP reductase synthesis (11, 12). In *S. typhimurium* glutamine starvation leads to increased levels of GMP reductase (61), suggesting a function of glutamine in the regulation of the *guaC* gene expression. This latter effect may be related to the fact that one of the products of the GMP reductase reaction is ammonia.

1. *GMP reductase* (EC 1.6.6.8)

GMP reductase catalyzes the reductive deamination of GMP:

$$GMP + NADPH + 2H^{+} \rightarrow IMP + NADP^{+} + NH_4{}^{+}$$

The enzymes from *S. typhimurium* (115, 161), *E. coli* (161), and *Aerobacter aerogenes* (20) have been studied. The purified *S. typhimurium* enzyme is specific for GMP and NADPH, and has an absolute requirement for a sulfhydryl compound such as glutathione (161). It consists of four apparently identical subunits with molecular weight of about 45 000 (115). The enzyme is inhibited by ATP and, much less, by AMP and IMP (115, 161). This inhibition may be completely reversed by GTP (Fig. 2.4). Thus it appears that conditions which lead to a lowering of the adenine nucleotide pool and an increase in the guanine nucleotide pool will increase the activity of GMP reductase.

Evidence has been obtained that GMP reductase of *S. typhimurium* is inactivated during purine starvation (115). The nature of this inactivation has not been elucidated.

F. Purine deoxyribonucleotide synthesis *de novo*

The endogenous synthesis of deoxynucleotides is catalyzed by ribonucleotide reductases, which are capable of reducing ribonucleotides to the corresponding 2′-deoxyribonucleotides. These enzymes are generally non-specific towards the nucleobase moiety, but highly specific towards the number of phosphates in the ribonucleotide substrate. The activity of the enzymes is regulated by both ribo- and deoxyribonucleoside triphosphate effectors (264).

At least two different types of ribonucleotide reductases have so far been found in microorganisms: (a) ribonucleoside *di*phosphate reductases, of which the *E. coli* enzyme has been most thoroughly studied: and (b) ribonucleoside *tri*phosphate reductases represented by the well characterized

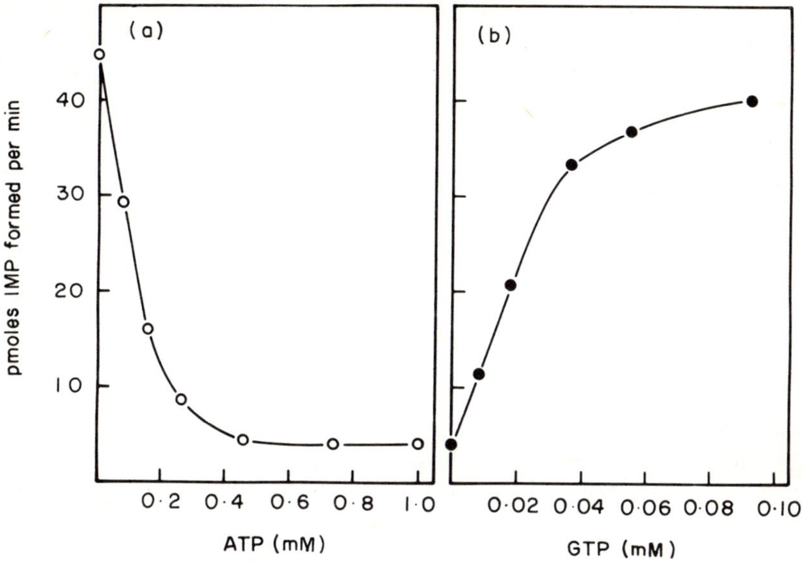

FIG. 2.4. Effects of ATP and GTP on the activity of partially purified GMP reductase of *S. typhimurium*. The GMP concentration is 2 mM throughout. (a) effect of increasing concentration of ATP in the absence of GTP; (b) effect of increasing concentrations of GTP in the presence of 1 mM ATP (115).

enzyme from *Lactobacillus leichmannii*. Besides differing in their ribonucleotide substrates (i.e. di- or triphosphates, respectively), the two types of enzymes differ in their cofactor requirements. The ribonucleoside diphosphate reductases contain firmly bound non-heme iron which is required for activity, whereas the ribonucleoside triphosphate reductases show an absolute requirement for coenzyme-B_{12}. Further details on the properties of these enzymes, as well as their distribution throughout the microbial world may be found in Chapter 4.

IV. SALVAGE PATHWAYS IN *E. COLI* AND *S. TYPHIMURIUM*

Purine bases and nucleosides are not intermediates in the *de novo* synthesis of purine nucleotides. Nevertheless they are readily incorporated into nucleic acids by enteric bacteria when supplied in the medium (295). Purine auxotrophic mutants blocked in the *de novo* synthesis of IMP are capable of satisfying their entire requirement for both adenine nucleotides and guanine nucleotides by anyone of the following compounds: adenine, adenosine, deoxyadenosine, guanine, guanosine, deoxyguanosine, hypoxanthine, inosine, and xanthine, although the efficiency by which the individual compounds support growth varies somewhat (Table 2.1).

Three main approaches have been used to study the utilization of exogenous purines: (1) incorporation of different radiolabeled purines and purine nucleosides into nucleic acid, and subsequent determination of the distribution of radioactivity in the adenine and guanine moieties of RNA and DNA; (2) isolation and characterization of the enzymes involved in purine salvage; and (3) characterization of mutants defective in each of these enzymes. Fig. 2.5 shows the salvage pathways operating in *E. coli* and *S. typhimurium*.

E. coli and *S. typhimurium* can use purine ribo- and deoxyribonucleosides as total carbon and energy sources. This capacity is dependent on efficient phosphorolytic cleavage of the N-glycosylic bond of the nucleosides yielding the free nucleobase and ribose 1-phosphate or deoxyribose 1-phosphate. Since nucleoside catabolism is treated separately in this book (Chapter 5), phosphorolysis of nucleosides will only be discussed here in relation to the utilization of purine nucleosides for nucleotide synthesis. The amino group of adenosine and of adenine after its conversion to adenosine, may serve as the sole nitrogen source (207).

TABLE 2.1

Growth rates of purine auxotrophic mutants of *E. coli*, *S. typhimurium* and *B. subtilis* on various purine compounds[a]

Compound added	*E. coli* *purE*	*S. typhimurium* *purE*	*B. subtilis* *pur*[b]
Adenine	43	66	42
Hypoxanthine	42	50	44
Guanine	76	60	63
Xanthine	102	76	158
Adenosine	42	64	40
Deoxyadenosine	46	53	47
Inosine	42	60	44
Deoxyinosine	42	54	46
Guanosine	68	68	79
Deoxyguanosine	68	59	80
Xanthosine	>400	>400	>400
2′ and 3′-AMP	220	94	272
5′-AMP	52	184	132
3′:5′ cyclic AMP	>400	>400	>400
5′-IMP	68	146	254

[a] Growth rates were determined in glucose minimal medium and given as doubling times in min. Purine concentrations were 0.1 mM.

[b] Blocked in the conversion of AIR to SAICAR, *see* Fig. 2.1.

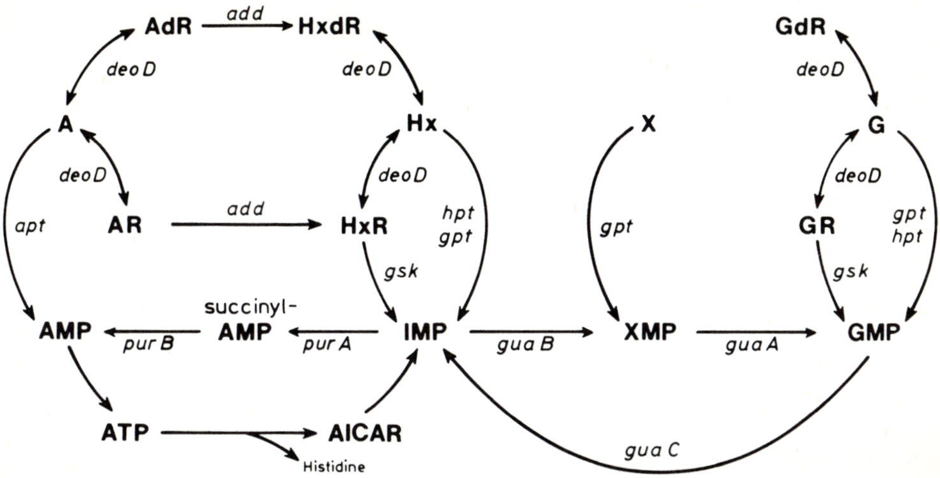

FIG. 2.5. Purine salvage and interconversion pathways of *E. coli* and *S. typhimurium*. The individual enzymes of the pathways are identified by their gene symbols (*see* Appendix). Nucleobases are identified as follows: A, adenine; G, guanine; Hx, hypoxanthine and X, xanthine. The corresponding ribonucleosides and deoxyribonucleosides are identified by R and dR respectively e.g. AR, adenosine and AdR, deoxyadenosine.

The utilization of extracellular purine bases and nucleosides requires translocation of the compounds through the cellular membranes. The mechanisms underlying these processes are the subject of Chapter 6, and will not be discussed here.

A. Adenine

Adenine is converted directly to AMP through phosphoribosylation with PRPP, catalyzed by adenine phosphoribosyltransferase (Fig. 2.5). Through the operation of the histidine biosynthetic pathway adenine nucleotides, i.e. ATP, may subsequently be converted to AICAR and hence to IMP and GMP (*see* p. 37). This pathway is blocked when histidine is present in the growth medium (160). It has been found, however, that adenine will satisfy the entire purine requirement of purine auxotrophs blocked in the *de novo* synthesis of IMP, i.e. a *purE* mutant, even when histidine is present in the growth medium, and that it will support growth of a purine auxotroph defective in adenine phosphoribosyltransferase activity (*purE, apt*) (Table 2.2) (101, 132, 207). These observations show the existence of an alternative pathway for converting adenine to IMP, which does not require histidine biosynthesis. The alternative pathway involves an intermediate formation of adenosine from adenine and ribose 1-phosphate, followed by deamination of adenosine to

TABLE 2.2
The role of purine nucleoside phosphorylase (*deoD*) and adenine phosphoribosyltransferase (*apt*) in adenine utilization.[a]

	Relevant genotype	Growth rates on		
		Adenine	Adenine + histidine	Hypoxanthine
E. coli	*purE*	51	54	48
	purE,deoD	116	>400	49
	purE,apt	74	88	50
	purE,deoD,apt	>400	>400	47
S. typhimurium	*purE*	70	102	54
	purE,deoD	>400	>400	54

[a] Growth rates were determined in glucose minimal medium and given as doubling times in min. Purines were added to 0.1 mM.

inosine. This was concluded from the observations that *purE* mutants with additional mutations inactivating either purine nucleoside phosphorylase (*deoD*) or adenosine deaminase (*add*) are unable to grow on adenine plus histidine (Table 2.2). At the same time these results excluded the presence of adenine and AMP deaminase activities.

From Table 2.2 it is apparent that the purine nucleoside phosphorylase-adenosine deaminase pathway operates more efficiently in *E. coli* than in *S. typhimurium* (better growth of a *purE* mutant on adenine plus histidine). In accordance with this *S. typhimurium apt* mutants cannot reutilize endogenously formed adenine and they excrete adenine into the growth medium (121, 123). An *E. coli apt* mutant excretes adenine only if the purine nucleoside phosphorylase-adenosine deaminase pathway is blocked by an additional mutation in the *deoD* or the *add* gene (133). The rate limiting step in *S. typhimurium* has been shown to be the ribosylation of adenine by purine nucleoside phosphorylase (101). The fact that adenine induces the synthesis of adenosine deaminase in *E. coli* but not in *S. typhimurium* (207, 227) may also contribute to the higher efficiency of this pathway in *E. coli*.

Another striking difference between *E. coli* and *S. typhimurium* is the observation (Table 2.2) that a *purE deoD* mutant of *S. typhimurium* cannot grow on adenine as purine source. This was shown to be due to insufficient conversion of ATP to IMP through the histidine pathway (101).

The purine analogs 2,6-diaminopurine and 6-methylaminopurine are metabolized by the same enzymes as adenine. By the action of adenine phosphoribosyltransferase they are converted to toxic nucleotides. Mutants defective in adenine phosphoribosyltransferase, i.e. *apt* mutants, are resistant to high concentrations (2 mM) of these analogs (122, 131).

Both analogs may also be converted by purine nucleoside phosphorylase to the corresponding ribonucleosides, which in turn are deaminated by adenosine deaminase to guanosine and inosine, respectively, and hence serve as purine sources (60, 125, 131, 152). When used in low, non toxic concentrations (0.1 mM), 2,6-diaminopurine and 6-methylaminopurine may serve as purine sources for purine auxotrophs, but the growth is poor. However, fast growing mutants are easily selected. In *S. typhimurium* such a selection yielded mutants with constitutive synthesis of purine nucleoside phosphorylase (*deoR* mutants) 160), whereas mutants with increased levels of adenosine deaminase were obtained in *E. coli* (207). In both cases the mutations result in an increased capacity to convert the analogs to guanosine and inosine, respectively.

B. Adenosine

Mutants deficient in the conversion of IMP to AMP, i.e. *purA* or *purB* mutants, have a growth requirement for adenine, which may be satisfied by adenosine. However, the growth yield on adenosine is only 10% of that observed on equimolar amounts of adenine. Since the growth yield on adenosine is identical to that on adenine in a *purA* mutant, deficient in adenosine deaminase (*add*), it may be concluded that the predominant route for adenosine metabolism is via deamination to inosine (101, 207). Inosine cannot satisfy the adenine requirement of a *purA* mutant, whereas it can satisfy the entire purine requirement of a *purE* mutant via its conversion to IMP. The beneficial effect of an *add* mutation on the growth yield of *purA* and *purB* mutants on adenosine as adenine source has been used for selection of *add* mutants (101, 116). Adenosine is rapidly deaminated in purine prototrophic strains as well, as illustrated in Fig. 2.6, which shows the degradation of adenosine in three different prototrophic strains. When adenosine deaminase is present (Fig. 2.6(a) and (b)) the majority of the added adenosine is deaminated. In the adenosine deaminase negative strain (Fig. 2.6(c)) adenosine is degraded equally fast, but in this case by direct phosphorolysis to adenine.

A triple mutant of the genotype *purA,add,deoD*, deficient in both adenosine deaminase and purine nucleoside phosphorylase has been constructed in *S. typhimurium*. This mutant cannot utilize adenosine, indicating that adenosine must be cleaved phosphorolytically to adenine before it can be converted to AMP (101). This further implies that *S. typhimurium* does not contain adenosine kinase activity. An *E. coli purB,add,deoD* mutant can grow slowly on adenosine (116, 133). This is probably due to hydrolytic cleavage of

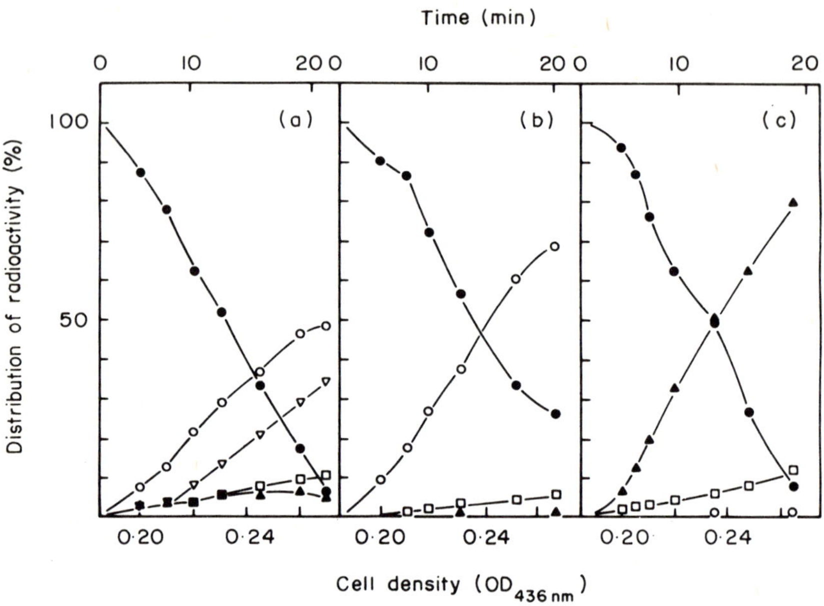

FIG. 2.6. Metabolism of adenosine in *E. coli.* [8-^{14}C] adenosine (50 nmole/ml) was added at time zero to cultures of: (a) wild type *E. coli*; (b) a purine nucleoside phosphorylase deficient *E. coli* mutant; and (c) an adenosine deaminase deficient *E. coli* mutant. At different times following addition of the labelled adenosine, samples of the cultures were analyzed for the distribution of label in the medium: ●——●, adenosine; ○——○, inosine; ▽——▽, hypoxanthine; ▲——▲, adenine. Label incorporated into whole cells was also determined, □——□.

adenosine to adenine, catalyzed by 5-methylthioadenosine nucleosidase, an enzyme which normally functions in S-adenosylmethionine metabolism (*see* p. 70).

An *E. coli purA,add,deoD* strain with an additional mutation resulting in a heat labile adenine phosphoribosyltransferase cannot grow on adenine or adenosine at the non permissive temperature. This shows that *E. coli* is devoid of adenosine kinase activity (152). In agreement with this, it has been found that adenosine cannot be converted to AMP in cell free extracts of an *E. coli* (*add,deoD,apt*) strain (unpublished).

C. Hypoxanthine

Hypoxanthine may be metabolized by two different pathways. Either it may be converted directly to IMP, catalyzed by a phosphoribosyltransferase, or it may be ribosylated to inosine, catalyzed by purine nucleoside phosphorylase (Fig. 2.5). *E. coli* and *S. typhimurium* contain two enzymes which both catalyze the synthesis of IMP from hypoxanthine and PRPP, i.e. hypoxanthine phosphoribosyltransferase and guanine phosphoribosyltransferase, encoded by the unlinked genes *hpt* and *gpt*, respectively (30, 73, 102, 116). This is clearly illustrated in Table 2.3, which shows that *purD,deoD,hpt* and *purD,deoD,gpt* mutants can grow on hypoxanthine as sole purine source, whereas a *purD,deoD,hpt,gpt* mutant cannot. Moreover, it appears from the growth rate of these mutants (Table 2.3) that hypoxanthine phosphoribosyltransferase contributes more to the conversion of hypoxanthine to IMP than guanine phosphoribosyltransferase.

TABLE 2.3
Growth rates of phosphoribosyltransferase negative, purine auxotrophic strains of *E. coli* on purine bases.[a]

	Growth rates on		
Relevant genotype	Hypoxanthine	Guanine	Xanthine
purD,deoD	55	60	77
purD,deoD,hpt	89	60	76
purD,deoD,gpt	54	265	>400
purD,deoD,hpt,gpt[b]	>400	>400	>400
purD,hpt,gpt[c]	>400	>400	>400

[a] Growth rates were determined in glucose minimal medium and given as doubling times in min. Purine concentrations were 0.1 mM.
[b] This strain will grow on adenine plus guanosine with a doubling time of 54 min.
[c] This strain will grow on adenine with a doubling time of 55 min.

Introduction of a *hpt* mutation leads to excretion of hypoxanthine even in a wild type strain. This shows that hypoxanthine is continuously formed in the cells and reutilized by the *hpt* coded enzyme. No purine excretion is observed in *gpt* mutants (unpublished).

The poor growth on hypoxanthine of a purine requiring *hpt*, *gpt* mutant of *E. coli* (Table 2.3) indicates that the conversion of hypoxanthine to IMP through the intermediate formation of inosine is of minor importance in purine salvage (21).

The purine analog 6-mercaptopurine is toxic to *E. coli* and *S. typhimurium* when added in fairly high concentrations (2 mM) to the growth medium. Mutants resistant to this analog are easily obtained, and most of them appear to be *hpt* mutants (30, 116, 168). These mutants still contain the *gpt* encoded enzyme, but apparently 6-mercaptopurine is a very poor substrate for this enzyme.

D. Inosine

The metabolism of the purine and ribose moieties of inosine has been studied in exponentially growing cultures of wild type *E. coli*, by following the incorporation of [U-^{14}C]- and [8-^{14}C]-inosine into the individual ribonucleoside triphosphates. The results showed that the purine moiety of inosine is rapidly incorporated into purine ribonucleotides. The ribose moiety of inosine is incorporated to the same extent into purine and pyrimidine ribonucleotides (111), suggesting that in wild type strains inosine is cleaved before it is utilized for purine nucleotide synthesis.

A *purE,deoD* mutant, in which inosine catabolism is prevented, will grow slowly on inosine as the sole purine source (Table 2.4) (101, 207). This indicates that *E. coli* and *S. typhimurium* contain inosine kinase activity, and that this activity is rate limiting for growth on inosine of such mutants. From the incorporation experiments it was concluded that inosine kinase activity does not contribute to inosine utilization in strains containing a functional purine nucleoside phosphorylase (111). Growth of the *purE,deoD* mutants on inosine is stimulated by simultaneous addition of adenine and histidine (Table 2.4). Under these conditions inosine is only required to satisfy the cellular requirement for guanine nucleotides.

Introduction of a mutation inactivating guanosine kinase (*gsk*) into a *purE,deoD* strain results in complete loss of the capacity to grow on inosine even in the presence of adenine and histidine (Table 2.4), indicating that one single protein is responsible for both inosine and guanosine kinase activity.

E. Guanine

Mutants deficient in either IMP dehydrogenase (*guaB* mutants) or GMP synthetase (*guaA* mutants) require guanine for growth. Guanine will also

TABLE 2.4

Growth rates of purine nucleoside phosphorylase negative, purine auxotrophic strains of *E. coli* and *S. typhimurium* on inosine.[a]

		Growth rates on		
	Relevant genotype	Inosine	Inosine + Adenine + Histidine	Hypoxanthine
S. typhimurium	*purE,deoD*	146	68	54
E. coli	*purE,deoD*	240	52	46
	purE,deoD,gsk	>400	>400	48

[a] Growth rates were determined in glucose minimal medium and given as doubling times in min. Purine concentrations were 0.1 mM.

support growth of a purine auxotroph blocked in the synthesis of IMP, although it is less efficient than hypoxanthine (Table 2.1).

Guanine may be metabolized by two different routes. Either it can be converted directly to GMP by phosphoribosyltransferases, or it may be ribosylated to guanosine by purine nucleoside phosphorylase.

In *E. coli* phosphoribosylation of guanine is catalyzed by both hypoxanthine phosphoribosyltransferase and by guanine phosphoribosyltransferase, although the latter enzyme seems to be more effective (Table 2.3). The contribution to guanine utilization by the second route, i.e. via conversion to guanosine followed by phosphorylation of guanosine to GMP, appears to be less significant, since a *purD,hpt,gpt* mutant, which contains a functional purine nucleoside phosphorylase, grows poorly on guanine (Table 2.3).

A somewhat different situation is found in *S. typhimurium*, where a *purE,gpt* mutant cannot grow on guanine (13). Apparently neither hypoxanthine phosphoribosyltransferase nor purine nucleoside phosphorylase is sufficiently active with guanine as substrate to ensure an adequate supply of purine nucleotides for growth. In a *S. typhimurium purE,gpt* strain, mutants capable of utilizing guanine for growth have been obtained (13). One of these were shown to contain a mutation in the *hpt* gene, presumably specifying a hypoxanthine phosphoribosyltransferase with altered specificity towards guanine.

The guanine analog 8-azaguanine is toxic to *S. typhimurium* when present in high concentrations (1 mM). Mutants resistant to 1 mM 8-azaguanine have been isolated and among them *gpt* mutants are very frequent (13, 73). A conditional mutant with a temperature sensitive GMP kinase has also been obtained by selecting for 8-azaguanine resistance (115). This shows that the analog is toxic only after conversion to 8-azaGDP or 8-azaGTP.

F. Guanosine

Incorporation of [U-^{14}C]- and [8-^{14}C]-guanosine into the individual nucleoside triphosphates has been studied in exponentially growing wild type cells of *E. coli* (111). The results differed from those obtained with inosine (p. 48) in that about 30% of the added guanosine is incorporated into GTP without prior cleavage to guanine.

Mutants defective in purine nucleoside phosphorylase cannot degrade guanosine to guanine. Cells of *E. coli* and *S. typhimurium* with *guaB* and *deoD* mutations grow with wild type rates with either guanosine or guanine as a source of GMP. These results show that these organisms contain guanosine kinase activity, and that this activity is not rate limiting for growth of the mutants on guanosine (101). Similarly a *purE,deoD* mutant will grow with normal growth rates on guanosine, provided the adenine requirement of the strain is satisfied by adenine and histidine. Mutants lacking guanosine kinase activity, i.e. *purE,deoD,gsk* mutants, cannot grow on guanosine + adenine + histidine, but growth is restored if guanosine is replaced by guanine (116).

It is therefore surprising, that a *purE,deoD* mutant cannot grow at all with guanosine as the sole source of purine, i.e. guanosine cannot supply the adenine requirement of such a strain (Table 2.5). In contrast inosine will serve as purine source (Table 2.4), even though the inosine kinase activity is only 5–10% of the guanosine kinase activity in these strains (116).

TABLE 2.5

Growth rates of *E. coli* purine auxotrophs with a defective purine nucleoside phosphorylase (*deoD*) on various purine sources.[a]

	Growth rates on		
Relevant genotype	Guanosine	Guanine	Hypoxanthine
purE	79	137	48
purE,deoD	>400	236	47
*purE,deoD,gsk**[b]	111	196	44
purE,deoD,gsk,purI*	78	48	47
purE,deoD,gsk,purF*	56	52	43
purE,deoD,gsk,purG*	86	60	44
purE,deoD,gsk,prs*	58	>400	87

[a] Growth rates were determined in glucose minimal medium and given as doubling times in min. Purine concentrations were 0.1 mM.

[b] *gsk** indicates a mutation in the *gsk* gene which results in the synthesis of a guanosine kinase with increased heat stability and lowered K_m for guanosine.

In order to study this phenomenon further, *purE,deoD* strains of both *E. coli* and *S. typhimurium* were used to select for mutants capable of using guanosine as the sole purine source (72, 107). Among the mutants isolated were some which showed increased activity of GMP reductase. Most mutants obtained appeared to have acquired two mutations: One which maps in or very close to the structural gene for guanosine kinase (*gsk*), and which results in the synthesis of an altered guanosine kinase (*gsk**) with lowered K_m for guanosine and increased heat-stability of the enzyme. The second mutation causes a block in purine *de novo* biosynthesis, prior to *purE*, i.e. either in *purF*, *purI*, *purG*, or in the structural gene for PRPP synthetase (*prs*). The *purF*, *purI*, or *purG* mutations alone did not improve growth of a *purE,deoD* strain on guanosine significantly, whereas a *gsk** mutation alone has an appreciable effect. Together, however, they ensure near normal growth rates (Table 2.5, and 61, 115).

It seems conceivable, that mutations which improve the guanosine kinase or the GMP reductase activities of the cells will improve growth on guanosine of the *purE,deoD* strains. It remains to be explained, how mutations in the early steps of the purine *de novo* pathway affect guanosine utilization. (In the present context PRPP synthetase may be regarded as the first enzyme in purine *de novo* biosynthesis.) One possibility is, that an intermediate of the *de novo* pathway accumulates in a *purE* strain and interferes with guanosine utilization, and that accumulation of this intermediate is prevented by additional, earlier blocks in the pathway. In keeping with this it has been found, that accumulation of intermediates of purine *de novo* biosynthesis in *S. typhimurium* causes a decrease in the GMP reductase activity of the cells (115). If indeed mutations in the early steps of purine *de novo* biosynthesis affect guanosine utilization by promoting GMP reductase activity of *purE* strains, one would likewise expect growth on guanine to be stimulated in such mutants. As shown in Table 2.5, this is indeed observed, whereas the *gsk** mutation has no such effect.

Lack of growth on guanosine as a purine source has also been observed in *purD,deoD* strains of *E. coli* (154). Selection of mutants growing on guanosine has so far yielded two different types: One mutation was shown to map between 10 and 15 min on the *E. coli* chromosome (*see* Appendix), and may therefore be similar to the *gsk** mutants. The other type was shown to contain a mutation in the *rpoB* gene encoding the β-subunit of RNA polymerase.

G. Xanthine and xanthosine

XMP is an intermediate in the endogenous synthesis of GMP from IMP (p. 34). Exogenous xanthine can satisfy the guanine requirement of a *guaB* mutant, but obviously not of a *guaA* mutant which is blocked in the conversion of XMP to GMP (197). Xanthine will also satisfy the total purine requirement of a purine auxotroph, although somewhat less efficiently than guanine (Table 2.1). As shown in Table 2.3 the utilization of xanthine is entirely dependent on a functional guanine phosphoribosyltransferase, whereas hypoxanthine phosphoribosyltransferase is not involved in the conversion of xanthine to XMP. This has been confirmed by *in vitro* studies of the specificities of the two phosphoribosyltransferases (111, 138).

Xanthosine, on the other hand, cannot satisfy a general purine requirement (Table 2.1), nor will it support growth of a *guaB* mutant. This indicates that xanthosine is not cleaved to xanthine, and that *E. coli* and *S. typhimurium* are devoid of xanthosine kinase activity. *E. coli* does, however, contain an inducible xanthosine phosphorylase, which requires millimolar concentrations of xanthosine for its induction. The properties of this enzyme, which is absent from *S. typhimurium*, are discussed in Chapter 5 (p. 207).

H. Purine deoxyribonucleosides

Purine auxotrophs of *E. coli* and *S. typhimurium* can utilize deoxyadenosine, deoxyinosine, or deoxyguanosine as sole sources of purines (Table 2.1). This capacity is lost by the introduction of a *deoD* mutation, which inactivates purine nucleoside phosphorylase. Since it has been established that the cells are devoid of purine deoxyribonucleoside kinase activities (126, 153), it follows that deoxyribonucleosides must be cleaved to the free bases in order to be utilized for nucleotide synthesis. Deoxyadenosine may, however, serve as a total nitrogen source in these organisms through deamination by adenosine deaminase (p. 237).

I. Purine nucleotides

Extracellular purine nucleotides are utilized by *E. coli* and *S. typhimurium* as purine sources (Table 2.1). Since nucleotides cannot penetrate the cytoplasmic membrane, they must be dephosphorylated before they can enter the cell (287). Studies on an *E. coli purA,ush* mutant which is defective in the periplasmic 5′-nucleotidase activity (*ush*), have shown that it cannot use 5′-AMP as an adenine source (33), indicating that the 5′-nucleotidase is required for the utilization of exogenous 5′-nucleoside monophosphates by *E. coli*. Wild type *S. typhimurium* is devoid of this 5′-nucleotidase (271). Nevertheless a *purE* mutant of *S. typhimurium* will grow on 5′-nucleoside monophosphates as purine sources (Table 2.1), albeit poorly. Obviously *S. typhimurium* contains one or more enzyme(s) capable of hydrolyzing 5′-nucleotides.

In both *E. coli* and *S. typhimurium* a number of periplasmic phosphatases have been identified, which will hydrolyze 5′-, 3′-, and 2′:3′-cyclic nucleotides to the corresponding nucleosides. Thus *E. coli*, besides the 5′-nucleotidase, contains a non-specific acid phosphatase (193), a 2′:3′-cyclic phosphodiesterase which also cleaves 3′-nucleotides (124), and an alkaline phosphatase (226). The latter enzyme is practically absent when cells are grown in media containing inorganic phosphate (226). *S. typhimurium* is devoid of alkaline phosphatase and 5′-nucleotidase (130, 280), but contain two different non-specific acid phosphatases (271) (cf. Table 3.3, p. 107).

K. Regulation of purine salvage

In *E. coli* and *S. typhimurium* the conversion of preformed purine bases to nucleotides occurs predominantly by phosphoribosylation catalyzed by phosphoribosyltransferases. Only very limited amounts of the purine bases are utilized through the intermediate formation of nucleosides.

The intracellular levels of the purine phosphoribosyltransferases do not

TABLE 2.6
Effects of amino acid starvation on the rate of uptake of purine nucleosides and bases in wild type and in *relA* strains of *E. coli* K12.

		Uptake[a] (nmoles/min/mg dry weight)		
Relevant genotype[b]	Valine present[c]	Adenine	Hypoxanthine[d]	Guanosine[e]
relA$^+$	−	3.7	3.2	1.3
	+	1.7	0.0	0.0
relA	−	1.7	2.4	0.5
	+	1.6	2.5	1.4

[a] [8-^{14}C]-labelled purine nucleosides and bases were added to exponentially growing cultures in glycerol minimal medium. With 5 sec intervals, samples were harvested and washed, and the total radioactivity accumulated within the cells determined.
[b] Identical results were obtained in *deoD* and *deoD*$^+$ strains.
[c] Valine (0.1 mM) was added 5 min before the labelled compound in order to induce isoleucine starvation (40).
[d] Similar results were obtained with guanine.
[e] Similar results were obtained with inosine.

change significantly following addition of purines to the growth medium, or during purine starvation of purine auxotrophic mutants (205). This seems to indicate that neither purine bases nor purine nucleotides control the synthesis of these enzymes. It has been reported (100) that growth of *E. coli* under conditions which inhibit *de novo* purine biosynthesis and render the cells dependent on exogenous purines, results in a 100-fold enrichment of adenine phosphoribosyltransferase. However, the final level of enzyme found under these conditions is 4–10 fold lower than found by others in crude extracts of wild type. *E. coli*, grown in the presence of in the absence of exogenous purines (116, 252).

The activities of hypoxanthine phosphoribosyltransferase and guanine phosphoribosyltransferase, but not of adenine phosphoribosyltransferase, are inhibited by ppGpp *in vitro* (99). That such a negative control mechanism is operating *in vivo* as well, is indicated by the observation (Table 2.6) that hypoxanthine and guanine are not utilized by a stringent (*relA*$^+$) strain of *E. coli* following amino acid starvation, whereas they are utilized normally by amino acid starved *relA* strains. Such a control system may be beneficial for the cells since it couples the utilization of exogenous purines to the net synthesis of RNA.

As a rule purine nucleosides are degraded before utilization for nucleotide synthesis. Adenosine and deoxyadenosine may be degraded phosphorolytically to adenine or deaminated to inosine and deoxyinosine, respectively, but more than 90% follows the latter pathway. Inosine and deoxyinosine are almost quantitatively cleaved to hypoxanthine by purine nucleoside phosphorylase, which is induced by purine nucleosides (Chapter 5, p. 218). Thus the conversion of adenosine, deoxyadenosine, inosine, and deoxyinosine to nucleotides proceeds through the intermediate formation of hypoxanthine. In *E. coli* the hypoxanthine formed furthermore induces the synthesis of adenosine deaminase (p. 220).

Guanosine is utilized somewhat differently for nucleotide synthesis. Although it is cleaved to guanine by purine nucleoside phosphorylase, 30% of it escapes phosphorolysis and is converted directly to GMP by guanosine kinase. As shown in Table 2.6, guanosine utilization seems to be negatively controlled by ppGpp in a way similar to that found for the utilization of hypoxanthine and guanine.

V. PURINE SALVAGE PATHWAYS IN OTHER MICROORGANISMS

A. Bacilli

Mutants of *B. subtilis* blocked in *de novo* synthesis of IMP will grow on either hypoxanthine, adenine, guanine, xanthine or on the corresponding nucleosides, although with variable growth rates (Table 2.1).

All the purine bases are converted to the corresponding ribonucleotides by phosphoribosyltransferases (18). A phosphoribosyltransferase specific for adenine has been characterized (17). The AMP formed in this way may be converted into IMP via the histidine pathway (174, 200). In addition *B. subtilis* contains adenine deaminase activity (unpublished), which enables the cells to convert adenine into IMP through hypoxanthine. It has been reported that *B. subtilis* contains AMP deaminase activity (200), but since this activity was demonstrated in whole cells it cannot be excluded that it was caused by the successive action of 5′-nucleotidase, adenosine phosphorylase and adenine deaminase.

The metabolism of purine nucleosides differs in several aspects from that found in *E. coli* and *S. typhimurium* (cf. Fig. 2.5). *B. subtilis* has the capacity to cleave purine, ribo- and deoxyribonucleosides phosphorolytically, but the reaction is catalyzed by two purine nucleoside phosphorylases rather than by one. One is specific for adenosine and deoxyadenosine, the second catalyzes the cleavage of inosine, deoxyinosine, guanosine, and deoxyguanosine (112). Mutants lacking adenosine phosphorylase activity but containing normal levels of inosine phosphorylase have been obtained (J. Neuhard, personal communication).

Adenosine deaminase activity has not been found in *B. subtilis* (Nygaard, unpublished observations). This implies that adenosine is catabolized exclusively through phosphorolysis.

B. cereus also contains two purine nucleoside phosphorylases (266). One of these has been extensively purified and shown to lack activity towards adenosine (62). In contrast to *B. subtilis*, *B. cereus* contains an adenosine deaminase and hence resembles enterobacteriaceae in its mode of metabolizing adenosine (184, 266, Fig. 2.6). The synthesis of adenosine deaminase in *B. cereus* is induced by adenine and repressed by inosine. Inosine also induces the synthesis of inosine/guanosine phosphorylase but not that of adenosine phosphorylase (266).

Purine deoxyribonucleoside kinases have been found both in *B. subtilis* (187, 228) and in *B. megaterium* (273). In *B. subtilis* two different enzymes are present, one that phosphorylates deoxyguanosine to dGMP, and another, which will phosphorylate deoxyadenosine and deoxycytidine. Mutants defective in the latter kinase have been characterized (187). Since the deoxyadenosine/deoxycytidine kinase of *B. subtilis* shows some activity toward adenosine *in vitro* (187), it is possible that *B. subtilis* also *in vivo* is capable of converting adenosine to AMP by direct phosphorylation.

B. subtilis can utilize purine nucleoside monophosphates as purine sources (Table 2.1). The cells produce both extra- and intracellular 3′-nucleotidases (54, 211, 248). Mutants with reduced extracellular nucleotidase activity cannot grow on 5′-AMP as adenine source (248). This indicates that the cell membrane of *B. subtilis* is impermeable for nucleotides (248).

A vast number of purine excreting mutants of *B. subtilis* have been isolated with the purpose of utilizing the excreted products commercially. The strains appear to carry multiple mutations affecting purine metabolism but they have not been completely characterized (109, 135, 169, 210).

B. Lactobacilli

In *Lactobacillus casei* purine bases are salvaged by phosphoribosyltransferases (138) and each of the naturally occurring purine bases may serve as the sole purine source (159). It has been established that *L. pentosus* (278) and probably also other *Lactobacillus* species are devoid of adenosine (deoxyadenosine) deaminase activity. Furthermore *L. acidophilus* and *L. leichmannii* contain purine deoxyribonucleoside kinase activities toward both deoxyadenosine and deoxyguanosine (41, 223). Most *Lactobacillus* species are unique in that they contain nucleoside deoxyribosyltransferases which catalyze the transfer of deoxyribosyl groups from a purine or pyrimidine deoxyribonucleoside to a purine or a pyrimidine base (103, 270). The same *Lactobacillus* species are generally devoid of purine nucleoside phosphorylases but contain purine ribonucleoside hydrolases (Chapter 5, p. 244). Thus it appears that these organisms are enzymatically equipped to scavenge deoxyribonucleosides for deoxyribonucleotide synthesis through the combined action of nucleoside deoxyribosyltransferases and deoxyribonucleoside kinases. In contrast purine ribonucleosides are utilized through hydrolytic cleavage to the free base which is then phosphoribosylated to the corresponding ribonucleoside monophosphate.

C. Neisseria

Purine utilization has been studied in some detail in wild type cells and in purine auxotrophs of *Neisseria meningitidis* by Jyssum (117, 118, 119, 120). From her studies, which included growth experiments, incorporation of radiolabeled purines, and enzyme analysis on cell free extracts, it appeared that: (a) purine bases can be utilized through phosphoribosylation; (b) adenosine is cleaved to adenine by a hydrolase; (c) other purine nucleosides are metabolically inert; (d) *N. meningitidis* lacks adenine deaminase, adenosine deaminase, AMP deaminase, and GMP reductase; (e) purine nucleotides will not serve as purine sources.

D. Fungi

1. Schizosaccharomyces pombe

Wild type cells of the fission yeast *S. pombe* will incorporate purine bases into nucleic acid (179). At the same time a storage pool of purine nucleosides/bases is formed endogenously (36, 37).

S. pombe possesses two purine phosphoribosyltransferases, one specific for adenine (*dap*) and the other with activities towards hypoxanthine, guanine and xanthine (*pur*) (42, 191). Mutants resistant to 2,6-diaminopurine, *dap* mutants, and mutants resistant to 8-aza-2-thioxanthine, *pur* mutants, have been isolated (42). Furthermore mutants lacking adenine deaminase activity have been characterized (1).

The main information about the purine salvage pathways in *S. pombe* has been obtained by studying the growth requirements of purine auxotrophic strains (43, 200). From such studies the following conclusions can be made: (a) Adenine may be metabolized either through the adenine phosphoribosyltransferase reaction or through deamination to hypoxanthine. The first enzyme is essential for the growth of an adenine requiring strain (Table 2.7, line 1 and 2). Adenine will satisfy a general purine requirement through either adenine phosphoribosyltransferase or adenine deaminase (Table 2.7, line 3–6). (b) Hypoxanthine utilization depends on a functional hypoxanthine phosphoribosyltransferase (Table 2.7, line 3 and 6). The same enzyme is required for the utilization of guanine and xanthine (Table 2.7, line 7 and 8). (c) Guanine and xanthine will not serve as general purine source, indicating that *S. pombe* is devoid of GMP reductase activity. However, guanine and xanthine will satisfy a specific guanine requirement of IMP dehydrogenase deficient mutants (Table 2.7, line 7 and 8). Guanine may also be deaminated to xanthine and thus serve as nitrogen source (222). (d) Adenosine will serve as a general

TABLE 2.7
Growth of purine auxotrophic strains of *S. pombe* on different purines.[a]

Relevant genotype[b]	Adenine	Hypoxanthine or inosine	Guanine or guanosine	Xanthine or xanthosine
ade	+	−	−	−
ade,dap[c]	−	−	−	−
hyp	+	+	−	−
hyp,dap	+	+	−	−
hyp,dea	+	+	−	−
hyp,pur	+	−	−	−
gua	−	−	+	+
gua,pur[c]	−	−	−	−

[a] Taken from (222).
[b] *ade* mutants are blocked in one of the two steps leading from IMP to AMP. *dap* mutants are devoid of adenine phosphoribosyltransferase activity. *dea* mutants are devoid of adenine deaminase activity. *gua* mutants are blocked in the step leading from IMP to XMP. *hyp* mutants are blocked in the *de novo* pathway, prior to the formation of CAIR (Fig. 2.1). *pur* mutants are devoid of hypoxanthine-guanine-xanthine phosphoribosyltransferase activity.
[c] The lethality of these genotypes has been deduced from the inability of spores from such mutants to germinate on any medium.

purine source, but only poorly as a specific adenine source, indicating that adenosine is predominantly metabolized through deamination to inosine (222). (e) Guanosine, inosine and xanthosine utilization involves cleaving of the nucleoside to the free base, since mutants deficient in hypoxanthine-guanine-xanthine phosphoribosyltransferase (*pur*) will not utilize nucleosides (Table 2.7, line 3–8). This is further supported by the observation that a *gua* mutant which is unable to cleave guanosine cannot utilize this nucleoside (222). The latter result indicates that *S. pombe* does not possess purine nucleoside kinase activity. (f) The utilization of purine nucleoside monophosphates depends on the presence of purine phosphoribosyltransferases (42, 106) suggesting that the nucleotides must be cleaved prior to utilization. *S. pombe* seems to be devoid of AMP deaminase activity (222).

Several purine analogs, carrying various substituents in position 6, support the growth of hypoxanthine requiring mutants, provided the cells possess adenine deaminase activity (1). However, 2,6-diaminopurine may serve as guanine source, even in adenine deaminase deficient strains. This conversion probably involves formation of 2,6-diaminopurine-ribonucleoside, followed by deamination to guanosine (*see* p. 45).

2. *Saccharomyces cerevisiae*

Wild type cells of *S. cerevisiae* will incorporate purine bases into nucleic acids (22, 23). Furthermore growth on minimal medium is stimulated by adenine and hypoxanthine. The utilization of exogenous purines involves initial entry into a pool of purine nucleosides and bases (189, 220).

The purine salvage pathways operating in *S. cerevisiae* appear to be similar to those reported for *S. pombe* but certain differences are apparent. Thus *S. cerevisiae* lacks adenine deaminase activity, suggesting that the conversion of adenine compounds occurs through deamination of adenosine. This is supported by the large pool of inosine found in cultures growing on adenine supplemented medium (22).

Guanine may serve as guanine source through phosphoribosylation as reported for *S. pombe*, but it is extensively deaminated to xanthine, which cannot be utilized as a purine source due to lack of xanthine phosphoribosyltransferase activity (4, 22, 63).

Wild type cells of *S. cerevisiae* are *insensitive* to 2,6-diaminopurine. Mutants which are sensitive have been isolated. They contain a three fold increased level of guanine phosphoribosyltransferase (155), suggesting that the adenine phosphoribosyltransferase of this organism cannot use the analog as substrate. A mutant which can grow on adenosine has been isolated in an adenine requiring strain. It contains increased adenosine kinase activity and has become sensitive to 3′-deoxyadenosine (5, 6).

3. Neurospora crassa

During the early stage of conidial germination RNA synthesis depends on a storage pool of inosine, as illustrated by studies on purine auxotrophs. The conidia of a mutant which is blocked in the *de novo* synthesis of IMP will germinate in the absence of purines, while conidia of a mutant blocked in the conversion of IMP to AMP will not, unless supplemented with adenine (216).

The adenine requirement of mutants blocked in the conversion of IMP to AMP can be satisfied by adenine, adenosine, deoxyadenosine, AMP, ATP or NAD (216). In mutants blocked in the *de novo* synthesis of IMP the same compounds, as well as hypoxanthine and inosine, will support growth (261). Equal growth rates are observed whether hypoxanthine, adenine or adenosine serve as the sole purine source (162, 163). Adenine may also be converted to hypoxanthine, xanthine and uric acid which are excreted into the medium (231).

Mutants deficient in IMP dehydrogenase require guanine or guanosine for growth (76). The finding that these two compounds do not support growth of a general purine auxotrophic mutant indicates that *Neurospora* cannot convert guanine compounds to adenine compounds (75). The production of xanthine from guanine is negligible in *Neurospora*.

Adenine, guanine and hypoxanthine phosphoribosyltransferase activities (164, 214, 232) as well as purine nucleoside hydrolase activity (170) have been identified in *Neurospora*. In addition intracellular 5′-nucleotidase activity has been observed (2).

Neurospora is resistant to many purine analogs (115), but 6-methylpurine will inhibit growth (214). Adenine, hypoxanthine or their respective ribonucleosides overcome the inhibition (214). Mutants resistant to 6-methylpurine have been isolated. One of these mutants is devoid of adenine phosphoribosyltransferase activity, while the level of hypoxanthine phosphoribosyltransferase is reduced 5-fold (214). It is not clear, whether this phenotype is the result of a single mutation.

4. *Physarum*

The purine salvage pathways of *Physarum polycephalum* have been studied in microplasmodia by measuring the incorporation of [^{14}C]-labelled purine bases and nucleosides into nucleotides, and by enzyme assays in cell free extracts (52, 56).

Although *Physarum* possesses purine phosphoribosyltransferases for both adenine, guanine, and hypoxanthine, they seem to lack transport systems for these bases. Exogenous guanine and hypoxanthine are not incorporated and only a very slow incorporation of adenine is observed.

Purine nucleosides, on the other hand, are readily incorporated into nucleotides and nucleic acids. Adenosine and deoxyadenosine labelled with [^{14}C] in the purine ring are incorporated into ATP, dATP, GTP, and dGTP, whereas labelled guanosine and deoxyguanosine are only incorporated into GTP and dGTP. Thus *Physarum* converts adenine nucleotides into guanine nucleotides, but not guanine nucleotides into adenine nucleotides. The finding that the purine ring of deoxyadenosine, deoxyinosine and deoxyguanosine is recovered in the *ribo*nucleoside triphosphates further shows that the deoxyribonucleosides are cleaved to the free bases before incorporation. Thus it seems that *endogenously* formed adenine, hypoxanthine and guanine may be utilized for nucleotide synthesis. In these experiments more dATP was derived from exogenously added deoxyadenosine than from adenosine, indicating that deoxyadenosine may be converted directly to dATP through the deoxyadenosine kinase reaction.

Enzyme measurements on crude extracts have verified that *Physarum* in addition to purine phosphoribosyltransferase activities contains purine ribonucleoside kinase, deoxyadenosine kinase and AMP deaminase activities. No adenosine deaminase, adenine deaminase or guanine deaminase activities could be demonstrated (56).

E. Purine auxotrophs

Many microorganisms, in particular parasites are unable to carry out *de novo* synthesis of purines and are therefore depending on the presence of the purine salvage pathways. In these cells no purine degradation occurs. The different organisms do, however, show variations in their ability to utilize exogenous purine for growth, suggesting that also the interconversion of nucleotides is deficient in the cells. All the parasitic protozoa investigated appear unable to synthesize purines *de novo* and are thus dependent on the host for a source of purines. This is of importance, since it might be expected to render the parasitic protozoa sensitive to the cytotoxic effect of purine analogs. The choice of analog will then be determined by the specific purine required by the protozoa and by the capacity of the salvage pathways.

1. *Mycoplasma*

The *Mycoplasma*, which are the smallest self replicating procaryotes, have a complex growth requirement (225). *Mycoplasma mycoides* and presumably also other *Mycoplasma* species have a specific requirement for guanine and growth is further stimulated by adenine. Besides the lack of *de novo* IMP synthesis *M. mycoides* is deficient in the conversion of IMP to GMP, hence the specific requirement for guanine.

By measuring the specific activity of purine nucleotides isolated from RNA following incorporation of [^{14}C]-labelled purines or purine nucleosides, and by determining the activity of the purine metabolizing enzymes in cell extracts, the major pathways of purine nucleotide synthesis in *Mycoplasma* have been defined (Fig. 2.7, 177, 178). If nucleosides are supplied as purine sources they undergo a phosphorolytic cleavage to nucleobases before utilization. Neither nucleoside kinase nor adenosine deaminase activities have been detected. Adenosine phosphorylase appears to be present in many *Mycoplasma* species and the activity of this enzyme seems to be the biochemical test of choice to detect *Mycoplasma* contamination in cell cultures (80, 85).

Several *Mycoplasma* species show 5′-nucleotidase activity, suggesting that also exogenous nucleotides may serve as purine sources (80).

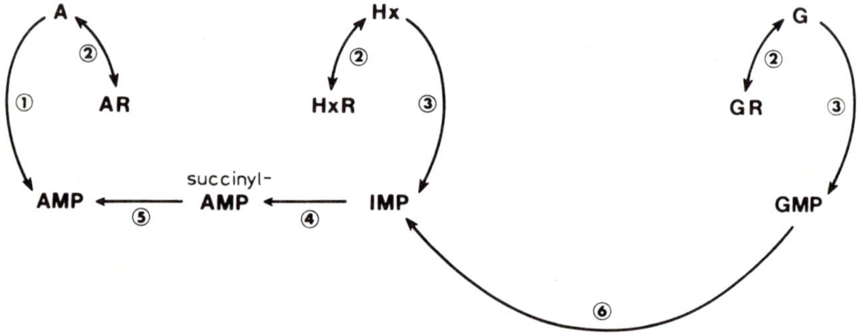

FIG. 2.7. Proposed pathways for purine salvage and interconversion in *Mycoplasma mycoides*. (1), adenine phosphoribosyltransferase; (2), purine nucleoside phosphorylase; (3), hypoxanthine/guanine phosphoribosyltransferase; (4), adenylosuccinate synthetase; (5), adenylosuccinate lyase; (6), GMP reductase.

2. Protozoa

Trypanosoma. In the life cycle of the *Trypanosoma* there are three main morphological stages of which the epimastigote form can be kept in culture (77). All three forms incorporate and interconvert radiolabelled purine bases and nucleosides.

In a detailed study of purine salvage in the culture form of *T. cruzi* it was found that adenine, hypoxanthine, guanine, xanthine and their respective ribonucleosides are equal in their ability to support growth. It was also shown in short time experiments (2h) with whole cells that the various purine bases and nucleosides mentioned above are converted to their respective nucleotide forms and that they are interconvertible (16).

On the basis of such results and of enzyme analysis in cell free extracts of all three forms of *T. cruzi* (78, 202) and on purine utilization studies in *T. brucei gambiense* (57) a pattern for the utilization of purine bases and nucleosides is proposed (Fig. 2.8). Neither adenine nor adenosine appears to be deaminated in these organisms, but since both compounds can be converted to guanine nucleotides a third deaminating enzyme, presumably AMP deaminase, must be present. Determination in the different forms of purine salvage enzymes furthermore indicated that there is little variation in purine metabolism during the life cycle of *T. cruzi* (78).

Leishmania. These protozoa grow intracellularly in macrophages in the amastigote form and in the promastigote form in the insect vector, the sand fly (77).

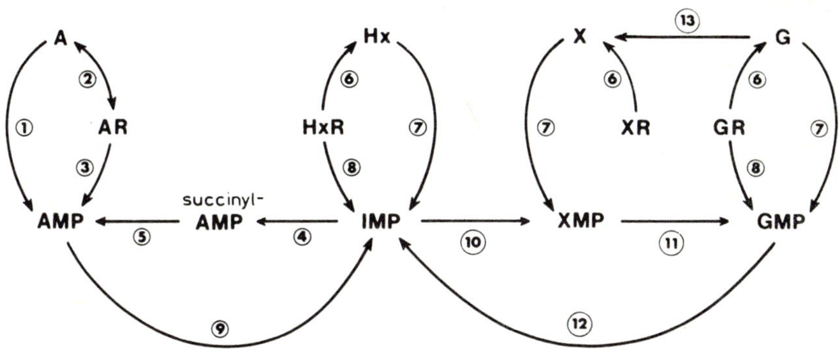

FIG. 2.8. Proposed pathways for the salvage and interconversion of purines in *Trypanosoma*. (1), adenine phosphoribosyltransferase; (2), adenosine phosphorylase; (3), adenosine kinase; (4), adenylosuccinate synthetase; (5), adenylosuccinate lyase; (6), nucleoside hydrolase(s); (7), purine phosphoribosyltransferase(s); (8), inosine (guanosine) kinase; (9), AMP deaminase; (10), IMP dehydrogenase; (11), GMP synthetase; (12), GMP reductase; (13), guanine deaminase.

Purine metabolism has been studied in the promastigote form, of *L. donovani* and *L. braziliensis* by growing cells in defined media with radiolabelled purines. Hypoxanthine is the preferred purine source. The major route of adenine and guanine metabolism was found to be deamination to hypoxanthine and xanthine respectively (166).

Leishmania species contain purine salvage and interconversion pathways (137, 140, 202, 269) similar to those of *Trypanosoma*, but with certain significant differences (Fig. 2.8). *Leishmania* possess adenine deaminase activity but is probably devoid of AMP deaminase activity (202). Furthermore at least four different enzymes which cleave nucleosides have been identified in *Leishmania*, including a specific purine 2′-deoxyribonucleoside hydrolase (136). There are also indications that a phosphotransferase rather than a kinase is responsible for the phosphorylation of inosine to IMP (137).

Crithidia. These parasites are insect trypanosomids, which can be readily cultivated (77). The pathways of purine salvage in *Crithidia* have not been defined in detail, but it is known that in *C. fasciculata* purine bases are converted into ribonucleotides by phosphoribosyltransferases (129). Furthermore it appears that deamination of adenine is obligatory for the conversion of adenine to IMP, since *C. fasciculata* will not grow on adenine in the presence of deoxycoformycin, an inhibitor of adenine deaminase. The growth is unaffected by deoxycoformycin, when hypoxanthine is the purine source (126).

There is evidence for the presence of ribonucleoside hydrolase activity towards inosine, guanosine and adenosine in *Crithidia* (202). No adenosine deaminase activity has been detected.

Thus the major differences in purine metabolism between *Crithidia* and *Trypanosomas* species (Fig. 2.8) seem to be that *Crithidia* contains adenine deaminase activity and lacks AMP deaminase.

Plasmodium. Most *Plasmodium* species parasitize mature erythrocytes. The life cycle of these organisms includes several stages and the morphology varies both with the species and with the stage of development. Most of the biochemical information of *Plasmodium* is derived from studies of parasites in the trophozoite form, growing within or removed from the erythrocyte. Only in this stage is the parasite accessible and abundant for the purpose of biochemical studies (77, 247).

Studies of the ability of isolated *P. lophurae* to utilize different purine compounds have indicated that hypoxanthine is the preferred purine source for the parasite (82, 267, 289). This would favor the host–parasite relationship for the *Plasmodium* since the maintenance of the adenine nucleotide pool of the host erythrocytes involves the continuous uptake of adenosine with concomitant formation of hypoxanthine (288, 289), which may subsequently act as purine source for the *Plasmodium*.

In agreement with this contention it has been reported that administration of allopurinol (p. 66), which is a potent inhibitor of xanthine oxidase, stimulates the multiplication of *P. berghei* in rats and mice (25). This effect most likely is due to the elevated plasma level of hypoxanthine, caused by the allopurinol treatment.

Coccidia. Purine metabolism has been studied in *Toxoplasma gondii* and *Eimeria tenella*. It is technically difficult to study incorporation of radiolabelled purines in these intracellular parasites (218, 276). In the case of *T. gondii* the problem has been partly circumvented by growing the parasite inside cultured Lesch-Nyhan fibroblasts, which lack hypoxanthine-guanine phosphoribosyltransferase, and thus are unable to utilize hypoxanthine and guanine (218). In this way it has been possible to show that *T. gondii* will incorporate hypoxanthine into both GMP and AMP of RNA, whereas guanine is mainly incorporated into GMP (218).

E. tenella contains two purine phosphoribosyltransferases, one which is specific for adenine and another which will use hypoxanthine, guanine and xanthine as substrates (277). The latter enzyme seems to be of crucial importance for the parasite since it has been observed that allopurinol, which specifically inhibits this enzyme (277), inhibits the growth of *E. tenella* in cultured chicken kidney ephithelial cells (230). The anticoccidial agent arprinocid [6-amino-9-(2-chloro-6-fluorobenzyl)-purine], which is a potent inhibitor of IMP dehydrogenase, selectively inhibits incorporation of [^{3}H] hypoxanthine into nucleic acids of *E. tenella* during its intracellular growth (276). The results indicate that the normal purine source of *E. tenella* is hypoxanthine.

Both *T. gondii* and *E. tenella* possess adenosine kinase activity, while only *T. gondii* mutants, resistant to adenine arabinoside, have been isolated. They contain a defective adenosine kinase (219, 241) and incorporation of [^{3}H] adenosine into nucleic acids of these mutants is only 25 percent of the incorporation seen in the wild type (241). The residual incorporation probably occurs via the successive action of adenosine deaminase, purine nucleoside phosphorylase and hypoxanthine phosphoribosyltransferase.

Tetrahymena. These protozoa are not parasites, but free living organisms in fresh water habitats.

T. pyriformis was the first microorganism shown to require an external purine source for growth. It lacks the *de novo* synthesis of purines and in addition is unable to convert IMP to GMP (93, 94). It therefore has a specific guanine requirement which may be satisfied with guanine, guanosine or GMP.

Other *Tetrahymena* species also have a specific requirement for guanine. Adenine and hypoxanthine both have a sparing effect on the guanine requirement, suggesting that *T. pyriformis* contains phosphoribosyltransferases and that adenine and hypoxanthine can both be converted into AMP (93). Neither adenine nor guanine are deaminated. The pathway responsible for the conversion of GMP to IMP has not been identified.

Guanosine and inosine are cleaved phosphorolytically by *T. pyriformis*, while adenosine cleaving activity has not been detected (94). Both adenosine kinase and adenosine deaminase activities have been observed (32, 95), whereas the organism seems to be devoid of AMP deaminase activity.

VI. INDIVIDUAL ENZYMES OF PURINE SALVAGE

A. Purine phosphoribosyltransferases

The major metabolic pathway for the conversion of purines to mononucleotides is a transfer of the phosphoribosyl group from 5-phosphoribosyl 1-pyrophosphate to the free purine base. The purine phosphoribosyltransferases, sometimes referred to as purine nucleotide pyrophosphorylases, thus catalyze the following reaction:

$$\text{purine} + \text{PRPP} \rightarrow \text{purine nucleoside 5'-monophosphate} + \text{pyrophosphate}$$

Enzymes capable of catalyzing this transfer are widely distributed in nature. They seem to be present in almost all organisms, although the specificity towards the purine bases may show some differences. In general one enzyme, the adenine phosphoribosyltransferase, acts on all 6-amino-purines, while either one enzyme or two enzymes with overlapping specificities for the different 6-oxo-purines have been identified in the various microorganisms investigated (18, 97, 98, 129, 138, 191, 269, 275).

1. *Adenine phosphoribosyltransferase* (EC 2.4.2.7)

Adenine phosphoribosyltransferase from *E. coli* (100, 252), *S. typhimurium* (121, 122), and *B. subtilis* (17) has been studied. The enzymes require Mg^{2+} or Mn^{2+} for activity and are specific for adenine. 2,6-diaminopurine will also serve as a substrate. The kinetics of the *E. coli* enzyme indicated that purine nucleotides, especially AMP, inhibit the reaction in competition with PRPP (98). K_m values for adenine and PRPP vary considerably in the different reports, (1.3–20 μM for adenine and 10–125 μM for PRPP). Electrophoresis of the purified enzyme (MW 40 000) in the absence of Mg^{2+} ions causes dissociation of the enzyme into several catalytically active bands, indicating a composite structure of the native enzyme (100).

From *Leishmania donovani* adenine phosphoribosyltransferase has been purified by affinity chromatography. This preparation showed no detectable activity towards hypoxanthine or guanine. The activity of the enzyme is stabilized by addition of Mg^{2+} ions and PRPP. The latter compound seems to affect the state of aggregation of the enzyme, since gel filtration in the presence of Mg^{2+} and PRPP indicated a molecular weight of 54 000, while omittance of PRPP resulted in a protein with a molecular weight of 25 000 (269).

The enzymes from *Schizosaccharomyces pombe* (191) and *Mycoplasma mycoides* (252) also appear to occur in multimeric forms and to be stabilized by Mg^{2+} and/or PRPP. Adenine phosphoribosyltransferase from *Crithidia fasciculata*, is unique in that it is activated most efficiently by Zn^{2+} ions (129).

2. Hypoxanthine-guanine-xanthine phosphoribosyltransferases

In *E. coli* (116, 138, 168, 176) and in *S. typhimurium* (30, 73, 97) two 6-oxopurine phosphoribosyltransferases have been found, i.e. hypoxanthine phosphoribosyltransferase (EC 2.4.2.8) and guanine phosphoribosyltransferase (EC 2.4.2.22). Both hypoxanthine and guanine will serve as substrates for the two enzymes, but with different efficiency. The guanine phosphoribosyltransferase will also use xanthine as a substrate. Both enzymes are inhibited by ppGpp (99). Other microorganisms, including *Lactobacilli*, yeasts and *Protozoa* have been shown to contain either one or two 6-oxopurine phosphoribosyltransferase(s) with overlapping specificities. Common traits in all these enzymes are the requirement for Mg^{2+} or Mn^{2+} for activity, K_m values for PRPP of about 50 μM, and K_m values for the purine bases less than 50 μM.

In *Lactobacillus casei* two enzymes have been identified, one which reacts with hypoxanthine and guanine, and another which reacts with xanthine, and to a lesser extent with guanine (138).

In *Schizosaccharomyces pombe* a single enzyme is responsible for the phosphorylation of hypoxanthine, guanine and xanthine. It is feedback inhibited by IMP, pyrophosphate and by guanine nucleotides (191).

From *Saccharomyces cerevisiae* a single enzyme has been purified (179, 236). It will use hypoxanthine ($K_m = 26$ μM), guanine (41 μM) but not xanthine (4) as substrates. *S. cerevisiae* is therefore unable to use xanthine as a purine source (p. 58). The purine analog 6-thioguanine is a very efficient substrate, but is not toxic to the cells (203). The enzyme is a dimer consisting of two, probably identical subunits with molecular weights of 29 500 (203), and kinetic analysis has indicated an ordered Bi Bi mechanism (4).

Two enzymes have been identified in *Leishmania donovani*, one which reacts with hypoxanthine and guanine, and one which reacts with xanthine but not with hypoxanthine. For the first enzyme a molecular weight of 110 000 was determined in the absence of substrates, while addition of PRPP caused dissociation of the enzyme into smaller subunits. The molecular weight of the xanthine phosphoribosyltransferase (54 000) was independent of the presence of PRPP (269).

In *Crithidia fasciculata* guanine and hypoxanthine phosphoribosyltransferases are separate enzymes with overlapping specificities (129). A single enzyme which reacts with hypoxanthine and guanine has been isolated from *Plasmodium chabaudi* (275). From *Eimeria tenella* a single enzyme has been purified which shows affinity for hypoxanthine, guanine, and xanthine. The activity is strongly inhibited by allopurinol (4-hydroxypyrazolo-[3,4-*d*]pyrimidine) a drug which is known to inhibit growth of this organism in tissue cultures (230).

B. Purine nucleoside kinases

1. *Adenosine kinase* (EC 2.7.1.20)

A number of microorganisms contain adenosine kinase which catalyzes the phosphorylation of adenosine:

$$\text{Adenosine} + \text{ATP} \rightarrow \text{AMP} + \text{ADP}$$

Deoxyadenosine is only weakly accepted as substrate. The enzyme has been purified from brewers yeast to apparent homogeneity (144). It requires Mg^{2+} for activity and is stabilized by adenosine and dithiothreitol. In the presence of adenosine the enzyme exists in the form of a single polypeptide, with a molecular weight of about 40 000.

E. coli does not contain adenosine kinase. Previous reports on the occurrence of adenosine kinase activity in *E. coli* (116) have later been shown to be based on artifacts.

2. *Guanosine/inosine kinase* (EC 2.7.1.73)

This enzyme has only been identified in *E. coli* and *S. typhimurium*. It catalyzes a phosphorylation of guanosine and inosine, but not of deoxyguanosine, deoxyinosine or xanthosine.

The enzyme has been partly purified from *E. coli* and shown to require Mg^{2+} and K^{+} ions for activity. K_m for guanosine is 16 μM (unpublished).

3. *Purine deoxyribonucleoside kinases*

Kinase activities towards purine deoxyribonucleosides are absent from *E. coli* and *S. typhimurium*, but several *bacilli* and *lactobacilli* have been shown to contain enzymes which will phosphorylate purine deoxyribonucleosides (41, 50, 187, 223, 273).

From *Lactobacillus acidophilus* two kinases have been separated. One enzyme is responsible for the phosphorylation of deoxycytidine ($K_m = 5\ \mu$M) and deoxyadenosine ($K_m = 5\ \mu M$), and another enzyme reacts preferentially with deoxyguanosine ($K_m = 4\ \mu M$). Both enzymes have molecular weights of about 50 000. For the deoxycytidine/deoxyadenosine kinase dGTP was found to be the preferred phosphate donor (41).

From *B. subtilis* two kinase activities with multiple specificities have been separated. One of these enzymes shows the following relative activities towards its substrates: deoxycytidine 100%, adenosine 19%, cytidine 10%, deoxyadenosine 8.5%, and deoxyguanosine $<5\%$. GTP is the preferred phosphate donor and dATP acts as feed back inhibitor. The K_m values for deoxyadenosine and deoxycytidine are equal ($5\ \mu M$) (187).

The second kinase has not yet been investigated, but appears to be specific for deoxyguanosine, as judged from measurements on crude extracts of a mutant which lacked the deoxycytidine/deoxyadenosine kinase (187).

C. Purine nucleoside phosphorylases and hydrolases

Purine nucleosides are cleaved by most microorganisms. The cleavage is catalyzed either by phosphorylases or by hydrolases (nucleosidases):

(i) purine(deoxy)ribonucleoside + phosphate $\rightleftharpoons$
purine + (deoxy)ribose 1-phosphate

(ii) purine ribonucleoside + $H_2O \rightarrow$ purine + ribose

Purine nucleoside phosphorylase (EC 2.4.2.1) catalyzes the reversible reaction (i) and the equilibrium favors nucleoside synthesis (113). Purine nucleoside phosphorylases from different sources differ in the specificity towards the purine bases. In *E. coli* (81), *S. typhimurium* (204) and in *Bacillus cereus* (266) the synthesis of purine nucleoside phosphorylase is induced by exogenous nucleosides.

In general nucleoside hydrolases catalyze the hydrolysis of ribonucleosides only. However, an enzyme which is specific for 2-deoxyribonucleoside has been isolated from *Leishmania donovani* (136).

For a more detailed account of the properties of the purine nucleoside phosphorylases and purine nucleoside hydrolases see Chapter 5.

As discussed in the previous sections purine nucleoside phosphorylases, and purine nucleoside hydrolases play an important role in the utilization of exogenous nucleosides. Purine nucleoside phosphorylases may also promote the utilization of exogenous adenine.

D. Nucleoside deoxyribosyltransferases

Nucleoside deoxyribosyltransferases or *trans*-N-deoxyribosylases (EC 2.4.2.6) catalyze the following phosphate independent reactions

(i) purine deoxyribonucleoside + purine* ⇄ purine* deoxyribonucleoside + purine

(ii) purine deoxyribonucleoside + pyrimidine ⇄ pyrimidine deoxyribonucleoside + purine

(iii) pyrimidine deoxyribonucleoside + pyrimidine* ⇄ pyrimidine* deoxyribonucleoside + pyrimidine

The enzymes have been found only in species of the subgenus *Thermobacterium* of *Lactobacillus* (9, 270).

From *L. helveticus* two distinct enzymes have been separated and purified, one catalyzing specifically reaction (i) and another catalyzing reactions (ii) and (iii) (38, 39).

The purine specific enzyme from *L. helveticus* which catalyzes reaction (i) shows K_m values for purine bases in the range of 0.04–0.09 mM and of about 0.4 mM for the purine deoxyribonucleosides. The enzyme has a molecular weight of 86 000 with a subunit molecular weight of 16 000 (103).

All *Lactobacillus* species which contain nucleoside deoxyribosyltransferases require exogenous deoxyribonucleosides for growth. When the cells are starved for deoxyribonucleosides the synthesis of the nucleoside deoxyribosyltransferases is derepressed (9).

E. Deaminases

1. Adenosine deaminase (EC 3.5.4.4)

$$\text{Adenosine} + H_2O \rightarrow \text{inosine} + NH_3$$

Adenosine deaminase catalyzes the deamination of adenosine and deoxyadenosine to inosine and deoxyinosine, respectively.

The enzyme has been purified from several sources (*see also* Chapter 5). The *E. coli* enzyme shows a broad specificity towards adenosine derivatives (207). Thus purine nucleosides with substituents in the 6-position are hydrolyzed, e.g. 2,6-diaminopurine ribonucleoside and 6-methylaminopurine ribonucleoside. Also altered pentose moieties are tolerated in the substrates, e.g. 3′-deoxyribose, 2′,3′-dideoxyribose and arabinose. In *E. coli* coformycin is a potent inhibitor of the enzyme (206).

Adenosine deaminase plays an important physiological role in the conversion of adenine to IMP (*see* p. 43). In *E. coli*, but not in *S. typhimurium*, the synthesis of the enzyme is induced by adenine and hypoxanthine (207, 227). Induction of adenosine deaminase by adenine derivatives is also observed in *B. cereus* (266).

2. *Adenine deaminase* (EC 3.5.4.2)

$$\text{Adenine} + H_2O \rightarrow \text{hypoxanthine} + NH_3$$

The enzyme is specific for adenine and no dialyzable cofactors are required. Adenine deaminase has been demonstrated in a number of microorganisms, but is absent from *E. coli* and *S. typhimurium* (101, 147).

Adenine deaminase from *Azotobacter vinelandii* and *Candida utilis* has been purified (83). K_m values for adenine were 10 μM and 30 μM respectively. Both enzymes as well as the enzyme from *Schizosaccharomyces pombe* will deaminate 2,6-diaminopurine and have also been found capable of removing a variety of substituents at position 6 of the purine ring (1, 83).

The enzymes from *Leishmania* species and from *Crithidia fasciculata* all have very high K_m values (0.7–2.2 mM). Coformycin and deoxycoformycin are potent inhibitors of these protozoan enzymes (127, 128).

VII. INTRACELLULAR FORMATION OF PURINE NUCLEOSIDES AND NUCLEOBASES

A. Degradation of RNA

Nucleosides and nucleobases are continuously generated in the cells as degradation products of nucleic acids and nucleotides. Ribosomal RNA and mature tRNA are normally quite stable in the cells, but the various mRNA species together with a small fraction of oligonucleotides which are formed in the maturation process of tRNA and ribosomal RNA are rapidly degraded by nucleolytic enzymes (254). It is at present not known whether the degradation products are nucleoside 2′-, 3′-, or 5′-monophosphates or possibly nucleoside 5′-diphosphates. The endogenously formed nucleoside 5′-phosphates may be directly utilized for nucleic acid synthesis after rephosphorylation, while nucleoside 2′- and 3′-monophosphates must be further degraded to nucleosides or to nucleobases prior to their utilization. Under certain conditions, i.e. starvation for nitrogen or carbon sources, also stable RNA is degraded (192). The majority of the degradation products are then excreted from the cells. Among the excretion products nucleobases, nucleosides, and nucleoside 3′- and 5′-phosphates have been identified (31, 124).

Several enzymes degrade nucleoside 2′-, 3′-, and 5′-monophosphates

(47, 48). In many cases the location of these enzymes has been shown to be external to the cytoplasmic membrane (237, 248, 271, 280, 287). Such a location might indicate that the main physiological function of these hydrolytic enzymes is to degrade nucleotides coming from the outside. This does not exclude that these nucleotidases may be involved in the hydrolysis of endogenous formed nucleotides. Several observations suggest that this might be the case: In *enterobacteriaceae* a cytoplasmic protein has been characterized which is a specific inhibitor of the periplasmic 5′-nucleotidase (193). In *Bacillus* ATP is a potent and specific inhibitor of the 5′-nucleotidase activities (54, 244). Furthermore, excretion of nucleic acid degradation products was measured during starvation of an *E. coli* mutant lacking 5′-nucleotidase. In this mutant the relative amount of uridine 5′-monophosphates in the excretion products was increased compared to the wild type. A mutation in the 3′-nucleotidase had no such effect. Although the results reported involved pyrimidine compounds solely, the specificity of the 5′-nucleotidase for both purine and pyrimidine nucleotides strongly suggests that hydrolysis of purine nucleotides follows a pattern similar to that of pyrimidine nucleotides (31).

Very little information exists about cytoplasmic nucleotidases in microorganisms. Only in *Bacillus* their presence is well documented (211, 248).

Some microorganisms contain a specific AMP nucleosidase (147) which catalyzes the hydrolysis of AMP to adenine and ribose 5-phosphate. The catabolism of AMP by this enzyme has been studied in detail in *Azotobacter vinelandii* and in *E. coli* (147, 239, 291). The results suggest that AMP nucleosidase is involved in the regulation of the energy charge of the cells (240). In *A. vinelandii* the adenine formed is deaminated to hypoxanthine (239), which is excreted, while in *E. coli* the adenine is excreted directly (247). An *A. vinelandii* strain which is defective in AMP nucleosidase contains a sevenfold elevated AMP pool, but grows normally (148).

B. Turnover of S-adenosylmethionine

Large amounts of adenine/adenosine are continuously produced from the metabolism of S-adenosylmethionine. This compound is the primary methyl-donor in numerous transmethylation reactions. The demethylated product in these reactions is S-adenosylhomocysteine which in *E. coli* and in other bacteria (274) is cleaved hydrolytically at the glycosylic bond yielding adenine and S-ribosylhomocysteine. This reaction (Fig. 2.9) is irreversible and is catalyzed by 5′-methylthioadenosine nucleosidase (49, 55), in the earlier literature sometimes referred to as S-adenosylhomocysteine nucleosidase. The enzyme exhibits low activity towards adenosine and deoxyadenosine (115). In yeast and in other fungi S-adenosylhomocysteine is cleaved by S-adenosylhomocysteine hydrolase to adenosine and homocysteine (Fig. 2.9).

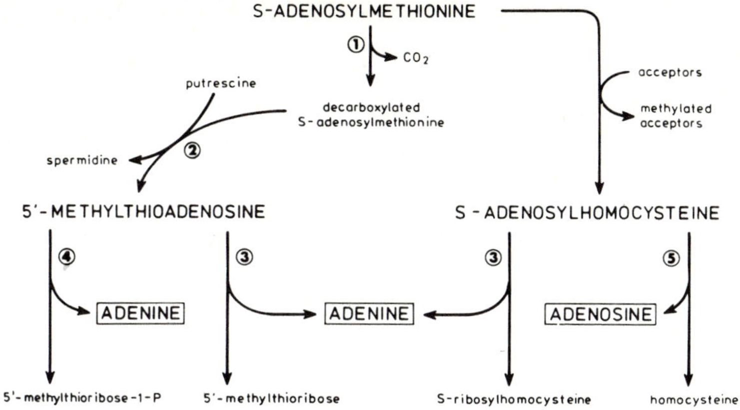

FIG. 2.9. Adenine and adenosine formation from S-adenosylmethionine. (1), adenosylmethionine decarboxylase; (2), aminopropyl transferase; (3), 5′-methylthioadenosine nucleosidase (S-adenosylhomocysteine nucleosidase); (4), 5′-methylthioadenosine phosphorylase; (5), S-adenosylhomocysteine hydrolase. Reaction (3) occurs in bacteria, reaction (5) occurs in yeast, and reaction (4), has so far, been shown to occur only in *Caldariella acidophila*.

This reaction is reversible with an equilibrium strongly favoring adenosine formation (274). S-adenosylhomocysteine hydrolase has not been found in bacteria.

In the synthesis of polyamines such as spermidine and spermine from putrescine, S-adenosylmethionine acts as a donor of the aminopropyl group after a decarboxylation has occurred (Fig. 2.9, 262, 263). The resulting 5′-methylthioadenosine is further cleaved by 5′-methylthioadenosine nucleosidase to adenine and 5′-methylthioribose (Fig. 2.9, 55). An enzyme which degrades 5′-methylthioadenosine phosphorolytically has been characterized in *Caldariella acidophila* (Fig. 2.9, 27).

It is known that S-adenosylhomocysteine and 5′-methylthioadenosine are toxic to mammalian cells because of their adverse effects on transmethylation processes (89). They are produced in considerable amounts in microorganisms and might be toxic here also. Thus, it is an advantage to the cells to be able to degrade S-adenosylhomocysteine as well as 5′-methylthioadenosine and to reutilize the degradation products.

C. Naturally occurring purine nucleoside analogs

A number of purine nucleoside analogs are formed by certain bacteria and fungi and excreted into the media. These analogs represent a diverse group of biologically active compounds and most of them show antibiotic activity (261).

Some of them are formed directly from *de novo* synthesized purines e.g. cordycepin (145, Fig. 2.10), while others are synthesized by pathways different from the purine *de novo* biosynthetic pathway, e.g. formycin B (208, Fig. 2.10). For detailed reviews see references 259, 260, 261. Some of the purine nucleoside analogs such as cordycepin must be metabolized by the purine salvage and interconversion pathway in order to form the biological active compounds.

A few of the analogs are known to inhibit specific enzymes of the purine salvage and interconversion pathways. Thus coformycin (Fig. 2.10) and its 2′-deoxyisomer are potent inhibitors of adenosine deaminase (206, 261) and adenine deaminase (128), while formycin B (Fig. 2.10) is a weak inhibitor of purine nucleoside phosphorylase (258). The effect of the latter compound is potentiated ten fold by halogen substitution at the 5′-position (258). Decoyinine and psicofuranine (Fig. 2.10) are antibiotics with an adenosine-like structure which specifically inhibits GMP synthetase (253, 259).

The toxicity of most purine nucleoside analogs is due to inhibition of either DNA, RNA, protein or cell wall synthesis. For further information on this subject see references 259, 260.

Certain purine compounds, the cytokinins, which stimulate growth and development of plants, have been isolated from the culture media of some

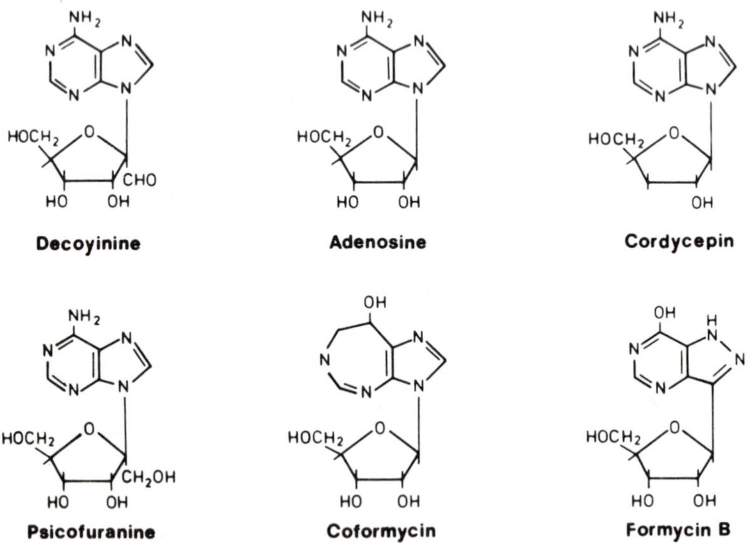

FIG. 2.10. Structural formula of adenosine and some purine nucleoside analogs which interferes with nucleoside and nucleotide metabolism.

microorganisms (260). The cytokinins contain modified adenine structures similar to those found in tRNA molecules and are believed to arise from the breakdown of tRNA in these microorganisms (74, 79). Many microorganisms which form symbiotic or parasitic associations with higher plants produce and excrete cytokinins, and it has been speculated that this may be of importance for the interaction between the microorganism and the higher plant (74).

VIII. INTERPLAY BETWEEN THE PURINE *DE NOVO* AND SALVAGE PATHWAYS

The purine salvage pathways are not essential for purine prototrophic microorganisms. Nevertheless they have a capacity to utilize preformed purine compounds. This is beneficial to the cells, because it enables them to reutilize purine bases and nucleosides which are produced endogenously through the turnover of nucleic acids, nucleotides, and S-adenosylmethionine. Furthermore it allows the cells to scavenge purine compounds which may be present in the growth medium. Since purine *de novo* synthesis is quite expensive for the cells, considerable savings may be obtained by salvage of available, preformed purine compounds, provided *de novo* synthesis decreases simultaneously.

Experiments with a variety of purine prototrophic microorganisms, e.g. *E. coli*, *S. typhimurium*, *Micrococcus sodonensis*, *Neurospora crassa*, and *Physarum polycephalum*, have shown that when exogenous purine bases and nucleosides are used for nucleic acid synthesis, the contribution from *de novo* pathway concomitantly decreases (8, 56, 205, 217, 251, 295). Thus, regulatory systems seem to exist, which can turn off purine *de novo* synthesis when preformed purine compounds are available. The extent to which the *de novo* pathway is suppressed by exogenous purines varies considerably in different microorganisms and with the purine compound available (Table 2.8).

In several microorganisms it has been observed that addition of purine

TABLE 2.8

Utilization of purine bases for RNA synthesis in *S. typhimurium* prototrophs.[a]

Additions (33 μM)	Percentage contribution from exogenous purines to RNA nucleotides: AMP	GMP
Adenine	86	59
Hypoxanthine	89	92
Guanine	25	105

[a] Cells were grown with [^{14}C]-labelled purines and the contribution to the overall synthesis from the exogenous source was calculated (295).

nucleosides and bases may result in growth inhibition. These effects are usually observed when high concentrations of the purine compounds are added, and they probably reflect unphysiological conditions. The extensive literature dealing with this subject has recently been reviewed (87, 88).

A. Regulation of the synthesis of the *de novo* enzymes

In a number of microorganisms, such as *A. aerogenes*, *E. coli*, *S. typhimurium*, and *B. subtilis*, starvation of purine auxotrophic mutants results in increased levels of some of the *de novo* biosynthetic enzymes. In purine prototrophic strains of the same organisms exogenous purines cause a repression of the synthesis of the purine *de novo* enzymes (107, 149, 181, 196). In *S. typhimurium* hypoxanthine or guanine repress the synthesis of the first four enzymes of the *de novo* pathway, while adenine has no effect (Table 2.9). Mutants which show high, non-repressible synthesis of some of the enzymes have been isolated in *S. typhimurium* (14), in *E. coli* (150), and in yeast (7, 46).

TABLE 2.9
Effect of exogenous supplied purines on the level of the four first enzymes of the purine *de novo* biosynthetic pathway in wild type *S. typhimurium* (105).

Addition to the medium (0.1 mM)	Enzyme activity (nmoles/min/mg protein)			
	PRPP-amidotransferase (*purF*)	GAR-synthetase (*purD*)	GAR-formyl-transferase	FGAR-amidotransferase (*purG*)
None	66	16	39	10
Adenine	56	19	39	11
Hypoxanthine	15	2	3	3
Guanine	24	8	9	5

From extracts of wild type of *E. coli* a protein has been isolated which will bind to DNA containing the *purE* gene. This binding was shown to be ATP-dependent. Under the same conditions the protein also binds to DNA containing the *purA* gene. Furthermore it was found that the protein binds to DNA containing the *purF* or *purI* genes, and that this binding was GTP-dependent (134).

In an attempt to identify a possible low-molecular weight effector which would trigger the repression of the synthesis of the *de novo* enzymes, a number of *S. typhimurium* mutants were grown in the presence of exogenous purines, and the individual purine nucleotide pools, as well as the levels of the first four enzymes of the *de novo* pathway were determined. The mutants and the purine additions were chosen so as to obtain large variations in the steady state concentrations of endogenous adenine and guanine nucleotides (105). No straightforward correlation was observed between the levels of the four enzymes and the endogenous adenine and guanine nucleotide pools. This suggests that purine nucleotides are probably not the repressing metabolites (105).

In the course of these studies it was observed that addition of guanine to a mutant defective in guanine phosphoribosyltransferase (*gpt*) results in repression of the synthesis of the purine *de novo* enzymes, accompanied by a reduction of the growth rate and a decrease in the ATP and GTP pools, as well as an increase in the PRPP pools (Table 2.10). Thus it seems that in a purine prototrophic *gpt* strain addition of guanine causes partial purine starvation.

Mutants defective in both guanine and hypoxanthine phosphoribosyltransferases (*gpt,hpt* mutants) contain fully repressed levels of the *de novo* enzymes (Table 2.10). When such mutants were grown in minimal medium, their low growth rate and purine nucleotide pools indicated partial purine starvation (Table 2.10). Such cells excrete hypoxanthine (*see* p. 47) due to their inability to reutilize endogenously produced hypoxanthine. These observations indicate that the purine nucleoside triphosphates are not directly involved in the repression mechanism. Rather it appears that the repressing metabolites may be free hypoxanthine and guanine (105).

The synthesis of the enzymes catalyzing the conversion of IMP to GMP, and of IMP to AMP, are also regulated by purine compounds in many microorganisms. Addition of guanine or guanosine represses the synthesis of the two enzymes involved in GMP synthesis from IMP in *E. coli* (197), *B. subtilis* (199), and *Neurospora* (36). Both adenine and guanine and their nucleoside derivatives can repress the synthesis of the two enzymes involved in AMP synthesis from IMP in *A. aerogenes* (196) and *B. subtilis* (199).

B. Regulation of the activity of the *de novo* enzymes

The *de novo* biosynthesis of IMP may be controlled by regulation of the activity of the first enzyme in the pathway, i.e. PRPP amidotransferase, by exogenous purines. From *in vitro* studies it is known that PRPP amidotransferase from different sources is inhibited by adenine and guanine nucleotides (p. 31). Strong evidence for the operation of a feedback control *in vivo* was obtained by Gots *et al.*, who followed the formation of the purine *de novo* intermediary compounds in *pur* mutants of *S. typhimurium* and *E. coli*. They observed that addition of preformed purines to cultures of these mutants resulted in an immediate strong decrease in the rate of formation of intermediates in the purine *de novo* synthesis (69, 70, 158). Similar observations have been reported for yeast (7, 46), and *Neurospora* (215). Another,

TABLE 2.10

Nucleotide pools, PRPP pools and purine *de novo* enzyme levels in mutants of *S. typhimurium*[a] (105).

Relevant genotype	Addition	PRPP-amido-transferase	GAR synthetase	GAR-formyl-transferase	FGAR-amido-transferase	ATP	GTP	PRPP	Doubling time (min)
		(nmoles/min/mg protein)				(nmoles/mg dry weight)			
wild type	None	66	16	39	10	5.7	2.8	1.1	54
	guanine (0.1 mM)	24	8	9	5	3.1	3.1	0.3	54
gpt	None	29	12	14	7	5.6	2.9	2.4	59
	guanine (0.1 mM)	15	nd[b]	4	nd[b]	3.1	1.6	4.5	77
hpt gpt	None	11	8	7	6	2.7	1.3	3.8	66

A Cells were grown in glucose minimal medium at 37°C.

[b] Not determined.

more indirect, piece of evidence for a feedback control is the reversal by thiamine of the inhibitory effect of adenine on the growth of several microorganisms. Since the initial five steps of purine and thiamine biosynthesis are identical (87, 183) the inhibiting effect of adenine may be explained by severe feedback inhibition of PRPP amidotransferase and hence of the biosynthesis of thiamine.

As discussed previously (Chapter 1) addition of purines to cultures of purine prototrophic strains of *E. coli* and *S. typhimurium* results in a strong reduction in the PRPP pools of the cells (Table 1.1, p. 11). This decrease may affect purine *de novo* synthesis preferentially, because the affinity of PRPP amidotransferase for PRPP is significantly lower than the affinities for PRPP of the purine phosphoribosyltransferase. A low PRPP pool therefore will preferentially be used for salvage of exogenous purines. Furthermore it has been shown that the feedback inhibition of PRPP amidotransferase by AMP and GMP is significantly potentiated when the concentration of PRPP is low (172, 173, 195, 234).

The conversion of IMP to AMP and GMP may also be the target of feedback control by exogenous purines. The first enzyme specific for AMP synthesis, adenylosuccinate synthetase, is inhibited by AMP, and the first enzyme specific for GMP synthesis, IMP dehydrogenase, is inhibited by GMP. Indirect evidence for the participation of these enzymes in the regulation of AMP and GMP synthesis *in vivo* is apparent from Table 2.8, which shows that addition of guanine inhibits the *de novo* synthesis of GMP more than that of AMP. The differential inhibition after adenine addition on the *de novo* synthesis of AMP and GMP is less pronounced.

C. Sporulation and spore germination

Many observations have indicated that purine compounds are involved in the initial process of sporulation. (a) Inosine and adenosine stimulates sporulation in *B. subtilis* (243) and *Myxococcus xanthus* (250), respectively. (b) The levels of purine nucleoside phosphorylase show a transient increase of 10–15 fold during the early period of sporulation of *B. cereus* (51). (c) In *B. subtilis*, PRPP amidotransferase, the first enzyme in the purine *de novo* pathway undergoes rapid oxygen-dependent inactivation as cells enter the stationary phase of growth (283). (d) In *B. subtilis*, all conditions of nutritional deficiency, known to induce sporulation result in a decrease of the GTP pool at the onset of sporulation (58, 157).

The pathway of the purine *de novo* synthesis is incomplete in the dormant spores. It has been found for *Neurospora* (216) that the conidia contain storage pools of purine nucleosides. The salvage of these purine compounds as precursors for nucleic acid synthesis plays a transient but essential role in the initial germination period before the *de novo* synthesis takes over. In *Aspergillus* it has furthermore been shown that addition of adenine (50 μM) increases the germination rate 10–15 fold (268).

In the first 20 minutes of germination of *B. megaterium* spores, *de novo* purine biosynthesis is absent. The RNA synthesis observed within a few minutes must therefore rely on preformed purine compounds stored in the dormant spore, partly as free nucleotides. However, more than 85% of the ribonucleotides needed for RNA synthesis during the first 30 minutes of germination are provided by degradation of dormant spore RNA (245, 246).

REFERENCES

1. Abbondandolo, A., Weyer, A., Heslot, H., and Lambert, M. (1971). Study of adenine aminohydrolase in the yeast, *Schizosaccharomyces pombe. J. Bacteriol.* **108**, 959–963.
2. Acharya, P., and Chhatpar, H. S. (1981). Purification and characterization of two forms of 5′AMP nucleotidase from *Neurospora crassa. Indian J. Exp. Biol.* **19**, 953–955.
3. Aida, K., Chung, S., Suzuki, J., and Yagi, I. (1965). On the purification and some properties of 5′-adenylic acid deaminating enzyme from *Microsporum audouini. Agric. Biol. Chem.* **29**, 508–514.
4. Ali, L. Z., and Sloan, D. L. (1982). Studies of the kinetic mechanism of hypoxanthine-guanine phosphoribosyltransferase from yeast. *J. Biol. Chem.* **257**, 1149–1155.
5. Anderson, J. M., and Roth, R. M. (1974). Cordycepin sensitivity in adenosine utilizing mutants of *Saccharomyces cerevisiae. Biochim. Biophys. Acta* **335**, 285–289.
6. Anderson, J. M., and Roth, R. (1976). Adenosine utilization in Cordycepin-sensitive mutants of *Saccharomyces cerevisiae. J. Bact.* **128**, 689–691.
7. Armitt, S., and Woods, R. A. (1970). Purine-excreting mutants of *Saccharomyces cerevisiae.* I. Isolation and genetic analysis. *Genet. Res.* **15**, 7–17.
8. Bagnara, A. S., and Finch, L. R. (1974). The effects of bases and nucleosides on the intracellular contents of nucleotides and 5-phosphoribosyl 1-pyrophosphate in *Escherichia coli. Eur. J. Biochem.* **41**, 421–430.
9. Beck, W. S., and Levin, M. (1963). Purification, kinetics, and repression control of bacterial trans-N-deoxyribosylase. *J. Biol. Chem.* **238**, 702–709.
10. Bell, R. M., Parsons, S. M., Dubravac, S. A., Redfield, A. G., and Koshland, D. E. (1974). Characterization of slowly interconvertible states of phosphoribosyladenosine triphosphate synthetase dependent on temperature, substrates and histidine. *J. Biol. Chem.* **249**, 4110–4118.
11. Benson, C. E., Brehmeyer, B. A., and Gots, J. S. (1971). Requirement of cyclic AMP for induction of GMP reductase in *Escherichia coli. Biochem. Biophys. Res. Comm.* **43**, 1089–1094.
12. Benson, C. E., and Gots, J. S. (1975). Regulation of GMP reductase in *Salmonella typhimurium. Biochim. Biophys. Acta* **403**, 47–57.
13. Benson, C. E., and Gots, J. S. (1975). Genetic modification of substrate specificity

of hypoxanthine phosphoribosyltransferase in *Salmonella typhimurium*. *J. Bacteriol.* **121**, 77–82.

14. Benson, C. E., and Gots, J. S. (1976). Occurrence of a regulatory deficiency in purine biosynthesis among *purA* mutants of *Salmonella typhimurium*. *Molec. Gen. Genet.* **145**, 31–36.
15. Berberich, M. A., and Gots, J. S. (1965). A structural gene mutation in *Salmonella typhimurium* resulting in repressibility of adenylosuccinase. *Proc. Natl. Acad. Sci. U.S.A.* **54**, 1254–1261.
16. Berens, R. L., Marr, J. J., Lafon, S. W., and Nelson, D. J. (1981). Purine metabolism in *Trypanosoma cruzi*. *Mol. Bio. chem. Parasitol.* **3**, 187–196.
17. Berlin, R. D. (1969). Adenylate pyrophosphorylase: Purification, reaction sequence, and inhibition by sodium ion. *Arch. Biochem. Biophys.* **134**, 120–129.
18. Berlin, R. D., and Stadtman, E. R. (1966). A possible role of purine nucleotide phyrophosphorylases in the regulation of purine uptake by *Bacillus subtilis*. *J. Biol. Chem.* **241**, 2679–2686.
19. Brenner, M., and Ames, B. N. (1971). The histidine operon and its regulation. *In* "Metabolic regulation" (H. J. Vogel, *Ed.*), Vol. 5, pp. 349–388. Academic Press, New York.
20. Brox, L. W., and Hampton, A. (1968). Inactivation of guanosine 5′-phosphate reductase by 6-chloro, 6-mercapto, and 2-amino-6-mercapto-9-*β*-D-ribofuranosylpurine 5′-phosphates. *Biochemistry* **7**, 398–405.
21. Brykun, J. A., and Sukhodolets, V. V. (1981). Phenotypical properties of the *Escherichia coli* K-12 purine auxotrophs with blocked synthesis of phosphoribosyltransferase for guanine, xanthine and hypoxanthine. *Genetika* **17**, 932–935.
22. Burridge, P. W., Woods, R. A., Henderson, J. F. (1977). Purine metabolism in *Saccharomyces cerevisiae*. *Can. J. Biochem.* **55**, 935–941.
23. Burridge, P. W., Woods, R. A., and Henderson, J. F. (1978). Altered purine metabolism in regulatory mutants of *Saccharomyces cerevisiae*. *J. Gen. Microbiol.* **107**, 403–406.
24. Buzzee, D. H., and Levin, A. P. (1968). Demonstration of an effector site for the enzyme inosine 5′-phosphate dehydrogenase. *Biochem. Biophys. Res. Comm.* **30**, 673–677.
25. Büngener, W. (1967) Adenosindeaminase und Nucleosidephosphorylase bei Malariaparasiten. *Z. Tropenmed. Parasitol.* **18**, 48–52.
26. Büngener, W. (1974). Influence of allopurinol on the multiplication of rodent malaria parasite. *Z. Tropenmed. Parasitol.* **25**, 309–312.
27. Carteni-Farina, M., Olivia, A., Romeo, G., Napolitano, G., DeRosa, M., Gambacarta, A., and Zappia, V. (1979). 5′-methylthioadenosine phosphorylase from *Caldariella acidophila*: Purification and properties. *Eur. J. Biochem.* **10**, 317–324.
28. Carter, C. E., and Cohen, L. H. (1956). The preparation and properties of adenylosuccinase and adenylosuccinic acid. *J. Biol. Chem.* **222**, 17–30.
29. Chapman, A. G., and Atkinson, D. E. (1977). Adenine nucleotide concentrations and turnover rates. Their correlation with biological activity in bacteria and yeast. *Adv. Microb. Physiol.* **15**, 253–306.
30. Chou, J. Y., and Martin, R. G. (1972). Purine phosphoribosyltransferases of *Salmonella typhimurium*. *J. Bacteriol.* **112**, 1010–1013.
31. Cohen, L., and Kaplan, R. (1977). Accumulation of nucleotides by starved *Escherichia cole* cells as a probe for the involvement of ribonucleases in ribonucleic acid degradation. *J. Bacteriol.* **129**, 651–657.

32. Conner, R. L., and Linden, C. (1970). Purine and pyrimidine 5′-nucleotide catabolism in *Tetrahymena pyriformis* W. *J. Protozool.* **17**, 659–662.
33. Cowman, A., and Beacham, I. R. (1980). Molecular cloning of the gene (ush) from *Escherichia coli* specifying periplasmic UDP-sugar hydrolase (5′-nucleotidase). *Gene* **12**, 281–286.
34. Cronan, J. E., and Godson, G. N. (1972). Mutants of *Escherichia coli* with temperature-sensitive lesions in membrane phospholipid synthesis: Genetic analysis of glycerol-3-phosphate acyltransferase mutants. *Molec. Gen. Genet.* **116**, 199–210.
35. Cronan, J. E., Ray, T. K., and Vagelos, P. R. (1970). Selection and characterization of an *E. coli* mutant defective in membrane lipid biosynthesis. *Proc. Natl. Acad. Sci. U.S.A.* **65**, 737–744.
36. Cummins, J. E., and Mitchison, J. M. (1964). A method for making autoradiographs of yeast cells which retain pool components. *Exp. Cell. Res.* **34**, 406–409.
37. Cummins, J. E., and Mitchison, J. M. (1967). Adenine uptake and pool formation in the fission yeast *Scizosaccharomyces pombe*. *Biochim. Biophys. Acta* **136**, 108–120.
38. Danzin, C., and Cardinaud, R. (1974). Deoxyribosyl transfer catalysis with trans-N-deoxyribosylase. *Eur. J. Biochem.* **48**, 255–262.
39. Danzin, C., and Cardinaud, R. (1976). Deoxyribosyl transfer catalysis with trans-N-deoxyribosylase. *Eur. J. Biochem.* **62**, 365–372.
40. DeFelice, M., Levinthal, M., Jaccarino, M., and Guardiola, J. (1979). Growth inhibition as a consequence of antagonism between related amino acids: effect of valine in *Escherichia coli* K-12. *Microbiol. Rev.* **43**, 42–58.
41. Deibel, R., and Ives, D. H. (1977). Resolution of two deoxynucleoside kinases from *Lactobacillus acidophilus* R-26 by blue sepharose affinity chromatography. *J. Biol. Chem.* **252**, 8235–8239.
42. DeGroodt, A., Whitehead, E. P., Heslot, H., and Poirier, L. (1971). The substrate specificity of purine phosphoribosyltransferases in *Schizosaccharomyces pombe*. *Biochem. J.* **122**, 415–420.
43. Demain, A. L. (1964). Nutrition of adenineless auxotrophs of yeasts. *J. Bacteriol.* **88**, 339–345.
44. DeWolf, W. E., Fullin, F. A., and Schramm, V. L. (1975). The catalytic site of AMP nucleosidase. Substrate specificity and pH effects with AMP and formycin 5′-phosphate. *J. Biol. Chem.* **25**, 10868–10875.
45. Dickinson, J. R. (1981). The *cdc* mutation in *Schizosaccharomyces pombe* is a temperature-sensitive defect in nucleoside diphosphokinase. *Eur. J. Biochem.* **119**, 341–345.
46. Dorfman, B. Z. (1969). The isolation of adenylosuccinate synthetase mutants in yeast by selection for constitutive behavior in pigmented strains. *Genetics* **61**, 377–389.
47. Drummond, G. J., and Yamamoto, M. (1971). Nucleoside cyclic phosphate diesterases. *In* "The enzymes" (P. D. Boyer, *Ed.*), Vol. IV, pp. 355–371. Academic Press, New York.
48. Drummond, G. J. (1971). Nucleotide phosphomonoesterases. *In* "The enzymes" (P. D. Boyer, *Ed.*), Vol. IV, pp. 337–354. Academic Press, New York.
49. Duerre, J. A. (1962). A hydrolytic nucleosidase acting on S-adenosylhomocysteine and on 5′-methylthioadenosine. *J. Biol. Chem.* **237**, 3737–3741.
50. Durham, J. P., and Ives, D. H. (1971). The metabolism of deoxyribonucleosides

in *Lactobacillus acidophilus*: Regulation of deoxyadenosine, deoxycytidine, deoxyguanosine, and deoxythymidine kinase activities by nucleotides. *Biochim. Biophys. Acta* **228**, 9–25.
51. Engelbrecht, H. L. (1972). Time course of purine nucleoside phosphorylase occurrence in sporulation of *Bacillus cereus*. *J. Bacteriol.* **111**, 33–36.
52. Evans, H. H. (1982). Nucleotide metabolism in *Physarum polycephalum*. *In* "Cell Biology of *Physarum* and *Didymium*" (H. C. Aldrich and J. W. Daniel, *Eds*), Vol. 2, pp. 183–216. Academic Press, New York.
53. Eyzaguirre, J., and Atkinson, D. E. (1974). Regulatory effects of purine ribonucleotides on *Escherichia coli* adenylosuccinate synthetase. *Arch. Biochem. Biophys.* **169**, 339–343.
54. Felicioli, R. A., Senesi, S., Marmocchi, F., Falcone, G., and Ipata, P. L. (1973). Nucleoside phosphomonoesterase during growth cycle of *Bacillus subtilis*. *Biochemistry* **12**, 547–552.
55. Ferro, A. J., Barrett, A., Shapiro, S. K. (1976). Kinetic properties and the effect of substrate analogues on 5′-methyl-thioadenosine nucleosidase from *Escherichia coli*. *Biochim. Biophys. Acta* **438**, 487–494.
56. Fink, K., and Nygaard, P. (1979). Purine metabolism in microplasmodia of *Physarum polycephalum*. *Biochim. Biophys. Acta* **563**, 269–277.
57. Fish, W. R., Marr, J. J., and Berens, R. L. (1982). Purine metabolism in *Trypanosoma brucei gambiense*. *Biochim. Biophys. Acta* **714**, 422–428.
58. Freese, E., Heinze, J. E., and Galliers, E. M. (1979). Partial purine deprivation causes sporulation of *Bacillus subtilis* in the presence of excess ammonia, glucose and phosphate. *J. Gen. Microbiol.* **115**, 193–205.
59. Gallant, J., Irr, J., and Cashel, M. (1971). The mechanism of amino acid control of guanylate and adenylate biosynthesis. *J. Biol. Chem.* **246**, 5812–5816.
60. Garber, B. B., and Gots, J. S. (1980). Utilization of 2.6-diaminopurine by *Salmonella typhimurium*. *J. Bacteriol.* **143**, 864–871.
61. Garber, B. B., Jochimsen, B. U., and Gots, J. S. (1980). Glutamine and related analogs regulate guanosine monophosphate reductase in *Salmonella typhimurium*. *J. Bacteriol.* **143**, 105–111.
62. Gardner, R., and Kornberg, A. (1967). Biochemical studies of bacterial sporulation and germination. V. Purine nucleoside phosphorylase of vegetative cells and spores of *Bacillus cereus*. *J. Biol. Chem.* **242**, 2383–2388.
63. Gardner, W. J. E., and Woods, R. A. (1979). Isolation and characterization of guanine auxotrophs in *Saccharomyces cerevisiae*. *Can. J. Microb.* **25**, 380–389.
64. Gilbert, H. J., Lowe, C. R., and Drabble, W. T. (1979). Inosine 5′-monophosphate dehydrogenase of *Escherichia coli*. Purification by affinity chromatography, subunit structure and inhibition by guanosine 5′-monophosphate. *Biochem. J.* **183**, 481–494.
65. Ginther, C. L., and Ingraham, J. L. (1974). Nucleoside diphosphokinase of *Salmonella typhimurium*. *J. Biol. Chem.* **249**, 3406–3411.
66. Ginther, C. L., and Ingraham, J. L. (1971). Cold-sensitive mutant of *Salmonella typhimurium* defective in nucleoside diphosphokinase. *J. Bacteriol.* **118**, 1020–1026.
67. Glaser, M., Nulty, W., and Vagelos, P. R. (1975). Role of adenylate kinase in the regulation of macromolecular biosynthesis in a putative mutant of *Escherichia coli* defective in membrane phospholipid biosynthesis. *J. Bacteriol.* **123**, 128–136.
68. Goelz, S. E., and Cronan, J. E. (1982). Adenylate kinase of *Escherichia coli*:

evidence for a functional interaction in phospholipid biosynthesis. *Biochemistry* **21**, 189–195.

69. Gollup, E. G., and Gots, J. S. (1959). Purine metabolism in bacteria. VI. Accumulations by mutants lacking adenylosuccinase. *J. Bacteriol.* **78**, 320–325.
70. Gots, J. S. (1957). Purine metabolism in bacteria. V. Feedback inhibition. *J. Biol. Chem.* **228**, 57–66.
71. Gots, J. S. (1971). Regulation of purine and pyrimidine metabolism. *In* "Metabolic pathways" Vol. 5, "Metabolic regulation" (H. J. Vogel, *Ed.*), pp. 225–255. Academic Press, New York.
72. Gots, J. S., Benson, C. E., Jochimsen, B., and Koduri, K. R. (1977). Microbial models and regulatory elements in the control of purine metabolism. *Ciba Found. Symp.* **48**, 23–41. Elsevier North-Holland Inc.
73. Gots, J. S., Benson, C. E., and Shumas, S. R. (1972). Genetic separation of hypoxanthine and guanine-xanthine phosphoribosyltransferase activity by deletion mutations in *Salmonella typhimurium*. *J. Bacteriol.* **112**, 910–916.
74. Greene, E. M. (1980). Cytokinin production by microorganisms. *Bot. Rev.* **46**, 25–74.
75. Greer, W. L., Pendyala, L., and Wellman, A. M. (1980). Guanine metabolism in *Neurospora*. *Can. J. Biochem.* **58**, 369–376.
76. Greer, W. L., and Wellman, A. M. (1980). Isolation and characterization of guanine auxotrophs in *Neurospora crassa*. *Can. J. Microbiol.* **26**, 1412–1415.
77. Gutteridge, W. E., and Coombs, G. H. (1977). Biochemistry of parasitic protozoa. MacMillan Press, London.
78. Gutteridge, W. E., and Davies, M. J. (1981). Enzymes of purine salvage in *Trypanasoma cruzi*. *FEBS Lett.* **127**, 211–214.
79. Hall, R. H. (1970). N^6-(Δ^2-isopentyl) adenosine: Chemical reactions, biosynthesis, metabolism and significance to the structure and function of tRNA. *In* "Progress in nucleic acid research and molecular biology" (J. N. Davidson and W. E. Cohn, *Eds.*), Vol. 10, pp. 57–86. Academic Press, New York.
80. Hamet, M., Bonissol, C., and Cartier, P. (1980). Enzymatic activities on purine and pyrimidine metabolism in nine mycoplasma species contaminating cell cultures. *Clin. Chim. Acta* **103**, 15–22.
81. Hammer-Jespersen, K., Munch-Petersen, A., Nygaard, P., and Schwartz, M. (1971). Induction of enzymes involved in the catabolism of deoxyribonucleosides and ribonucleosides in *Escherichia coli* K12. *Eur. J. Biochem.* **19**, 533–538.
82. Hansen, B. D., Sleeman, K., and Pappas, P. W. (1980). Purine base and nucleoside uptake in *Plasmodium berghei* and host erythrocytes. *J. Parasitology* **66**, 205–212.
83. Hartenstein, R. C., and Fridovich, J. (1967). Adenine aminohydrolase. An investigation of specificity. *J. Biol. Chem.* **242**, 740–746.
84. Hartman, S. C. (1970). Purines and pyrimidines. *In* "Metabolic Pathways" (D. M. Greenberg, *Ed.*), Vol. V, pp. 1–68. Academic Press, New York.
85. Hatanaka, M., Giudice, R. D., and Long, C. (1975). Adenine formation from adenosine by mycoplasmas: Adenosine phosphorylase activity. *Proc. Natl. Acad. Sci. U.S.A.* **72**, 1401–1405.
86. Henderson, J. F. (1972). Regulation of purine biosynthesis. *Am. Chem. Soc. Washington.*
87. Henderson, J. F. (1980). Inhibition of microbial growth by naturally occurring purine bases and ribonucleosides. *Pharmacol. Therapy* **8**, 605–627.
88. Henderson, J. F., Scott, F. W., and Lowe, J. K. (1980). Toxicity of naturally occurring purine deoxyribonucleosides. *Pharmacol. Therapy* **8**, 573–604.

89. Herschfield, M. S. (1979). Apparent suicide inactivation of human lymphoblast S-adenosylhomocysteine hydrolase by 2′-deoxyadenosine and adenine arabinoside. *J. Biol. Chem.* **254**, 22–25.
90. Heyde, E., and Morrison, J. F. (1976). Studies on inosine monophosphate dehydrogenase. An association-dissociating system. *Biochim. Biophys. Acta* **429**, 635–644.
91. Heyde, E., and Morrison, J. F. (1976). Studies on inosine monophosphate dehydrogenase. Isotope exchange at equilibrium. *Biochim. Biophys. Acta* **429**, 661–671.
92. Heyde, E., Nagabhushanam, A., Vonarx, M., and Morrison, J. F. (1976). Studies on inosine monophosphate dehydrogenase. Steady state kinetics. *Biochim. Biophys. Acta* **429**, 645–660.
93. Hill, D. L. (1972). Purine, pyrimidine, and nucleic acid metabolism. *In* "The Biochemistry and Physiology of Tetrahymena" Vol. VI, pp. 125–162. Academic Press, New York.
94. Hill, D. L., and Chambers, P. (1967). The purine and pyrimidine metabolism of *Tetrahymena pyriformis. J. Cell. Physiol.* **69**, 321–330.
95. Hill, D. L., Straight, S., and Allan, P. W. (1970). Use of *Tetrahymena pyriformis* to evaluate the effect of purine and pyrimidine analogs. *J. Protozool.* **17**, 619–623.
96. Hiraga, S., and Sugino, Y. (1966). Nucleoside monophosphatkinases of *Escherichia coli* infected with an RNA phage. *Biochim. Biophys. Acta* **114**, 416–418.
97. Hochstadt, J. (1978). Hypoxanthine phosphoribosyltransferase and guanine phosphoribosyltransferase from enteric bacteria. *In* "Meth. Enzymol." (P. A. Hoffee and M. E. Jones, *Eds.*), Vol. LI, pp. 549–558. Academic Press, New York.
98. Hochstadt, J. (1978). Adenine phosphoribosyltransferase from *Escherichia coli. In* "Meth. Enzymol." (P. A. Hoffee and M. E. Jones, *Eds.*), Vol. LI, pp. 558–567. Academic Press, New York.
99. Hochstadt-Ozer, J., and Cashel, M. (1972). The regulation of purine utilization in bacteria. V. Inhibition of purine phosphoribosyltransferase activities and purine uptake in isolated membrane vesicles by guanosine tetraphosphate. *J. Biol. Chem.* **247**, 7067–7072.
100. Hochstadt-Ozer, J., and Stadtman, E. R. (1971). The regulation in bacteria of purine utilization. I. Purification of adenine phosphoribosyltransferase from *Escherichia coli* K12 and control of activity by nucleotides. *J. Biol. Chem.* **246**, 5294–5303.
101. Hoffmeyer, J., and Neuhard, J. (1971). Metabolism of exogenous purine bases and nucleosides by *Salmonella typhimurium. J. Bacteriol.* **106**, 14–24.
102. Holden, J. A., Harriman, P. D., and Wall, J. D. (1976). *Escherichia coli* mutants deficient in guanine-xanthine phosphoribosyltransferase. *J. Bacteriol.* **126**, 1141–1148.
103. Holguin, J., and Cardinaud, R. (1975). Trans-N-deoxyribolase: Purification by affinity chromatography and characterization. *Eur. J. Biochem.* **54**, 505–514.
104. Holmes, R. K., and Singer, M. F. (1973). Purification and characterization of adenylate kinase as an apparent adenosine triphosphate-dependent inhibitor of ribonuclease II in *Escherichia coli. J. Biol. Chem.* **248**, 2014–2021.
105. Houlberg, U. (1981). Regulation of purin biosynthesen/regulation af *pur* gen expression i *Salmonella typhimurium.* Thesis, University Copenhagen.
106. Housset, P., Nagy, M. (1977). Study of the role of purine phosphoribosyltransferases in the uptake of adenine and guanine by *Schizosaccharomyces pombe* cells. *Eur. J. Biochem.* **73**, 99–105.
107. Hove-Jensen, B., and Nygaard, P. (1982). Phosphoribosylpyrophosphate

synthetase of *Escherichia coli*: Identification of a mutant enzyme. *Eur. J. Biochem.* **126**, 327–332.
108. Ishii, K., and Shiio, J. (1970). Regulation of purine ribonucleoside synthesis by end product inhibition. III. Effect of purine nucleotides on succino-AMP synthetase of *Bacillus subtilis*. *J. Biochem.* **68**, 171–176.
109. Ishii, K., and Shiio, J. (1972). Improved inosine production and derepression of purine nucleotide biosynthetic enzymes in 8-azaguanine resistant mutants of *Bacillus subtilis*. *Agric. Biol. Chem.* **36**, 1511–1522.
110. Ito, Y., Tomasselli, A. G., and Noda, L. H. (1980). ATP., AMP phosphotransferase from bakers yeast. *Eur. J. Biochem.* **105**, 85–92.
111. Jensen, K. F. (1977). Purine nucleoside phosphorylases in bacteria. Thesis, University Copenhagen.
112. Jensen, K. F. (1978). Two purine nucleoside phosphorylases in *Bacillus subtilis*. Purification and some properties of the adenosine-specific phosphorylase. *Biochim. Biophys. Acta* **525**, 346–356.
113. Jensen, K. F., and Nygaard, P. (1975). Purine nucleoside phosphorylase from *Escherichia coli* and *Salmonella typhimurium*. Purification and some properties *Eur. J. Biochem.* **51**, 253–265.
114. Jha, K. K. (1971). Genes conferring resistance to 8-aza adenine in *Neurospora crassa* and the variability of resistant alleles in the aza-1 locus with respect to excretion of purines. *Molec. Gen. Genet.* **114**, 156–167.
115. Jochimsen, B. (1979). GMP reductase-guanosine kinase: Regulation. Thesis, University Copenhagen.
116. Jochimsen, B., Nygaard, P., and Vestergaard, T. (1975). Location on the chromosome of *Escherichia coli* of genes governing purine metabolism. *Molec. Gen. Genet.* **143**, 85–91.
117. Jyssym, S. (1974). Purine metabolism in *Neisseria meningitidis*. 1. Utilization of exogenous adenine. *Acta. Path. Microbiol. Scand.* **Section B, 82**, 508–520.
118. Jyssum, S. (1974). Purine metabolism in *Neisseria meningitidis*. 2. Utilization of exogenous adenosine, guanosine and inosine. *Acta Path. Microbiol. Scand.* **Section B, 82**, 885–894.
119. Jyssum, S. (1975). Purine metabolism in *Neisseria meningitidis*. 3. Utilization of exogenous hypoxanthine, guanine and xanthine. *Acta Path. Microbiol. Scand.* **Section B, 83**, 397–406.
120. Jyssum, S., and Jyssym, K. (1975). Purine metabolism in *Neisseria meningitidis*. 4. Interconversion of purine ribonucleotides. *Acta Path. Microbiol. Scand.* **Section B, 83**, 407–415.
121. Kalle, G. P., and Gots, J. S. (1961). Mechanism of resistance to 2,6-diaminopurine in *Salmonella typhimurium*. *Biochim. Biophys. Acta* **51**, 130–137.
122. Kalle, G. P., and Gots, J. S. (1961). Alterations in purine nucleotide pyrophosphorylases and resistance to purine analogues. *Biochim. Biophys. Acta* **53**, 166–173.
123. Kalle, G. P., and Gots, J. S. (1962). Metabolic abnormalities of 2,6-diaminopurine resistant mutants of *Salmonella typhimurium*. *Proc. Soc. Exp. Biol. Med.* **109**, 277–281.
124. Kaplan, R., Cohen, L., and Yagil, E. (1975). Acid-soluble degradation products of ribonucleic acid in *Escherichia coli* and the role of nucleotidases in their catabolism. *J. Bact.* **124**, 1159–1164.
125. Karlström, O. (1968). Mutants of *Escherichia coli* defective in ribonucleoside and deoxyribonucleoside catabolism. *J. Bacteriol.* **95**, 1069–1077.

126. Karlström, O. (1970). Failure of *Escherichia coli* B to incorporate added deoxycytidine, deoxyadenosine, and deoxyguanosine into DNA. *Eur. J. Biochem.* **17**, 68–71.
127. Kidder, G. W., Dewey, V. C., and Nolan, L. L. (1977). Adenine deaminase of a eucaryotic animal cell, *Crithidia fasciculata*. *Archiv. Biochem. Biophys.* **183**, 7–12.
128. Kidder, G. W., and Nolan, L. L. (1979). Adenine aminohydrolase: occurrence and possible significance in trypanosomid flagellates. *Proc. Natl. Acad. Sci. U.S.A.* **76**, 3670–3672.
129. Kidder, G. A., Nolan, L. L., and Dewey, V. C. (1979). The purine phosphoribosyltransferases of *Crithidia fasciculata*. *J. Parasitology* **65**, 520–525.
130. Kier, L. D., Weppelman, R., and Ames, B. N. (1977). Resolution and purification of three periplasmic phosphatases of *Salmonella typhimurium*. *J. Bacteriol.* **130**, 399–410.
131. Kocharian, S. M., Chukanova, T. I., and Sukhodolets, V. V. (1977). Mutations of resistance to 2,6-dianimopurine and 6-methylpurine affecting adenine phosphoribosyltransferase in *Escherichia coli*. *Genetika* **13**, 1821–1830.
132. Kocharian, S. M., Livshitz, V. A., and Sukhodolets, V. V. (1975). Genetic study of *Escherichia coli* K-12 mutants resistant to 2,6-diaminopurine. *Genetika* **11**, 79–89.
133. Kocharian, S. M., and Sukhodolets, V. V. (1976). Phenotypic manifestion of 2,6-diamino purine-resistant mutations (*apt*) in genomes of *Escherichia coli* K-12 purine auxotrophs. *Genetika* **12**, 100–108.
134. Koduri, R. K., and Gots, J. S. (1980). A DNA-binding protein with specificity for *pur* genes in *Escherichia coli*. *J. Biol. Chem.* **255**, 9594–9598.
135. Komatsu, K. J., and Kodaira, R. (1973). Accumulation of nucleic acid-related substances by microorganisms. V. Role of purine nucleoside phosphorylase on the accumulation of guanosine by mutants of a *Bacillus* strain. *J. Gen. Appl. Microb.* **19**, 263–271.
136. Koszalka, G. W., and Krenitsky, T. A. (1979). Nucleosidases of *Leishmania donovani*. *J. Biol. Chem.* **254**, 8185–8193.
137. Krenitsky, T. A., Koszalka, G. W., Tuttle, J. V., Adamczyk, D. L., Elion, G. B., and Marr, J. R. (1980). Purine salvage enzymes in man and *Leishmania donovani*. *In* "Purine Metabolism in Man" Part III. Adv. Exp. Med. Biol." (A. Rapado, R. W. E. Watts, and C. H. M. M. DeBruyn, *Eds.*), Vol. 122B, pp. 51–56.
138. Krenitzky, T. A., Neil, S. M., and Miller, R. L. (1970). Guanine and xanthine phosphoribosyltransfer activities of *Lactobacillus casei* and *Escherichia coli*. Their relationship to hypoxanthine and adenine phosphoribosyltransfer activities. *J. Biol. Chem.* **245**, 2605–2611.
139. Krishnaiah, K. V. (1975). Inosine acid 5′-monophosphate dehydrogenase of *Escherichia coli*: Purification by affinity chromatography and some properties. *Archiv. Biochem. Biophys.* **170**, 567–575.
140. Königk, E. (1978). Purine nucleotide metabolism in promastigotes of *Leishmania tropica*: Inhibitory effect of allopurinol and analogues of purine nucleosides. *Tropenmed. Parasit.* **29**, 435–438.
141. Lambden, P. R., and Drabble, W. T. (1973). Inosine 5′-monophosphate dehydrogenase of *Escherichia coli* K-12: the nature of the inhibition by guanosine 5′-monophosphate. *Biochem. J.* **133**, 607–608.
142. Lambden, P. R., and Drabble, W. T. (1973). The *gua* operon of *Escherichia coli* K-12: Evidence for polarity from *GuaB* to *GuaA*. *J. Bacteriol.* **115**, 992–1002.

143. Lee, B. H., and Hartman, S. C. (1974). Preferential utilization of glutamine for amination of xanthosine 5′-phosphate to guanosine 5′-phosphate by purified enzymes from *Escherichia coli*. *Biochem. Biophys. Res. Comm.* **60**, 918–925.

144. Leibach, T. K., Spiess, G. I., Neudecker, T. J., Pescke, G. J., Puchwein, G., and Hartmann, G. R. (1971). Purification and properties of adenosine kinase from dried brewers yeast. *Z. Physiol. Chem.* **352**, 328–344.

145. Lennon, M. B., and Suhadolnik, R. J. (1976). Biosynthesis of 3′-deoxyadenosine by *Cordyceps militaris*. Mechanism of reduction. *Biochim. Biophys. Acta* **425**, 532–536.

146. Leung, H. B., and Schramm, V. L. (1978). The role of adenosine monophosphate nucleosidase in the regulation of adenine nucleotide levels in *Azotobacter vinelandii* during aerobi-anaerobic transition. *Archiv. Biochem. Biophys.* **190**, 46–56.

147. Leung, H. B., and Schramm, V. L. (1980). Adenylate degradation in *Escherichia coli*. The role of AMP nucleosidase and properties of the purified enzyme. *J. Biol. Chem.* **255**, 10867–10874.

148. Leung, H. B., and Schramm, V. L. (1981). A mutant AMP nucleosidase. *J. Biol. Chem.* **256**, 12823–12829.

149. Levin, A. P., and Magasanik, B. (1961). The effect of purines on the formation of two enzymes involved in purine biosynthesis. *J. Biol. Chem.* **236**, 184–188.

150. Levine, R. A., and Taylor, M. W. (1981). Selection for purine regulatory mutants in an *E. coli* hypoxanthine phosphoribosyltransferase-guanine phosphoribosyltransferase double mutant. *Molec. Gen. Genet.* **181**, 313–318.

151. Lichtenstein, J., Barner, H. D., and Cohen, S. S. (1960). The metabolism of exogenously supplied nucleotides by *Escherichia coli*. *J. Biol. Chem.* **235**, 457–465.

152. Livshits, V. A. (1976). Effect of 2,6-diaminopurine resistant mutations on the utilization of adenine and adenosine by adenine-requiring *Escherichia coli* K-12 strains. *Genetika* **12**, 180–182.

153. Livshits, V. A., and Sukhodolets, V. V. (1973). Mutants of *Escherichia coli* K-12. Defective in purine nucleoside phosphorylase and in *de novo* synthesis of purine nucleotides. *Genetika* **9**, 66–75.

154. Livshits, V. A., and Sukhodolets, V. V. (1973). On the role of adenine nucleotides in the regulation purine ribonucleoside utilization by mutants of *Escherichia coli* K-12 defective in purine nucleoside phosphorylase. *Genetika* **9**, 102–111.

155. Lomax, C. A., and Woods, R. A. (1969). Mutant of yeast sensitive to 2,6-diaminopurine. *J. Bacteriol.* **100**, 817–822.

156. Lomax, C. A., and Woods, R. A. (1973). A complex genetic locus controlling purine biosynthesis in yeast. *Molec. Gen. Genet.* **120**, 139–149.

157. Lopez, J. M., Marks, C. L., and Freese, E. (1979). The decrease of guanine nucleotides initiates sporulation of *Bacillus subtilis*. *Biochim. Biophys. Acta* **587**, 238–252.

158. Love, S. H., and Gots, J. S. (1955). Purine metabolism in bacteria. III. Accumulation of a new pentose-containing arylamine by a purine requiring mutant of *Escherichia coli*. *J. Biol. Chem.* **212**, 647–654.

159. Magasanik, B. (1962). *In* "The Bacteria" Vol. *3* "Biosynthesis of purine and pyrimidine nucleotides" (J. C. Gunsalus, and R. Y. Stanier, *Eds.*), pp. 295–334. Academic Press, New York.

160. Magasanik, B., and Karibian, D. (1960). Purine nucleotide cycles and their metabolic role. *J. Biol. Chem.* **235**, 2672–2681.

161. Mager, J., and Magasanik, B. (1960). Guanosine 5′-phosphate reductase and its role in the interconversion of purine nucleotides. *J. Biol. Chem.* **235**, 1474–1478.
162. Magill, J. M., and Magill, C. W. (1973). Correlation of growth inhibition patterns to nucleoside transport models in *Neurospora crassa*. *J. Bacteriol.* **116**, 1071–1074.
163. Magill, J. M., and Magill, C. W. (1975). Purine base transport in *Neurospora crassa*. *J. Bacteriol.* **124**, 149–154.
164. Magill, C. W., Sabina, R. L., Garber, T. L., and Magill, J. M. (1982). Guanine uptake and metabolism in *Neurospora crassa*. *J. Bacteriol.* **149**, 941–947.
165. Markham, G. D., and Reed, G. H. (1977). Adenylosuccinate synthetase from *Azotobacter vinelandii*: Purification, properties and steady-state kinetics. *Archiv. Biochem. Biophys.* **184**, 24–35.
166. Marr, J. J., Berens, R. L., and Nelson, D. J. (1978). Purine metabolism in *Leishmania braziliensis*. *Archiv. Biochem. Biophys.* **544**, 360–371.
167. Martin, R. G. (1963). The first enzyme in histidine biosynthesis: The nature of feedback inhibition by histidine. *J. Biol. Chem.* **238**, 257–268.
168. Martin, W. R., and Yang, R. R. (1972). Inosine and guanine phosphoribosyltransferase in *Escherichia coli*. *Biochem. Biophys. Res. Comm.* **48**, 1641–1648.
169. Matsui, H., Sato, K., Enei, H., and Hirose, Y. (1979). Production of guanosine by psicofuranine and decoyinine resistant mutants of *Bacillus subtilis*. *Agric. Biol. Chem.* **43**, 1739–1744.
170. Mattoo, A. K., Dandekar, A. M., Trivedi, J. P., Majmudar, G. H., and Patel, D. M. (1979). Regulation of *Neurospora crassa* nucleosidase by some environmental factors. *Z. Allg. Mikrobiol.* **19**, 253–260.
171. Mehra, R. K., and Drabble, W. T. (1981). Dual control of the *gua* operon of *Escherichia coli* K12 by adenine and guanine nucleotides. *J. Gen. Microbiol.* **123**, 27–37.
172. Messenger, L. J., and Zalkin, H. (1979). Glutamine phosphoribosylpyrophosphate amidotransferase from *Escherichia coli*. *J. Biol. Chem.* **254**, 3382–3392.
173. Meyer, E., and Switzer, R. L. (1979). Regulation of *Bacillus subtilis* glutamine phosphoribosylpyrophosphate amidotransferase activity by end products. *J. Biol. Chem.* **254**, 5397–5402.
174. Midorikawa, Y., Akiya, T., Kuninaka, A., and Yoshino, H. (1972). Conversion of hypoxanthine derivatives to guanine derivatives by *Bacillus subtilis*. *Agric. Biol. Chem.* **36**, 1529–1535.
175. Miller, R. L., and Bieber, A. L. (1969). Substrate binding specificity and properties of inosine monophosphate: pyrophosphate phosphoribosyltransferase (EC 2.4.2.8) from brewers yeast. *Biochemistry* **8**, 603–608.
176. Miller, R. L., Ramsey, G. A., Krenitsky, T. A., and Elion, G. B. (1972). Guanine phosphoribosyltransferase from *Escherichia coli*. Specificity and properties. *Biochemistry* **11**, 4723–4731.
177. Mitchell, A., and Finch, L. R. (1977). Pathway of nucleotide biosynthesis in *Mycoplasma mycoides* subsp. *mycoides*. *J. Bacteriol.* **130**, 1047–1054.
178. Mitchell, A., Sin, J. L., and Finch, L. R. (1978). Enzymes of purine metabolism in *Mycoplasma mycoides* subsp. *mycoides*. *J. Bacteriol.* **134**, 706–712.
179. Mitchison, J. M., Cummins, J. E., Gross, R. R., and Creanor, J. (1969). The uptake of bases and their incorporation into RNA during the cell cycle of *Schizosaccharomyces pombe* in normal growth and after a step-down. *Exp. Cell. Res.* **57**, 411–422.

180. Moat, A. G., and Friedman, H. (1960). The biosynthesis and interconversion of purines and their derivatives. *Bact. Rev.* **24**, 309–339.
181. Momose, H., Nishikawa, H., and Shiio, J. (1966). Regulation of purine nucleotide synthesis in *Bacillus subtilis*. I. Enzyme repression by purine derivatives. *J. Biochem.* **59**, 325–331.
182. Morton, D. P., and Parsons, S. M. (1977). Synergistic inhibition of ATP phosphoribosyltransferase by guanosine tetraphosphate and histidine. *Biochem. Biophys. Res. Comm.* **74**, 172–177.
183. Moyed, H. S. (1964). Inhibition of the biosynthesis of the pyrimidine portion of thiamine by adenosine. *J. Bacteriol* **88**, 1024–1029.
184. Mura, U., Sgarella, F., and Ipata, P. L. (1978). Utilization of exogenous purine compounds in *Bacillus cereus*. Translocation of the ribose moity of inosine. *J. Biol. Chem.* **253**, 7905–7909.
185. Murakami, K. (1979). AMP deaminase from bakers yeast. Kinetic and molecular properties. *J. Biochem.* **86**, 1331–1336.
186. Murakami, K., and Yoshino, M. (1980). Ion-dependent activation of AMP nucleosidase from *Azotobacter vinelandii*. *Biochim. Biophys. Acta* **6**, 153–159.
187. Møllgaard, H. (1980). Deoxyadenosine/deoxycytidine kinase from *Bacillus subtilis*. Purification, characterization, and physiological function. *J. Biol. Chem.* **255**, 8216–8220.
188. Nagy, M. (1970). Regulation of the biosynthesis of purine nucleotides in *Schizosaccharomyces pombe*. I. Properties of the phosphoribosylpyrophosphate: glutamine amidotransferase of the wild strain and of a mutant desensitized towards feedback modifiers. *Biochim. Biophys. Acta* **198**, 471–481.
189. Nagy, M. (1979). Studies on purine transport and on purine content in vacuoles isolated from *Saccharomyces cerevisiae*. *Biochim. Biophys. Acta* **558**, 221–232.
190. Nagy, M., Djembo-Taty, M., and Heslot, H. (1973). Regulation of the biosynthesis of purine nucleotides in *Schizosaccharomyces pombe*. III. Kinetic studies of adenylosuccinate synthetase. *Biochim. Biophys. Acta* **309**, 1–10.
191. Nagy, M., and Ribet, A. M. (1977). Purification and comparative study of adenine and guanine phosphoribosyltransferases from *Schizosaccharomyces pombe*. *Eur. J. Biochem.* **77**, 77–85.
192. Nath, K., and Koch, A. L. (1971). Protein degradation in *Escherichia coli*. II. Strain differences in the degradation of protein and nucleic acid resulting from starvation. *J. Biol. Chem.* **246**, 6956–6967.
193. Neu, H. C. (1967). The 5′-nucleotidase of *Escherichia coli*. *J. Biol. Chem.* **242**, 3905–3911.
194. Newell, P. C., and Tucker, R. G. (1968). Biosynthesis of the pyrimidine moiety of thiamine. A new route of the pyrimidine biosynthesis involving purine intermediates. *Biochem. J.* **106**, 279–287.
195. Nierlich, D. P., and Magasanik, B. (1965). Regulation of purine ribonucleotide synthesis by end product inhibition. The effect of adenine and guanine ribonucleotides on the 5′-phosphoribosyl-pyrophosphate amidotransferase of *Aerobacter aerogenes*. *J. Biol. Chem.* **240**, 358–365.
196. Nierlich, D. P., and Magasanik, B. (1971). Control by feedback repression of the enzymes of purine biosynthesis in *Aerobacter aerogenes*. *Biochim. Biophys. Acta* **230**, 349–361.
197. Nijkamp, H. J. J., and Haan, P. G. (1967). Genetic and biochemical studies of the guanosine 5′-monophosphate pathway in *Escherichia coli*. *Biochim. Biophys. Acta* **145**, 31–40.

198. Nijkamp, H. J. J., and Oskamp, A. A. G. (1968). Regulation of the biosynthesis of guanosine 5′-monophosphate: evidence for one operon. *J. Mol. Biol.* **35**, 103–109.
199. Nishikawa, H., Momose, H., and Shiio, J. (1967). Regulation of purine nucleotide synthesis in *Bacillus subtilis*. II. Specificity of purine derivatives for enzyme repression. *J. Biochem.* **62**, 92–98.
200. Nishikawa, H., Momose, H., and Shiio, J. (1968). Pathway of purine nucleotide synthesis in *Bacillus subtilis*. *J. Biochem.* **63**, 149–155.
201. Nisonson, I., Tannenbaum, M., and Neu, H. C. (1969). Surface localization of *Escherichia coli* 5′-nucleotidase by electron microscopy. *J. Bacteriol.* **100**, 1083–1090.
202. Nolan, L. L., and Kidder, G. W. (1980). Inhibition of growth and purine metabolizing enzymes of *Trypanosomid flagellates* by N^6-methyladenine. *Antimicr. Agents Chemotherapy* **17**, 567–571.
203. Nussbaum, R. L., and Caskey, C. T. (1981). Purification and characterization of hypoxanthine-guanine phosphoribosyltransferase from *Saccharomyces cerevisiae*. *Biochemistry* **20**, 4584–4590.
204. Nygaard, P. (1973). Nucleoside-catabolizing enzymes in *Salmonella typhimurium*. Induction by ribonucleosides. *Eur. J. Biochem.* **36**, 267–272.
205. Nygaard, P. (1977). Functioning of purine salvage pathways in *Escherichia coli* K-12. *In* "Purine Metabolism in Man" (M. M. Müller, E. Kaiser, and J. E. Seegmiller, *Eds.*), Vol. II, pp. 186–195. Plenum Press, New York.
206. Nygaard, P. (1978). Adenosine deaminase from *Escherichia coli*. *In* "Meth. Enzymol." (P. A. Hoffee, and M. E. Jones, *Eds.*), Vol. LI, pp. 508–512. Academic Press, New York.
207. Nygaard, P. (In press) Multiple functions of purine nucleoside phosphorylase and adenosine deaminase in *Escherichia coli*. *In* "Molecular and cellular Regulation of Enzyme Activity" Proceedings of Symposium in Halle, GDR.
208. Ochi, K., Yashima, S., Eguchi, Y., and Matsushitu, K. (1979). Biosynthesis of formycin. *J. Biol. Chem.* **254**, 8819–8824.
209. Oeshger, M. P., and Bessman, M. J. (1966). Purification and properties of guanylate kinase from *Escherichia coli*. *J. Biol. Chem.* **241**, 5452–5460.
210. Ogata, K. (1975). The microbial production of nucleic acid-related compounds. *Adv. Appl. Microbiol.* **19**, 209–247.
211. Ozaki, H., and Shiio, J. (1979). Two cytoplasmic 5′-nucleotidases of *Bacillus subtilis* K. *J. Biochem.* **85**, 1083–1089.
212. Patel, N., Moyed, H. S., and Kane, J. F. (1975). Xanthosine-5′-phosphate amidotransferase from *Escherichia coli*. *J. Biol. Chem.* **250**, 2609–2613.
213. Patel, N., Moyed, H. S., and Kane, J. F. (1977). Properties of xanthosine 5′-monophosphate-amidotransferase from *Escherichia coli*. *Arch. Biochem. Biophys.* **178**, 652–661.
214. Pendyala, L., Smyth, J., and Wellman, A. M. (1979). Nature of 6-methylpurine inhibition and characterization of two 6-methylpurine resistant mutants of *Neurospora crassa*. *J. Bacteriol.* **137**, 248–255.
215. Pendyala, A., and Wellman, A. M. (1975). Effect of histidine on purine nucleotide synthesis and utilization in *Neurospora crassa*. *J. Bacteriol.* **124**, 78–85.
216. Pendyala, L., and Wellman, A. M. (1975). Endogenous purine metabolism in the conidia of wild type and certain adenine mutants of *Neurospora crassa*. I. The nature of the reserve pools, and pool utilization during adenine starvation. *Biochim. Biophys. Acta* **385**, 194–206.

217. Pendyala, L., and Wellman, A. M. (1980). Purine biosynthesis and its regulation in *Neurospora crassa*. *Biochim. Biophys. Acta* **607**, 350–360.
218. Pfefferkorn, E. R., and Pfefferkorn, L. C. (1977). *Toxoplasma gondii*: Specific labeling of nucleic acids of intracellular parasites in Lesch-Nyhan cells. *Exp. Parasitol.* **41**, 95–104.
219. Pfefferkorn, E. R., and Pfefferkorn, L. C. (1978). The biochemical basis for resistance to adenine arabinoside in a mutant of *Toxoplasma gondii*. *J. Parasitol.* **64**, 486–492.
220. Pickering, W. R., and Woods, R. A. (1972). The uptake and incorporation by wild type *Saccharomyces cerevisiae* and a mutant resistant to 4-aminopyrazolo (3,4-d) pyrimidine. *Biochim. Biophys. Acta* **264**, 45–58.
221. Pourquié, J. (1969). Regulation of the biosynthesis of purine nucleotides in *Schizosaccharomyces pombe*. II. Kinetic studies of IMP dehydrogenase. *Biochim. Biophys. Acta* **185**, 310–315.
222. Pourquié, J., and Heslot, H. (1971). Utilization and interconversion of purine derivatives in the fission yeast *Schizosaccharomyces pombe*. *Genet. Res.* **18**, 33–44.
223. Powell, J. W., and Wachsman, J. T. (1973). Evidence for four deoxynucleoside kinase activities in extracts of *Lactobacillus leichmannii*. *Appl. Microbiol.* **25**, 869–872.
224. Ratliff, R. L., Weaver, R. H., Lardy, H. A., Kuby, S. A. (1964). Nucleoside triphosphate-nucleoside diphosphate transphosphorylase (nucleoside diphosphokinase). *J. Biol. Chem.* **239**, 301–309.
225. Razin, S. (1978). The mycoplasmas. *Microb. Rev.* **42**, 414–470.
226. Reid, T. W., and Wilson, I. B. (1971). *E. coli* alkaline phosphatase. *In* "The Enzymes (P. D. Boyer, *Ed.*), Vol. IV, pp. 373–415. Academic Press, New York.
227. Remy, C. N., and Love, S. H. (1968). Induction of adenosine deaminase in *Escherichia coli*. *J. Bacteriol.* **96**, 76–85.
228. Rima, B. K., and Takahashi, J. (1978). Deoxyribonucleoside-requiring mutants of *Bacillus subtilis*. *J. Gen. Microbiol.* **107**, 139–145.
229. Rudolph, F. B., and Fromm, H. J. (1969). Initial rate studies of adenylosuccinate synthetase with product and competitive inhibitors. *J. Biol. Chem.* **244**, 3832–3839.
230. Ryley, J. F., and Wilson, R. G. (1972). Growth factor antagonism studies with *Coccidia* in tissue cultures. *Zeit. Parasitenkund.* **40**, 31–34.
231. Sabina, R. L., Hanks, A. R., Magill, J. M., and Magill, C. W. (1979). Role of purine base excretion in regulation of purine pools. *Molec. Gen. Genet.* **172**, 31–38.
232. Sabina, R. L., Magill, J. M., and Magill, C. W. (1976). Regulation of hypoxanthine transport in *Neurospora crassa*. *J. Bacteriol.* **128**, 598–603.
233. Sakamoto, N., Wesley Hatfield, G., and Moyed, H. S. (1972). Physical properties and subunit structure of xanthosine 5′-phosphate aminase. *J. Biol. Chem.* **247**, 5880–5887.
234. Satyanaragana, T., and Kaplan, J. G. (1971). Regulation of the purine pathway in bakers yeast activity and feedback inhibition of phosphoribosylpyrophosphate amidotransferase. *Archiv. Biochem. Biophys.* **142**, 40–47.
235. Schafer, M. P., Hannon, W. H., and Levin, A. P. (1974). *In vivo* and *in vitro* complementation between *guaB* and *in vivo* complementation between *guaA* auxotrophs of *Salmonella typhimurium*. *J. Bacteriol.* **117**, 1270–1279.
236. Schmidt, R., Wiegand, H., and Reichert, U. (1979). Purification and characterization of the hypoxanthine-guanine phosphoribosyltransferase from *Saccharomyces cerevisiae*. *Eur. J. Biochem.* **93**, 355–361.

237. Schrader, W. P., and Anderson, J. S. (1978). Membrane-bound nucleotidase of *Bacillus cereus*. *J. Bacteriol.* **133**, 576–583.
238. Schramm, V. L., and Fullin, F. A. (1978). Kinetics of adenosine monophosphate nucleosidase inactivation by phosphate and protection by substrate and allosteric activator. *J. Biol. Chem.* **253**, 2161–2167.
239. Schramm, V. L., and Lazonik, F. C. (1975). The pathway of adenylate catabolism in *Azotobacter vinelandii*. *J. Biol. Chem.* **250**, 1801–1808.
240. Schramm, V. L., and Leung, H. (1973). Regulation of adenosine monophosphate levels as a function of adenosine triphosphate and inorganic phosphate. *J. Biol. Chem.* **248**, 8313–8315.
241. Schwartzman, J. D., and Pfefferkorn, E. R. (1982). Purine synthesis and salvage in mutant host cells and parasites. *Exp. Parasitol.* **53**, 77–86.
242. Sedmak, J., and Ramaley, R. (1971). Purification and properties of *Bacillus subtilis*. Nucleoside diphosphokinase. *J. Biol. Chem.* **246**, 5365–5372.
243. Sekar, V., Wilson, S. P., and Hageman, J. H. (1981). Induction of *Bacillus subtilis* sporulation by nucleosides. Inosine appears to be sporogen. *J. Bacteriol.* **145**, 489–493.
244. Senesi, S., Falcone, G., Ipata, P. L., and Felicioli, R. A. (1974). Inhibition of phosphodiesterase from *Bacillus subtilis* by nucleoside triphosphates. *Biochemistry* **13**, 5008–5011.
245. Setlow, P. (1974). Energy and small-molecule metabolism during germination of *Bacillus* spores. Spores VI (P. Gerhardt, R. N. Costilow, and H. L. Sadoff, *Eds.*), pp. 443–450. Am. Soc. Microbiol.
246. Setlow, P., and Kornberg, A. (1970). Biochemical studies of bacterial sporulation and germination. XXIII. Nucleotide metabolism during spore germination. *J. Biol. Chem.* **245**, 3645–3651.
247. Sherman, I. W. (1979). Biochemistry of *Plasmodium* (Malarial parasites). *Microbiol. Rev.* **43**, 453–495.
248. Shiio, J., and Ozaki, H. (1978). Cellular distribution and some properties of 5′-nucleotidases in *Bacillus subtilis*. K. *J. Biochem.* **83**, 409–421.
249. Shimada, K., Fukumaki, Y., and Takagi, Y. (1976). Expression of the guanine operon of *Escherichia coli* as analyzed by bacteriophage lambda-induced mutations. *Molec. Gen. Genet.* **147**, 203–208.
250. Shimkets, L. J., and Dworkin, M. (1981). Excreted adenosine is a cell density signed for the initiation of fruiting body formation in *Myxococcus xanthus*. *Develop. Biol.* **84**, 51–60.
251. Shobe, C. R., and Campbell, J. N. (1973). Adenosine metabolism in *Micrococcus sodonensis*. *Can. J. Microbiol.* **19**, 1275–1284.
252. Sin, J. L., and Finch, L. R. (1972). Adenine phosphoribosyltransferase in *Mycoplasma mycoides* and *Escherichia coli*. *J. Bacteriol.* **112**, 439–444.
253. Slechta, L. (1960). Studies on the mode of action of psicofuranine. *Biochem. Pharmacol.* **5**, 96–102.
254. Smith, J. D. (1976). Transcription and processing of transfer RNA precursors. *In* "Progress in Nucleic Acid Research and Molecular Biology" (W. E. Cohn, *Ed.*), Vol. 16, pp. 25–73. Academic Press, New York.
255. Spector, T., and Beacham, L. M. (1975). Guanosine monophosphate synthetase from *Escherichia coli* B-96. *J. Biol. Chem.* **250**, 3101–3107.
256. Spector, T., Miller, R. L., Fyfe, J. A., and Krenitsky, T. A. (1974). GMP synthetase from *Escherichia coli* B-96. Interaction with substrate analogs. *Biochim. Biophys. Acta* **370**, 585–591.
257. Stayton, M. M., and Fromm, H. J. (1979). Guanosine 5′-diphosphate-3′-

diphosphate inhibition of adenylosuccinate synthetase. *J. Biol. Chem.* **254**, 2579–2581.

258. Stoeckler, J. D., Cambor, C., Kuhns, V., Chu, S. H., and Parks, R. E. (1982). Inhibitors of purine nucleoside phosphorylase C(8) and C(5′) substitutions. *Biochem. Pharmacol.* **31**, 163–171.
259. Suhadolnik, R. J. (1970). "Nucleoside Antibiotica." Wiley, New York.
260. Suhadolnik, R. J. (1979). "Nucleosides as biological probms". Wiley, New York.
261. Suhadolnik, R. H. (1979). Naturally occurring nucleoside and nucleotide antibiotics. *In* "Progress in Nucleic Acid Research and Molecular Biology" (W. E. Cohn, *Ed.*), Vol. 22, pp. 193–291. Academic Press, New York.
262. Tabor, C. W., and Tabor, H. (1976). 1,4-diaminobutane (putrescine), spermidine, and spermine. *Ann. Rev. Biochem.* **45**, 285–306.
263. Tabor, H., and Tabor, C. W. (1972). Biosynthesis and metabolism of 1,4-diaminobutane, spermidine, spermine, and related amines. *Adv. Enzymol. Relat. Areas Mol. Biol.* **36**, 204–268.
264. Thelander, L., and Reichard, P. (1979). Reduction of ribonucleotides. *Ann. Rev. Biochem.* **48**, 133–158.
265. Theze, J., and Margarita, D. (1972). Etude de l'adénylate chez *E. Coli* K12 et chez un mutant thermosensible. *Ann. l'Institut Pasteur.* **123**, 157–169.
266. Tozzi, M. G., Sgarella, F., and Ipata, P. L. (1981). Induction and repression of enzymes involved in exogenous purine compound utilization in *Bacillus cerus. Biochim. Biophys. Acta* **678**, 460–466.
267. Tracy, S. M., and Sherman, J. W. (1972). Purine uptake and utilization by the avian malaria parasite *Plasmodium lophurae. J. Protozool.* **19**, 541–549.
268. Tsay, Y., and Hanaoka, F. (1972). Transient effect of exogenous adenine on germination of *Aspergillus oryzae* conidia. *J. Gen. Apl. Microbiol.* **18**, 209–223.
269. Tuttle, J. V., and Krenitsky, T. A. (1980). Purine phosphoribosyltransferase from *Leishmania donovani. J. Biol. Chem.* **255**, 909–916.
270. Uerkvitz, W. (1971). Trans-N-deoxyribosylase from *Lactobacillus helveticus.* Crystallization and properties. *Eur. J. Biochem.* **23**, 387–395.
271. Uerkvitz, W., and Beck, C. F. (1981). Periplasmic phosphatases in *Salmonella typhimurium* LT2. *J. Biol. Chem.* **256**, 382–389.
272. Vogels, G. D., and v. d. Drift, C. (1976). Degradation of purines and pyrimidines by microorganisms. *Bact. Rev.* **40**, 403–468.
273. Wachsman, J. T., and Morgan, D. D. (1971). Deoxynucleoside kinases of *Bacillus megaterium. J. Bacteriol.* **105**, 787–792.
274. Walker, R. D., and Duerre, J. A. (1975). S-adenosylhomocysteine metabolism in various species. *Can. J. Biochem.* **53**, 312–319.
275. Walter, R. D., and Königk, E. (1974). Hypoxanthine-guanine phosphoribosyltransferase and adenine phosphoribosyltransferase from *Plasmodium chabaudi*, purification and properties. *Z. Trop. Parasitol.* **25**, 227–235.
276. Wang, C. C., Simashkevich, P. M., and Stotish, R. L. (1979). Mode of anticoccidial action of arprinocid. *Biochem. Pharmacol.* **28**, 2241–2248.
277. Wang, C. C., and Simashkevich, P. M. (1981). Purine metabolism in the protozoan parasite *Eimeria tennella. Proc. Natl. Acad. Sci. U.S.A.* **78**, 6618–6622.
278. Wang, T. P., and Lampen, J. O. (1951). The cleavage of adenosine, cytidine, and xanthosine by *Lactobacillus pentosus. J. Biol. Chem.* **192**, 339–347.
279. Watanabe, T., Ohe, T., and Tsujisaka, Y. (1976). Changes in the metabolic pathways of hypoxanthine in *Streptomyces. J. Gen. Appl. Microbiol.* **22**, 13–23.
280. Weppelman, R., Kier, L. D., and Ames, B. N. (1977). Properties of two

phosphatases and a cyclic phosphodiesterase of *Salmonella typhimurium*. *J. Bacteriol.* **130**, 411–419.
281. Westby, C. A., and Gots, J. S. (1969). Genetic blocks and unique features in the biosynthesis of 5′-phosphoribosyl-N-formylglycineamide in *Salmonella typhimurium*. *J. Biol. Chem.* **244**, 2095–2102.
282. Woodward, D. O. (1978). Adenylosuccinate AMP-lyase (*Neurospora crassa*). *In* "Meth. Enzymol." (P. A. Hoffee and M. E. Jones, *Eds.*), Vol. LI, pp. 202–213. Academic Press, New York.
283. Wong, J. Y., Meyer, E., and Switzer, R. L. (1977). Glutamine phosphoribosylpyrophosphate amidotransferase from *Bacillus subtilis*. *J. Biol. Chem.* **252**, 7424–7426.
284. Wu, T. W., and Scrimgeour, K. G. (1973). Properties of inosinic acid dehydrogenase from *Bacillus subtilis*. II. Kinetic properties. *Can. J. Biochem.* **51**, 1391–1398.
285. Wu, T. W., and Scrimgeour, K. G. (1973). Properties of inosinic acid dehydrogenase from *Bacillus subtilis*. I. Purification and physical properties. *Can. J. Biochem.* **51**, 1380–1390.
286. Wyngaarden, J. B., and Greenland, R. A. (1963). The inhibition of succinoadenylate kinosynthetase of *Escherichia coli* by adenosine and guanosine 5′-monophosphate. *J. Biol. Chem.* **238**, 1054–1057.
287. Yagil, E., and Beacham, J. R. (1975). Uptake of adenosine 5′-monophosphate by *Escherichia coli*. *J. Bacteriol.* **121**, 401–405.
288. Yamada, K. A., and Sherman, I. W. (1981). Purine metabolizing enzymes of *Plasmodium lophurae* and its host cell, the duckling (*Anas domesticus*) erythrocyte. *Mol. Biochem. Parasitol.* **2**, 349–358.
289. Yamada, K. A., and Sherman, I. W. (1981). Purine metabolism by the avian malarial parasitic *Plasmodium lophurae*. Molecular *Biochem. Parasitol.* **3**, 253–264.
290. Yoshino, M., Murakami, K., and Tsushima, K. (1979). AMP deaminase from bakers yeast. Purification and some regulatory properties. *Biochim. Biophys. Acta* **570**, 157–166.
291. Yoshino, M., Tsukada, T., Murakami, K., and Tsushima, K. (1980). Adenine nucleotide metabolism in *Azotobacter vinelandii*. Two metabolic pathways of AMP degradation. *Arch. Microbiol.* **128**, 222–230.
292. Zalkin, H., and Truitt, C. D. (1977). Characterization of the glutamine site of *Escherichia coli* guanosine 5′-monophosphate. *J. Biol. Chem.* **252**, 5431–5436.
293. Zappia, V., Zydek-Cwick, C. R., and Schlenk, F. (1969). The specificity of S-adenosylmethionine derivatives in methyl transfer reactions. *J. Biol. Chem.* **244**, 4499–4509.
294. Zielke, C. L., and Suelter, C. H. (1971). Purine, purine nucleoside, and purine nucleotide aminohydrolase. *In* "The Enzymes" (P. D. Boyer, *Ed.*), Vol. IV, pp. 47–78. Academic Press, New York.
295. Zimmerman, E. F., and Magasanik, B. (1964). Utilization and interconversion of purine bases and ribonucleosides by *Salmonella typhimurium*. *J. Biol. Chem.* **239**, 293–300.

3

Utilization of preformed pyrimidine bases and nucleosides

JAN NEUHARD

I. Introduction . . . 96
II. Pathways in *Salmonella typhimurium* and *Escherichia coli* . 97
A. Uracil metabolism . . . 98
B. Cytosine metabolism . . . 101
C. Uridine metabolism . . . 101
D. Cytidine metabolism . . . 102
E. Deoxyuridine metabolism . . . 104
F. Deoxycytidine metabolism . . . 105
G. Metabolism of exogenous pyrimidine nucleotides . . 105
III. Pathways in other microorganisms . . . 106
A. *Bacillus subtilis* . . . 106
B. Lactic acid bacteria . . . 109
C. *Pseudomonads* . . . 111
D. The *moraxella* group . . . 112
E. *Mycoplasmas* . . . 112
F. Fungi . . . 113
G. Protozoa . . . 117
H. Summary . . . 117
IV. Enzymes involved in pyrimidine salvage . . . 118
A. Uracil phosphoribosyltransferase . . . 118
B. Cytosine deaminase . . . 120
C. Cytidine deaminase . . . 120
D. Enzymes cleaving the glycosylic bond of pyrimidine nucleosides. . . . 121
E. Hydroxylases (dioxygenases) . . . 125
F. Uracil-5-carboxylic acid decarboxylase . . . 127
G. Pyrimidine ribonucleoside kinase . . . 127
H. Thymidine kinase . . . 128
I. Deoxycytidine kinase . . . 129
K. Nucleoside phosphotransferases . . . 130
L. CMP/dCMP kinase . . . 131
M. Ribonucleoside monophosphate glycosylases . . . 132
V. Physiological aspects of pyrimidine salvage . . . 133
A. Utilization of exogenous pyrimidine compounds . . 133

METABOLISM OF NUCLEOTIDES, NUCLEOSIDES AND NUCLEOBASES IN MICROORGANISMS
ISBN: 0-12-510580-0

B. Interrelation between pyrimidine *de novo* and salvage pathways 134
C. Endogenous turnover of pyrimidine compounds . . 135

I. INTRODUCTION

The *de novo* synthesis of pyrimidine nucleotides does not involve nucleobases or nucleosides as intermediates (Fig. 3.1). The ability of pyrimidine requiring mutants of a large number of microorganisms to utilize pyrimidine bases and nucleosides for growth shows that metabolic sequences, i.e. *salvage pathways*, capable of converting these compounds into pyrimidine nucleotides are of widespread occurrence.

The extensive use of radiolabeled pyrimidines for *in vivo* studies of nucleic acid synthesis in many microorganisms has further established the existence of salvage pathways, also in pyrimidine prototrophic organisms. In order to select and use the most suitable radioactive precursor for the labeling of nucleic acids *in vivo*, a detailed knowledge of pyrimidine salvage pathways is essential.

The salvage pathways also define the sensitivity of an organism to those pyrimidine analogs which require conversion to nucleotides for cytotoxicity. This is because most analogs are metabolized by the same enzymes as the natural nucleobases and nucleosides. Consequently, toxic pyrimidine analogs may be used for isolating mutants blocked in various steps of the salvage pathways. Studies of such mutants have provided information about the

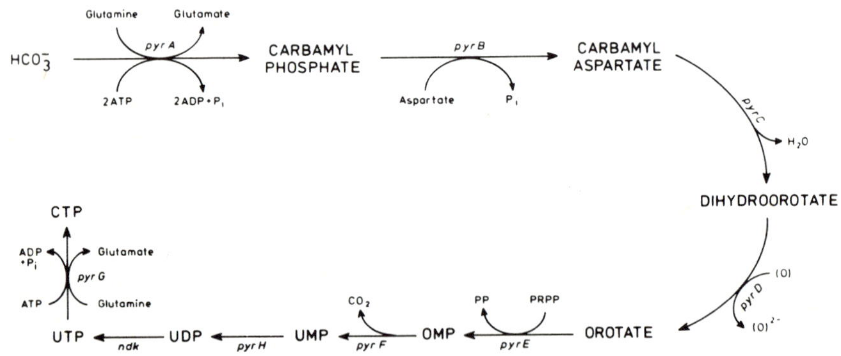

FIG. 3.1. *De novo* synthesis of UTP and CTP. The individual enzymes are identified by their gene symbols as follows: *ndk*, nucleoside diphosphokinase; *pyrA(car)*, carbamoylphosphate synthase; *pyrB*, aspartate carbamoyltransferase; *pyrC*, dihydroorotase; *pyrD*, dihydroorotate oxidase; *pyrE*, orotate phosphoribosyltransferase; *pyrF*, orotidine 5′-phosphate decarboxylase; *pyrG*, CTP synthetase; *pyrH*, UMP kinase.

physiological importance of the individual salvage enzymes. The analogs most commonly used are pyrimidines containing a fluoro-substitution on the 5-carbon of the ring.*

As defined above, pyrimidine salvage includes transport of exogenous nucleobases and nucleosides through the cell membrane(s). Since a separate chapter of this book is devoted to a detailed discussion of these processes (Chapter 6), they will not be discussed here.

Also the special aspects of the utilization of thymine and thymidine as precursors for dTTP and hence DNA synthesis are discussed separately in this book (Chapter 4), and will only be mentioned briefly in the present discussion.

II. PATHWAYS IN *SALMONELLA TYPHIMURIUM* AND *ESCHERICHIA COLI*

The pyrimidine salvage pathways in *Escherichia coli* and *Salmonella typhimurium* appear to be almost identical. The pyrimidine ring of each of the compounds uracil, cytosine, uridine, cytidine, deoxyuridine, and deoxycytidine is incorporated into the four pyrimidine nucleotides of RNA and DNA (26, 104), and each of these pyrimidine bases and nucleosides will satisfy the total pyrimidine requirements of pyrimidine auxotrophic strains (19, 104). Consequently, any one of the six corresponding 5-fluoropyrimidine bases and nucleosides inhibits growth of pyrimidine prototrophic strains (19, 63).

In addition to the bases and nucleosides certain pyrimidine nucleotides may also serve as precursors for nucleic acids, when added to the growth medium. Since nucleotides cannot permeate the cell membranes they must be cleaved to the corresponding nucleosides by periplasmic phosphatases, before they can be further metabolized (104, 170).

The capacity of both *S. typhimurium* and *E. coli* to utilize this variety of exogenous pyrimidine compounds is dependent on a network of salvage reactions, which is summarized in Fig. 3.2. Each of the enzyme reactions shown in the figure has been studied *in vitro* and, with the exception of the ribonucleoside monophosphate glycosylase reaction, mutants defective in the individual enzymes have been isolated and characterized. The genes coding for these enzymes have all been localized on the *S. typhimurium* and *E. coli* chromosomes, as shown on p. 307 in the Appendix.

* Abbreviations used: 5-fluoro substituted uracil, cytosine, uridine, cytidine, deoxyuridine, and deoxycytidine analogs are denoted FU, FC, FUR, FCR, FUdR, and FCdR, respectively. PRPP is 5-phosphoribosyl-1-pyrophosphate. CRP stands for the catabolite repressor protein, i.e. the cyclicAMP-binding protein. T, thymine; TdR, thymidine; U, uracil; UR, uridine; UdR, deoxyuridine; C, cytosine; CR, cytidine; CdR, deoxycytidine.

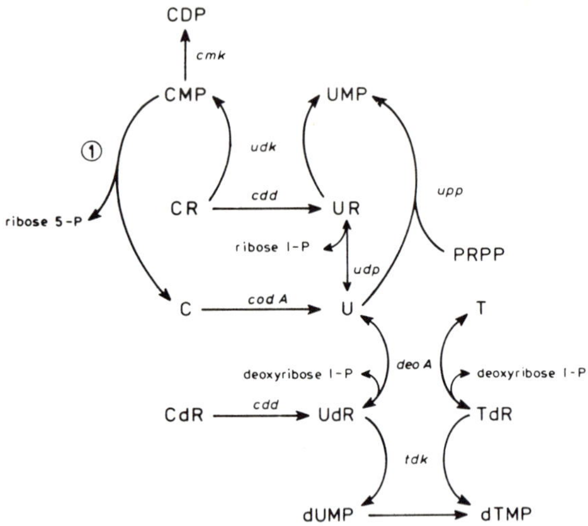

FIG. 3.2. Pyrimidine salvage pathways in *Escherichia coli* and *Salmonella typhimurium*. The individual reactions are identified by the corresponding gene designations: *cdd*, cytidine/deoxycytidine deaminase; *cmk*, CMP/dCMP kinase; *codA*, cytosine deaminase; *deoA*, thymidine phosphorylase; *tdk*, thymidine kinase; *udk*, uridine/cytidine kinase; *udp*, uridine phosphorylase; *upp*, uracil phosphoribosyltransferase; (1), ribonucleotide glycosylase.

Thymine and thymidine cannot serve as a total pyrimidine source but may serve as a source of dTMP (cf. Chapter 4).

A. Uracil metabolism

Uracil phosphoribosyltransferase catalyzes the formation of UMP from uracil and PRPP (Fig. 3.2 and p. 118). Mutant strains carrying mutations in the structural gene for this enzyme (*upp*) are unable to incorporate uracil into nucleic acids (29). Thus, a pyrimidine requiring *upp* mutant cannot grow with 100 μM uracil as the only pyrimidine source; it requires either uridine or cytidine for growth (Table 3.1). In addition pyrimidine prototrophic *upp* mutants are resistant to FU (Table 3.1); a phenotype which has been used in the selection of such mutants (Table 3.2).

An alternative route for converting uracil to UMP may be demonstrated. It involves the formation of uridine and requires the sequential action of uridine phosphorylase and uridine kinase. The phenotype of the *upp* mutants indicates, however, that this route is of minor importance under normal growth conditions. If a pyrimidine requiring *upp* mutant is supplied with either 1 mM uracil or with 100 μM uracil plus an exogenous source of ribose

TABLE 3.1

Effects of mutations in pyrimidine salvage pathways on the utilization of pyrimidines and the sensitivity towards 5-fluoropyrimidines in *Salmonella typhimurium* and *Escherichia coli*.[a]

	Growth											
Relevant genotype	Pyrimidine source[b]						Analog present[c]					
	U	C	UR	CR	UdR	CdR	FU	FC	FUR	FCR	FUdR	FCdR
upp	−	−	+	+	−	−	+	+	−	−	−	−
codA	+	−	+	+	+	+	−	+	−	−	−	−
udk	+	+	+	+	+	+	−	−	−	−	−	−
udp	+	+	+	+	+	+	−	−	−	−	−	−
upp,udk[d]	−	−	−	−	−	−	+	+	+	+	−	−
udp,udk	+	+	+	+	+	+	−	−	−	n.d.[e]	−	−
Δ cdd	+	+	+	+	+	−	−	−	−	−	−	+
Δ cdd,udk	+	+	+	−	+	−	−	−	−	+	−	+
Δ cdd, Δ cod	+	−	+	−	+	−	−	+	−	−	−	+
Δ cdd,upp	−	−	+	−	−	−	+	+	−	−	−	+
deoA	+	+	+	+	+	+	−	−	−	−	−	−
deoA,udp	+	+	+	+	−	−	−	−	−	−	−	−
deoA,udp,tdk	+	+	+	+	−	−	−	−	−	−	+	+

[a] The data are compiled from a large number of published and unpublished observations (19).
[b] These data were obtained with strains containing an additional mutation rendering them pyrimidine requiring.
[c] All strains were pyrimidine prototrophic.
[d] This strain contained a *pyrC* mutation which confers a temperature sensitive pyrimidine requirement. When pyrimidine sources were tested the strain was grown at 42°C (Pyr^-). When analog sensitivity was tested the strain was grown at 30°C (Pyr^+).
[e] n.d. = not determined.

1-phosphate in the form of a purine ribonucleoside it will grow albeit at a somewhat reduced rate. Thus, an increase in the internal concentration of either substrate for uridine phosphorylase will stimulate the flow of uracil to UMP through uridine. For the same reason a pyrimidine prototrophic *upp* mutant is sensitized to FU if a purine ribonucleoside is added to the medium. This has been used for selecting mutants defective in uridine phosphorylase, i.e. *udp* mutants (141, and Table 3.2).

Two physiological conditions have been reported to affect the capacity of a upp^+ strain to convert uracil and FU to ribonucleotides. The first is amino acid starvation of stringent strains of *E. coli*, which causes a 10-fold reduction in the incorporation of ^{14}C-uracil into UTP and CTP. Amino acid starvation of relaxed mutants (*rel*) has no such effect (52, 102). The second is partial guanine-starvation which results in resistance towards FU in *S. typhimurium* (84). *In vitro*, uracil phosphoribosyltransferase requires GTP for activity, and is inhibited by ppGpp (p. 118). Thus, either amino acid starvation, which causes ppGpp accumulation in stringent strains, or partial guanine starvation, which results in low internal GTP pools, may be expected to decrease the *in*

TABLE 3.2

Selection of *Escherichia coli*, *Salmonella typhimurium*, and *Bacillus subtilis* mutants defective in pyrimidine salvage enzymes by the use of 5-fluoro pyrimidine analogs.

Gene	Required genetic background	Resistance to	Additional selective phenotype	
			Analog sensitivity	Carbon source[a]
cdd[b]	thy^+	FCdR + U	$FUdR^s$	
cmk	*cdd*	FCR		
	pyr^+,*cdd*,*upp*	FCR	FUR^s	
codA[c]	pyr^+,pur^+	FC	FU^s	
dck[d]	thy^+,*cdd*	FCdR	$FUdR^s$	
tdk	thy^+,*tpp*	FUdR + U		
	thy^+,*tpp*,*udp*	FUdR		
	thy^+,pyr^+,*upp*	FU + PudR[e]		TdR^+,$PudR^+$
tpp	thy^+,pyr^+,*upp*	FU + PudR		TdR^-,$PudR^-$
udk	*cdd*	FCR		
	pyr^+,*upp*	FUR		
	pyr^+,*upp*	FU + PuR[f]		UR^+,$PudR^+$
udp	pyr^+,*upp*	FU + PuR		UR^-,$PudR^+$
upp	pyr^+	FU		

[a] The indicated nucleosides are used as the sole source of carbon.
[b] This selection does not work in *B. subtilis* due to the presence of deoxycytidine kinase.
[c] *B. subtilis* does not contain cytosine deaminase.
[d] *E. coli* and *S. typhimurium* are devoid of deoxycytidine kinases.
[e] PudR stands for purine deoxyribonucleosides.
[f] PuR stands for purine ribonucleosides.

vivo activity of uracil phosphoribosyltransferase, and thereby account for the observed depression in uracil and FU incorporation.

B. Cytosine metabolism

The only known route for cytosine utilization is via deamination to uracil catalyzed by cytosine deaminase. Mutants carrying mutations in the structural gene for cytosine deaminase (*codA*) are resistant to FC but sensitive to FU, and they cannot use cytosine as the only pyrimidine source (Tables 3.1 and 3.2, 126). This shows that cytosine can only be converted to uridine nucleotides via free uracil; but it does not show whether cytosine may serve directly as a precursor of CMP. Mutants unable to synthesize CTP due to a mutation in the CTP synthetase gene *pyrG* require cytidine for growth. This requirement cannot be satisfied by cytosine or by a combination of cytosine and a purine ribonucleoside (126). Thus cytosine cannot be converted directly to cytidine or to CMP.

The synthesis of cytosine deaminase is regulated in a very complex manner. It is repressed by addition of purines to the growth medium, and it is derepressed by starvation for pyrimidines and for nitrogen (cf. p. 120). As a result, cod^+ strains which have been cultured in a medium containing purines are phenotypically resistant to FC but sensitive to FU (L. Andersen, 1979. Thesis, University of Copenhagen).

Cytosine deamination by wild-type cells of *S. typhimurium* and *E. coli* is too slow to support any appreciable growth with cytosine present as the only nitrogen-source. However, mutants which grow equally well on cytosine or ammonia as nitrogen sources are readily isolated (*codR*). Such mutants contain 5–20 fold increased levels of cytosine deaminase (L. Andersen, 1979. Thesis, University of Copenhagen).

C. Uridine metabolism

Uridine may be utilized by two alternative routes: the first consists in the direct phosphorylation of uridine to UMP, with conservation of the glycosylic bond of the nucleoside. This reaction is catalyzed by uridine kinase. The second route involves phosphorolytic cleavage of uridine to uracil and ribose 1-phosphate catalyzed by uridine phosphorylase. Uracil is subsequently converted to UMP by uracil phosphoribosyltransferase or, if produced in surplus, excreted into the medium.

Studies with strains carrying mutations in the genes coding for uridine phosphorylase (*udp*), uridine kinase (*udk*), and uracil phosphoribosyltransferase (*upp*), alone or in different combinations, have shown that both pathways for uridine utilization are operating *in vivo* (Table 3.1 and Fig. 3.2). (a) Mutants blocked in the first pathway, i.e. *udk* mutants are sensitive to FUR and can satisfy their entire pyrimidine demand by uridine. (b) Mutants blocked in the second pathway due to mutations in either the *udp* or the *upp* genes are likewise sensitive to FUR and can satisfy their pyrimidine requirement with uridine. (c) Double mutants of the genotype *udk,upp*, in which both routes are blocked by mutation are FUR resistant and unable to utilize uridine as a source of UMP (Table 3.1). (d) Double mutants of the genotype *udp,udk*, however, are *sensitive* to FUR and capable of utilizing uridine as pyrimidine source, although at a somewhat reduced rate (Table 3.1). This indicates that the cells contain another enzyme activity besides uridine phosphorylase, which can convert uridine to uracil. The identity of this alternative uridine cleaving activity has not been established. However, it cannot be excluded that the responsible enzyme is thymidine phosphorylase (cf. Table 5.3).

The growth yield from uridine of pyrimidine-requiring *upp* mutants gives a direct measure of that fraction of a given amount of uridine, which is converted to UMP via uridine kinase *in vivo*. Any phosphorolytic cleavage of uridine to uracil will decrease the maximal yield, since uracil cannot be converted to UMP by such mutants. The growth yield from the same amount of uridine using a *pyr,upp*$^+$ strain gives the maximal yield. From the results of such growth yield experiments with both *S. typhimurium* (Fig. 3.3(a)) and *E. coli* (Fig. 3.3(b)) it appears that a large fraction of the added uridine escapes uridine kinase and is cleaved to uracil. In *S. typhimurium* only 25% of the added uridine is converted to UMP via uridine kinase, whereas the value in *E. coli* is about 50%. This difference between the two organisms is due to the fact that uridine phosphorylase is induced by uridine in *S. typhimurium* but not in *E. coli* (p. 220). Introduction of an additional mutation inactivating uridine phosphorylase results in a significant increase in the growth yield from uridine (Fig. 3.3(a)). However, even in the absence of uridine phosphorylase, a significant fraction of the added uridine is converted to uracil. As discussed above this residual uridine cleaving activity may be accounted for by thymidine phosphorylase.

If UMP is used instead of uridine as the limiting pyrimidine source, the growth yields obtained with *pyr,upp* and *pyr,upp*$^+$ strains are the same (Fig. 3.3(c)). Since UMP can only be utilized following dephosphorylation to uridine (104), and the dephosphorylation reaction is rate limiting for growth (compare Fig. 3.3(b) and 3.3(c)), these results show that if uridine is fed in limiting concentrations, it is quantitatively converted to UMP via the uridine kinase reaction. None of it is cleaved to uracil by uridine phosphorylase.

D. Cytidine metabolism

E. coli and *S. typhimurium* contain inducible cytidine/deoxycytidine de-

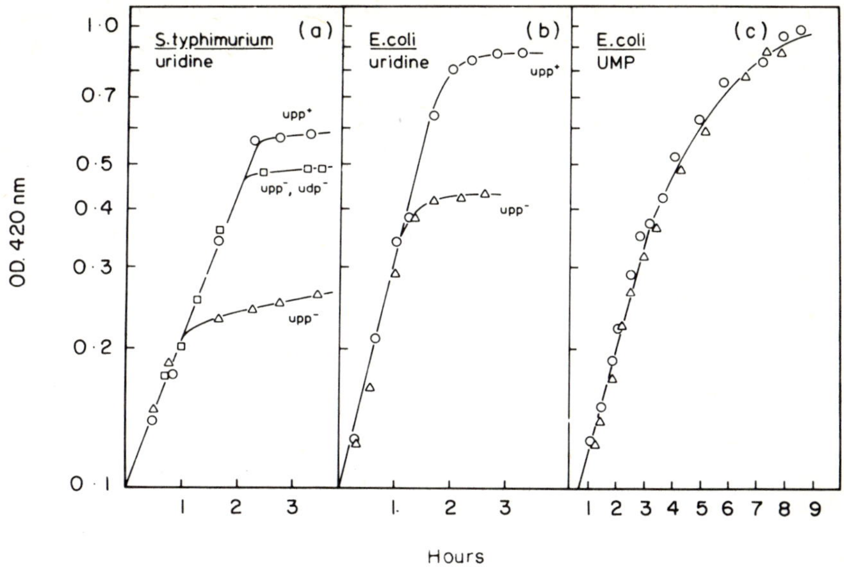

FIG. 3.3. Growth yield of various pyrimidine requiring mutants on: a, 6 μg uridine/ml; b, 8 μg uridine/ml; c, 12 μg 5′-UMP/ml as the sole sources of pyrimidines. ○——○, *pyrF*; △——△, *pyrF,upp*; □——□, *pyrF,upp,udp*. Note the difference in the abcissa in a and b vs. c. Glucose is used throughout as carbon source.

aminases which, when fully expressed, enable them to grow with cytidine as the only nitrogen source (Chapter 5). The product of the deamination, i.e. uridine, is further metabolized as described in the preceding section. In addition, both species contain cytidine kinase which converts cytidine directly to CMP. Any CMP formed is subsequently phosphorylated in two steps to CTP. The first phosphorylation is catalyzed by CMP/dCMP kinase (p. 131) and the second by nucleoside diphosphokinase (Chapter 4, and reference 20).

Mutants which lack cytidine deaminase, i.e. *cdd* mutants, are unable to grow on cytidine as nitrogen source, and are resistant to FCdR but sensitive to FCR (Table 3.1). Mutants lacking cytidine kinase are easily obtained in strains carrying *cdd* mutations, by selecting for resistance to FCR (Table 3.2). Such mutants have simultaneously lost the ability to phosphorylate uridine to UMP, indicating that both activities reside in the same protein, uridine/cytidine kinase (19, 63).

Pyrimidine auxotrophic strains containing a *cdd* mutation can synthesize all required cytosine nucleotides from cytidine, but they should not be able to convert cytidine to uridine nucleotides. However, they do grow with cytidine as the only pyrimidine source (Table 3.1), although with somewhat reduced growth rates (92, 125). This indicates the existence of a minor pathway from cytidine to uridine nucleotides not involving cytidine deaminase. The nature of this alternative route was investigated in *E. coli*, using a pyrimidine-requiring strain from which the *cdd*-gene was deleted (Δcdd). This strain grows slowly on cytidine as pyrimidine source, but not at all with deoxycytidine. The introduction into this strain of additional mutations in either one of the genes *cod*, *udk*, or *upp* completely abolishes growth on cytidine as pyrimidine source (Table 3.1; Josephsen and Neuhard, unpublished). Thus the slow conversion of cytidine to uridine nucleotides observed in Δcdd mutant involves both cytidine phosphorylation and cytosine deamination, and may proceed as follows:

$$\text{cytidine} \xrightarrow[udk]{} \text{CMP} \longrightarrow \text{cytosine} \xrightarrow[codA]{} \text{uracil} \xrightarrow[upp]{} \text{UMP}$$

In support of this formulation enzymes capable of hydrolyzing CMP to cytosine and ribose 5-phosphate (ribonucleotide glycosylases) have been found in both *E. coli* and *S. typhimurium* (p. 132). Due to lack of mutants defective in ribonucleotide glycosylase it has not yet been possible to establish its proposed function in cytidine metabolism.

Mutant strains which are unable to synthesize CTP owing to a mutation in the *pyrG* gene encoding CTP synthetase require cytidine for growth (126). This requirement can only be satisfied if the strain is defective in cytidine deaminase, i.e. a *pyrG* mutation is incompatible with a cdd^+ allele (17). Thus, exogenous cytidine is predominantly metabolized by cytidine deaminase in a cdd^+ strain. Little, if any, is phosphorylated to CMP.

E. Deoxyuridine metabolism

Deoxyuridine may undergo two different metabolic transformations. It may be cleaved phosphorolytically to uracil and deoxyribose 1-phosphate and thereby serve as a total source of both pyrimidines and of carbon, and it may be phosphorylated to dUMP by thymidine kinase.

Deoxyuridine cleavage is catalyzed by the inducible enzyme thymidine phosphorylase (p. 122). Although mutants defective in thymidine phosphorylase, i.e. *deoA* mutants, are unable to grow on deoxyuridine as the only carbon and energy source (Chapter 5), they have retained the capacity to use deoxyuridine as a total source of pyrimidines (Table 3.1). This residual capacity of a *deoA* mutant to degrade deoxyuridine is dependent on a functional uridine phosphorylase. A strain which lacks both thymidine phosphorylase and uridine phosphorylase cannot catabolize deoxyuridine (15) and is unable to grow on deoxyuridine as pyrimidine source (Table 3.1, 19).

Mutants unable to phosphorylate deoxyuridine to dUMP has been isolated by selecting for FUdR resistance in strains unable to degrade it to FU, i.e. in *deoA,udp* strains (Table 3.2). Such mutants have acquired an additional mutation in the structural gene for thymidine kinase (*tdk*) (Table 3.1, 18). Mutants in *tdk* may also be selected by resistance to FUdR in *deoA,udp*$^+$ strains provided the toxicity of FU is suppressed either by addition of uracil or by a *upp* mutation (Table 3.2). However, the *deoA* mutation is essential for such selections in order to protect FUdR from being cleaved too rapidly (19). An alternative method of selecting mutants lacking thymidine kinase is based on the observation that *upp* mutants are sensitized to FU, if a purine deoxyribonucleoside is simultaneously added as a source of deoxyribose 1-phosphate (32). Under such conditions thymidine phosphorylase catalyzes the formation of FUdR from FU and deoxyribose 1-phosphate. Starting with a strain lacking uracil phosphoribosyltransferase one may obtain mutants resistant to FU in the presence of purine deoxyribonucleosides owing to additional mutations in either one of the genes coding for thymidine kinase, thymidine phosphorylase, and purine nucleoside phosphorylase (*deoD*) (Table 3.2).

F. Deoxycytidine metabolism

E. coli and *S. typhimurium* can use deoxycytidine as a total source of pyrimidines, nitrogen, and carbon. Strains lacking cytidine deaminase activity, due to a mutation in the *cdd* gene, have lost all these capacities, but they have retained the ability to utilize deoxyuridine as the only pyrimidine and carbon source (19). Accordingly *cdd* mutants are resistant to the toxic compounds FCdR, but sensitive to FUdR (Tables 3.1 and 3.2) (90, 126). It has been shown that a *cdd* mutant of *E. coli* fails to incorporate radioactivity from [U-^{14}C]-deoxycytidine into DNA (91), and that no deoxycytidine kinase activity can be detected in crude extracts of *E. coli* and *S. typhimurium* (83, 91). On basis of these observations it can be concluded that deoxycytidine utilization, in contrast to what was found for cytidine, is entirely dependent on a functional cytidine deaminase. *E. coli* and *S. typhimurium* are not able to cleave deoxycytidine to cytosine, nor can they phosphorylate it to deoxycytidine nucleotides.

G. Metabolism of exogenous pyrimidine nucleotides

The cell envelope of most microorganisms is impermeable to nucleotides. Thus nucleotides must be hydrolyzed to the corresponding nucleosides before they can be utilized as precursors for nucleic acid synthesis (104). Hydrolysis of exogenous nucleotides to nucleosides is presumably catalyzed by phosphatases (nucleotidases) located in the periplasmic space. This implies that nucleotides must penetrate the outer membrane before they reach the hydrolyzing enzymes. Mutants defective in the synthesis of specific outer membrane proteins involved in pore formation are unable to utilize exogenous nucleotides, although they contain normal levels of the periplasmic phosphatases (1, 13, 14).

E. coli and *S. typhimurium* contain a number of periplasmic phosphatases which will hydrolyse nucleoside monophosphates (3, 4, 51, 95, 123, 144, 156, 170). They are listed in Table 3.3 together with their specificities towards nucleotides. As shown, both species contain periplasmic "nucleotidases" capable of hydrolyzing both 3′-, 5′-, and cyclic 2′,3′-ribonucleotides as well as 5′-deoxyribonucleotides. This capacity for nucleotide hydrolysis explains why pyrimidine requiring mutants can grow on any of the corresponding uracil or cytosine nucleotides as sole pyrimidine sources. Accordingly, *S. typhimurium* mutants lacking both the 2′,3′-cyclic phosphate phosphodiesterase (3′-nucleotidase), and the two non-specific acid phosphatases have lost the ability to use 3′-UMP and 5′-UMP as total pyrimidine source (170), and *E. coli* mutants lacking alkaline phosphatase and 5′-nucleotidase activities cannot utilize 5′-UMP and 5′-dUMP (171).

Exogenous RNA can be used as a total pyrimidine source by pyrimidine requiring mutants of *S. typhimurium* (176). This requires participation of the periplasmic RNase I as shown by the lack of growth of a pyrimidine-requiring RNase I deficient mutant on exogenous RNA as the only pyrimidine source (176). Since 2′-,3′-cyclic phosphate phosphodiesterase is the only known periplasmic nucleotidase which will hydrolyze 2′,3′-cyclic nucleotides, the products of the RNase I reaction, one would expect that a mutant lacking this enzyme (*pde* in *S. typhimurium* and *cpdB* in *E. coli*) would be unable to use exogenous RNA as pyrimidine source. This was confirmed using a *pyrF,pde* mutant of *S. typhimurium* (Neuhard, unpublished experiments).

III. PATHWAYS IN OTHER MICROORGANISMS

A. *Bacillus subtilis*

1. *Metabolism of uracil compounds*

B. subtilis can utilize uracil, uridine, and deoxyuridine as total pyrimidine sources, and it is sensitive to the 5-fluoro analogs FU, FUR, and FUdR. Mutants defective in uracil phosphoribosyltransferase, i.e. *upp* mutants,* are resistant to FU but are sensitive to both FUR and FUdR (Neuhard, unpublished experiments). Thus, *B. subtilis* must contain kinases capable of phosphorylating uridine and deoxyuridine to UMP and dUMP respectively. Since *upp* mutants are sensitized to FU when supplied with purine ribo- or

* Mutations giving rise to FU resistance in *B. subtilis* have been located in four different loci on the chromosome, i.e. furA, furC, furE and furF (67). It is at present not known whether one of these genes is identical with the *upp* gene.

TABLE 3.3

Periplasmic nucleotidases in *Escherichia coli* and *Salmonella typhimurium*.

Enzyme	Organism	Structural gene (map location)	Nucleotide specificity	Reference
Alkaline phosphatase	*E. coli*	*phoA* (9′)	5′-ribo- and 5′-deoxyribonucleotides 2′- and 3′-ribonucleotides	144
5′ nucleotidase (UDP sugar hydrolase)	*E. coli*	*ush* (9′)	5′-ribo- and 5′-deoxyribonucleotides	14, 123, 124
2′,3′-cyclic phosphate phosphodiesterase (3′-nucleotidase)	*E. coli* and *S. typhimurium*	*cpdB* (94′) *pde* (94′)	2′,3′-cyclic ribonucleotides 3′-ribonucleotides	3, 4, 12, 170, 178
non-specific acid phosphatase I	*E. coli* and *S. typhimurium*	*phoN* (93′)[a] *phoP* (25′)	5′-ribo- and 5′-deoxyribonucleotides 3′-ribonucleotides	51, 96, 178
non-specific acid phosphatase II	*S. typhimurium*	*aphA* (unknown)	5′-ribo- and 5′-deoxyribonucleotides 3′-ribonucleotides	170

[a] Mutations in *S. typhimurium* causing lack of non-specific acid phosphatase have been located in two genes, *phoN* and *phoP* of which *phoN* has been suggested to be the structural gene coding for the enzyme, while the *phoP* gene probably is a regulatory gene (96).

deoxyribonucleosides as a source of endogenous pentose 1-phosphate, *B. subtilis* must contain both uridine and thymidine/deoxyuridine phosphorylase activities. In the presence of an exogenous source of ribose- or deoxyribose 1-phosphate the phosphorylases are stimulated in the direction of nucleoside synthesis. Accordingly, mutants defective in either thymidine phosphorylase or thymidine kinase are obtained by selecting for resistance to FU plus a purine deoxyribonucleoside in a *upp* strain (Table 3.2). Since thymidine phosphorylase deficient mutants lack uridine phosphorylase activity as well, it appears that both phosphorylase activities reside in one protein, i.e. pyrimidine nucleoside phosphorylase (*pdp*) (149).

Mutants defective in uridine kinase activity (*udk*) have been isolated by selecting for FUR resistance in strains lacking uracil phosphoribosyltransferase. Mutants defective in uridine kinase also lack cytidine kinase activity. Thus in *B. subtilis* like in enteric bacteria, one single enzyme catalyzes the phosphorylation of both uridine and cytidine (Neuhard, unpublished).

2. *Metabolism of cytosine compounds*

In contrast to enteric bacteria, *B. subtilis* cannot utilize cytosine as a precursor for nucleic acid synthesis (145, 174), and it is resistant to FC. Thus, cytosine is metabolically inert in *B. subtilis*, suggesting either that this organism is devoid of a cytosine deaminase, or that it lacks a system to transport cytosine through the cytoplasmic membrane.

Cytidine or deoxycytidine will serve as a total source of pyrimidines, and the corresponding 5-fluoro analogs FCR and FCdR inhibit growth. *B. subtilis* contain a cytidine/deoxycytidine deaminase which appears *not* to be inducible. Mutants lacking cytidine deaminase (*cdd* mutants) cannot fulfil their total pyrimidine requirement by cytidine or deoxycytidine (145). Thus the metabolic routes for converting exogenous cytidine and deoxycytidine to uracil ribonucleotides require cytidine deaminase. This is in contrast to enteric bacteria which contain an additional pathway for converting cytidine to uracil involving ribonucleotide glycosylase and cytosine deaminase (p. 104).

Cytidine and deoxycytidine may also be phosphorylated to CMP and dCMP respectively, and thereby serve as precursors of nucleic acids. This explains why *cdd* mutants are sensitive to both FCR and FCdR (145). Mutants unable to phosphorylate cytidine to CMP are readily obtained by selecting for FCR resistance in a *cdd* strain (Table 3.2). Such mutants (*udk*) lack both cytidine and uridine kinase activities, but can still phosphorylate deoxycytidine to dCMP (145) (Neuhard and Møllgaard, unpublished). Although a *udk,cdd* mutant cannot incorporate [^{14}C]-cytidine into nucleic acids it can still utilize [^{14}C]-uridine as a nucleic acid precursor (145). This latter result is readily explained, since the mutant still contains normal levels of pyrimidine nucleoside phosphorylase and uracil phosphoribosyltransferase, which will convert uridine to UMP via free uracil.

Mutants defective in deoxycytidine kinase (*dck*) are obtained by selecting for FCdR resistance in a *cdd* strain (Table 3.2). Such mutants are unable to utilize deoxycytidine as a precursor for DNA synthesis (145). Interestingly, *dck* mutants are also defective in deoxyadenosine kinase activity, suggesting that the deoxycytidine and deoxyadenosine kinase activities are encoded by the same gene (116). This has been confirmed by the isolation of a single protein with both kinase activities (116).

In addition to *dck* mutants, mutants defective in dCMP deaminase (*dcd*) are also obtained when selecting for FCdR resistance. Such mutants are unable to deaminate FdCMP to the potent thymidylate synthase inhibitor FdUMP (117, 146). It is somewhat surprising that a *cdd,dcd* mutant is resistant to FCdR, since it is still capable of converting FCdR to FdCMP. This indicates that *B. subtilis* either cannot phosphorylate FdCMP to FdCTP, or that FdCTP is not toxic. *B. subtilis* do contain kinases which will phosphorylate dCMP to dCDP and dCTP, but the specificities of these kinases towards the 5-fluoro analogues have not been investigated.

B. Lactic acid bacteria

1. Lactobacillus

A number of *Lactobacillus* species have an absolute growth requirement for one deoxyribonucleoside *and* a purine *and* a pyrimidine base (22). In the absence of deoxyribonucleosides the cells cannot synthesize the deoxyribonucleotide precursors for DNA, and hence they grow filamentous and die (21). In *L. leichmannii* and *L. lactis* the deoxyribonucleoside requirement can be met by addition of vitamin B_{12} which acts by supplying the coenzyme B_{12} required for the ribonucleotide reductases of these species (cf. Chapter 4).

In accordance with the growth requirements *L. leichmannii* and *L. acidophilus* contain kinase activities capable of phosphorylating the four deoxyribonucleosides to the corresponding deoxyribonucleotides (50, 140). In addition, deoxyribonucleoside requiring species possess *trans*-N-deoxyribosylases which catalyze the reversible, phosphate independent, transfer of deoxyribosyl groups from deoxyribonucleosides to any free purine or pyrimidine base (22). These same *Lactobacillus* species are devoid of enzymes capable of cleaving the glycosylic bond of deoxyribonucleosides either hydrolytically or phosphorolytically (88, 169, 175).

Several *Lactobacillus* species have a growth requirement for pyrimidines even in the presence of an adequate supply of deoxyribonucleosides or vitamin B_{12} (22, 175). Both uracil and uridine will satisfy this requirement. The utilization of uracil for UMP synthesis is catalyzed by uracil phosphoribosyltransferases (44, 107) and *L. leichmannii* mutants defective in this enzyme have been isolated (65). Studies on the metabolism of uridine led to the discovery that most *Lactobacillus* are devoid of nucleoside phosphorylases but contain nucleoside hydrolases specific for *ribo*nucleosides (101, 166). These enzymes are able to cleave both cytidine and uridine to ribose and the corresponding nucleobases, which may either participate in *trans*-N-deoxyribosylation or, in the case of uracil, serve as a precursor of UMP. It is not known whether *Lactobacillus* contain cytosine deaminase.

Based on enzyme assays on crude extracts of *L. pentosus* it was claimed that this organism lacks deoxycytidine deaminase activity (175). However, it has later been postulated, although with no experimental documentation, that both *L. pentosus* as well as a large number of other *Lactobacillus* species contain cytidine deaminase activity (152). Thus, at present it is uncertain to what an extent *Lactobacillus* have the capacity of deaminating cytosine, cytidine, and deoxycytidine.

In general it appears that *Lactobacillus* species which require deoxyribonucleosides for growth have evolved salvage pathways designed to scavenge exogenous deoxyribonucleosides for DNA synthesis. They contain *trans*-N-deoxyribosylases, deoxyribonucleoside kinases, and they are lacking deoxyribonucleoside cleaving enzymes. At the same time *Lactobacillus* have evolved a different mode of catabolizing cytidine and uridine, i.e. via hydrolysis catalyzed by ribonucleoside hydrolases (cf. Chapter 5).

Although very few studies have been carried out with *Lactobacillus* species with *no* deoxyribonucleoside requirements, it has been shown that *L. casei*, which belongs to this group, does not contain *trans*-N-deoxyribosylases and ribonucleoside hydrolases but contains nucleoside phosphorylase activity (22, 88).

2. *Diplococcus*

Diplococcus pneumoniae is sensitive to all 5-fluoro pyrimidine analogs (FCR was not tested) indicating the presence of extensive pyrimidine salvage in this organism (16). The biochemical basis of this sensitivity was studied by following the metabolism of [^{14}C]-labeled FU, FUR, and FUdR *in vivo*, combined with *in vitro* measurements of certain enzyme activities using the natural substrates. It appears that *D. pneumoniae* contains uridine kinase, uracil phosphoribosyltransferase, thymidine/deoxyuridine kinase, and cytosine deaminase, while it lacks nucleoside phosphorylases and hydrolases as well as *trans*-N-deoxyribosylases (16).

Firshein and Hasselbacher who studied the utilization of deoxyribonucleosides in pneumococci found that crude extracts of both virulent and avirulent strains of *D. pneumoniae* contain a potent deoxycytidine deaminase (cytidine was not tested), as well as kinase activities toward both thymidine and purine deoxyribonucleosides. However, they failed to detect deoxycytidine kinase activity and deoxyribonucleoside cleaving activity. They further found that almost all radioactivity found in DNA of the cells following incorporation of deoxycytidine, deoxyuridine and thymidine *in vivo*, was localized in the thymine moieties (56). Although their data suggest the absence of deoxycytidine kinase activity in *D. pneumoniae*, the presence of deoxycytidine deaminase may very well have masked the detection of such an activity both *in vivo* and *in vitro*.

C. Pseudomonads

Pyrimidine salvage has been studied in considerable detail in *Pseudomonas acidovorans* by Kelln and Warren (94). Their findings may be summarized as follows: (a) Pyrimidine ribo- and deoxyribonucleosides cannot serve as total carbon and energy sources. (b) *P. acidovorans* is resistant to the 5-fluoro analogs FUR, FCR, and FCdR, but sensitive to FU, FC, and FUdR. (c) Pyrimidine requiring mutants cannot utilize pyrimidine ribo- and deoxyribonucleosides as total sources of pyrimidines; they require either cytosine or uracil for growth. (d) Mutants resistant to FU due to lack of uracil pnosphoribosyltransferase activity are also resistant to FC (e) Mutants defective in thymidylate synthase are obligate thymidine requiring, they cannot grow on thymine (cf. Chapter 4). (f) Thymidine and deoxyuridine can be used to label the thymine residues of *P. acidovorans* DNA specifically. (g) Enzyme assays confirmed the presence of cytosine deaminase and thymidine/deoxyuridine kinase but failed to detect any cytidine/deoxycytidine deaminating and nucleoside cleaving activities.

This indicates that *P. acidovorans* can convert exogenous cytosine to uracil by cytosine deaminase, and can utilize uracil for UMP synthesis via phosphoribosylation. Besides the ability to phosphorylate deoxyuridine and thymidine to dUMP and dTMP, *P. acidovorans* is devoid of pyrimidine nucleoside metabolizing activities.

From the few reports available on pyrimidine salvage in other pseudomonads it appears that most of them contain cytosine deaminases and are devoid of cytidine/deoxycytidine deaminases (152). In contrast to *P. acidovorans* a number of other pseudomonads has been found to be able to catabolize ribonucleosides hydrolytically (152, 167).

D. The moraxella group

Mutants of *Neisseria meningitidis* defective in the *de novo* synthesis of pyrimidine nucleotides are obligate uracil requirers. Uridine, deoxyuridine, cytosine, cytidine, and deoxycytidine cannot be used as pyrimidine sources. Consistent with this, prototrophic strains are sensitive to FU but resistant to all other 5-fluoropyrimidine bases and nucleosides. Whereas radioactive uracil is readily incorporated into both RNA and DNA no radioactivity from cytosine, uridine, cytidine, thymine, or thymidine is incorporated into nucleic acids. In confirmation of these findings the only salvage pathway enzyme which could be detected in cell-free extracts was uracil phosphoribosyltransferase (87).

Only limited information is available on pyrimidine salvage in other members of the moraxella group. Of fifteen *Moraxella* strains tested only three contain thymidine kinase and thymidine phosphorylase, all the rest are unable to metabolize thymine and thymidine. Similarly, *Acinetobacter calcoaceticus* seems to be deficient in both thymidine phosphorylase and thymidine kinase activities. In addition it appears that uridine is also metabolically inert in *A. calcoaceticus* (86, 186).

Thus, the moraxella group represents some of the most restrictive bacteria with respect to pyrimidine salvage. The only preformed pyrimidine compound which can be utilized for nucleic acid synthesis is uracil.

E. Mycoplasmas

The growth requirements of *Mycoplasma* in defined medium are quite complex. *Mycoplasma mycoides* does not contain *de novo* pathways for pyrimidine and purine nucleotide synthesis and cannot synthesize thymidine nucleotides endogenously. Thus, it will not grow unless the medium contains uracil or uridine, thymine, and guanine. By labeling cells with [^{14}C]-pyrimidine bases or nucleosides and [^{32}P]-orthophosphate and determining the ratio ^{14}C/^{32}P in the individual nucleoside monophosphates derived from RNA, Mitchell and Finch (114) were able to define the pathways involved in the synthesis of ribonucleotides from different precursors. Their results indicate that *M. mycoides* is devoid of cytosine deaminase but does contain uracil phosphoribosyltransferase, uridine/cytidine kinase, a potent uridine cleaving activity, probably uridine phosphorylase, and cytidine deaminase. Except for cytidine deaminase the presence of the other activities were confirmed by enzyme assays on cell free extracts (114). The failure to detect cytidine deaminase activity in the extract is somewhat contradictory to the observation that addition of [^{14}C]-cytidine to growing cultures of *M. mycoides* results in extracellular accumulation of [^{14}C]-uracil and [^{14}C]-uridine (113, 114).

In addition to the above mentioned enzyme activities the growth requirement for thymine shows the presence of thymidine phosphorylase and thymidine kinase. The capacity of *M. mycoides* to metabolize deoxyuridine and deoxycytidine has not been studied. However, the presence of thymidine phosphorylase and thymidine kinase suggests that deoxyuridine may either be cleaved phosphorolytically to uracil and deoxyribose 1-phosphate or be phosphorylated to dUMP. It has not been established whether *M. mycoides* contains kinases for deoxycytidine, deoxyadenosine, and deoxyguanosine.

F. Fungi

1. Yeast

Utilization of nucleobases. Cytosine and uracil are utilized effectively by *Saccharomyces cerevisiae* as pyrimidine sources, and both FC and FU are highly toxic to pyrimidine prototrophic strains. Furthermore cytosine will serve as the only nitrogen source for *S. cerevisiae*. Strains specifically defective in cytosine utilization, i.e. resistant to FC but sensitive to FU, contain mutations in either of two genetically distinct loci, *cyt-p* and *cyt-da*, which code for cytosine permease and cytosine deaminase, respectively (57, 85). By selecting for resistance to FU, mutants defective in uracil phosphoribosyltransferase, i.e. *ups* mutants, and mutants lacking uracil permease, i.e. *ura-p* mutants, were obtained. The *ura-p* mutants remain sensitive to FC while the *ups* mutants are resistant to this analog. The abilities of the different mutants to utilize exogenous pyrimidines as pyrimidine sources are shown in Table 3.4. From the table it can be seen that the conversion of exogenous cytosine to UMP requires cytosine permease, cytosine deaminase, and uracil phosphoribosyltransferase, whereas the utilization of uracil requires a functional uracil permease and uracil phosphoribosyltransferase.

Utilization of nucleosides. Uridine is an efficient pyrimidine source for pyrimidine auxotrophic strains of *S. cerevisiae*, whereas cytidine is only utilized poorly. Resistance to FUR has been used to screen for mutants in uridine metabolism (57, 85). Characterization of such mutants revealed that at least three different genetic loci are involved in uridine utilization: the structural gene for uridine kinase (*urid-k*), the structural gene for a specific uridine hydrolase (*urid-rh*), and a gene encoding a uridine permease (*urid-p*). All three genes appear to be unlinked to each other and to the genes encoding proteins involved in nucleobase metabolism (57, 85). Mutants defective in uridine permease cannot utilize uridine as sole pyrimidine source, but they grow normally on cytosine and uracil (Table 3.4). Mutants lacking uridine kinase or uridine hydrolase can still convert uridine to UMP, either via free uracil (in the *urid-k* mutants) or by phosphorylation (in the *urid-rh* mutants)

TABLE 3.4

Effects of mutations in the pyrimidine salvage pathway on the utilization of pyrimidines in *Saccharomyces cerivisiae*.

		Pyrimidine source			
		Cytosine		Uracil	Uridine
Genotype[a]	Defective enzyme	0.1 mM	10 mM	0.1 mM	0.4 mM
wild type	none	+ +[b]	+ +	+ +	+ +
cyt-p	cytosine permease	−	+	+ +	+ +
cyt-da	cytosine deaminase	−	−	+ +	+ +
ura-p	uracil permease	+ +	+ +	−	+ +
ups	uracil phosphoribosyltransferase	−	−	−	+ +
urid-p	uridine permease	+ +	+ +	+ +	−
urid-k	uridine kinase	+ +	+ +	+ +	+ +
urid-rh	uridine hydrolase	+ +	+ +	+ +	+ +
urid-k *urid-rh*	uridine kinase and uridine hydrolase	+ +	+ +	+ +	−

[a] All mutants contain an additional mutation, *ura-2*, rendering the strains pyrimidine auxotrophic.
[b] + +, denotes full growth; +, denotes slow growth; −, denotes no growth.

(Table 3.4). Double mutants of the genotype *urid-k*, *urid-rh*, however, are unable to grow on uridine as pyrimidine source (Table 3.4).

S. cerevisiae contains cytidine deaminase activity (85). The observation that cytidine is a poor pyrimidine source may indicate that deamination of cytidine is a slow process, or that cytidine is not readily taken up by the cells.

S. cerevisiae cannot utilize thymine or thymidine as a source of thymidine nucleotides due to the lack of thymidine phosphorylase and thymidine kinase activities (cf. Chapter 4). This may explain why yeast is quite resistant to FUdR (57).

2. *Other fungi*

A number of fungi besides yeast lack thymidine kinase and are unable to utilize thymidine as a specific DNA precursor (59). *Neurospora crassa*, *Aspergillus nidulans*, and the yeast *Rhodotorula glutinis* have evolved a different pathway for the utilization of thymidine and thymine for nucleic acid synthesis (161, 162, 180). This involves oxidation of thymidine/deoxyuridine to thymine ribonucleoside/uridine catalyzed by a 2′-hydroxylase, followed by hydrolytic cleavage of the glycosylic bond to thymine/uracil and ribose. Subsequently thymine is oxidized by thymine 7-hydroxylase to uracil 5-carboxylic acid via the intermediate formation of 5-hydroxymethyluracil and 5-formyluracil. Finally uracil 5-carboxylic acid is decarboxylated to uracil

(Fig. 3.4). In combination with cytidine/deoxycytidine deaminase this route also explains the utilization of deoxycytidine as a general pyrimidine precursor for both RNA and DNA synthesis (Fig. 3.4; 162).

Pyrimidine requiring mutants of *N. crassa* can use uracil, uridine or cytidine for growth. However, they will not grow on thymidine or thymine as sole pyrimidine source unless growth is initiated by the addition of small amounts of uridine or cytidine. Mutants have been isolated which will grow on thymidine alone. They appear to be regulatory mutants containing increased

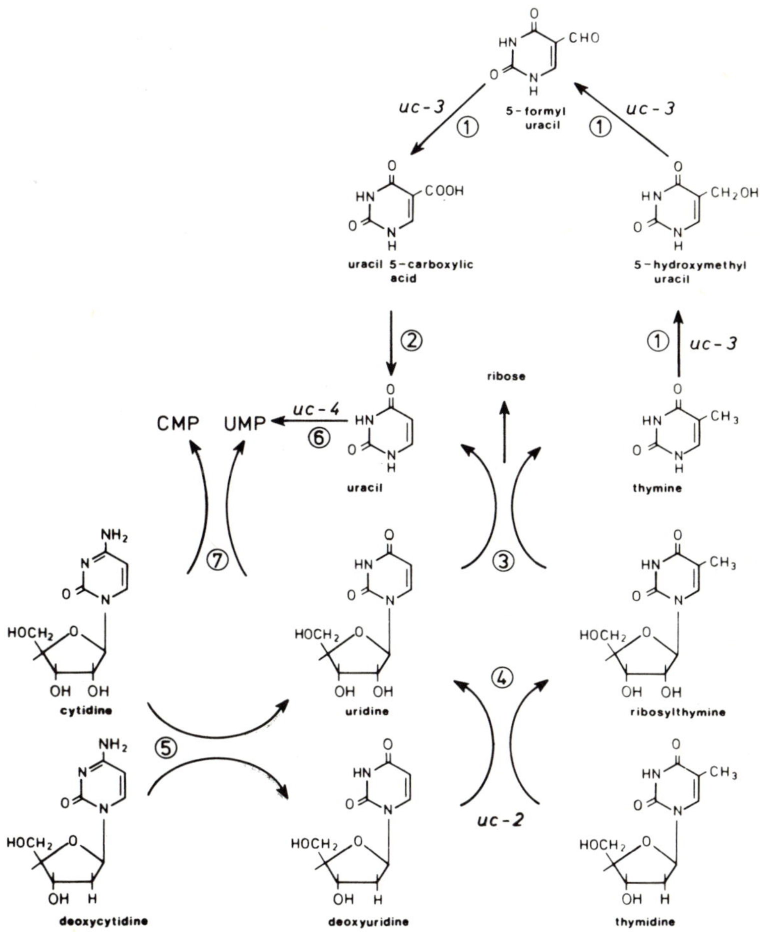

FIG. 3.4. Pyrimidine salvage pathways in *Neurospora crassa*. 1, thymine 7-hydroxylase; 2, uracil 5-carboxylic acid decarboxylase; 3, uridine hydrolase; 4, pyrimidine deoxyribonucleoside 2′-hydroxylase; 5, cytidine/deoxycytidine deaminase; 6, uracil phosphoribosyltransferase; 7, uridine/cytidine kinase.

levels of thymine 7-hydroxylase and uracil 5-carboxylic acid decarboxylase (58, 179).

Mutants with decreased 2′-hydroxylase activity (*uc*-2 mutants) have been isolated and characterized (162, 179). They can utilize thymine, but not thymidine or deoxyuridine as a pyrimidine source. Also a mutant defective in thymine 7-hydroxylase, (*uc*-3), has been characterized (162, 179). Its phenotype is somewhat peculiar since it can still utilize 5-formyluracil, but not 5-hydroxymethyluracil or thymine, as a pyrimidine source (Fig. 3.4). This was shown to be due to the presence in *N. crassa* of another activity, which in the presence of NAD^+ will oxidize 5-formyluracil to uracil 5-carboxylic acid (162).

In *N. crassa* uridine may be utilized for UMP synthesis by two routes, i.e. by phosphorylation catalyzed by uridine kinase or via hydrolytic cleavage followed by phosphoribosylation of the uracil thus formed. Both pathways are operating simultaneously since a pyrimidine requiring mutant defective in uracil phosphoribosyltransferase (*uc*-4) shows a reduced growth yield on uridine as the sole pyrimidine source (179).

The observation that FUdR in higher concentrations is toxic to many fungi despite the absence of both thymidine phosphorylase and thymidine kinase (57, 134) suggests that FUdR may be oxidized to FUR which is subsequently hydrolyzed to FU. This should be taken into account in studies involving the use of FUdR for selection of mutants of such organisms (134).

5-Fluorocytosine has been used as a potent antifungal agent (25). The high sensitivity of certain fungi towards FC depends on the presence of an efficient system for uptake and deamination of cytosine to uracil. This implies that resistance to this drug may readily develop due to accumulation of mutants defective in either cytosine permease or cytosine deaminase (71).

The metabolism of exogenous pyrimidine bases and nucleosides in the cellular slime mold *Physarum polycephalum* differs significantly from what is found in other fungi (55). Exogenous uracil, cytosine and thymine are not metabolized, and FU and FC do not inhibit growth even when present in very high concentration in the medium. The inability to utilize cytosine is due to lack of cytosine deaminase, whereas the failure to convert uracil to UMP *in vivo* seems to be caused by inability to transport uracil into the cell. Cell-free extracts of *P. polycephalum* do show uracil phosphoribosyltransferase activity.

Exogenous nucleosides are readily taken up and metabolized by microplasmodia of *P. polycephalum*, and FUR, FCR, FUdR, and FCdR are potent inhibitors of growth. The labeling pattern of nucleoside triphosphates by exogenous nucleosides combined with enzyme analysis on cell-free extracts revealed that both pyrimidine ribonucleosides and pyrimidine deoxyribonucleosides are phosphorylated to the corresponding nucleotides by kinases. In addition cytidine and deoxycytidine may undergo deamination, and uridine, deoxyuridine, and thymidine may be cleaved hydrolytically to uracil and pentoses (55).

The thymidine kinase activity of *Physarum* undergoes periodic fluctuations during the nuclear division cycle, reaching a sharp maximum during the early S-phase, immediately following mitosis (62).

G. Protozoa

The ciliated protozoa *Tetrahymena* lacks the pyrimidine and purine *de novo* pathways and has therefore an absolute growth requirement for an exogenous pyrimidine and purine base. The pyrimidine requirement may be satisfied with uracil, uridine or cytidine, but not with cytosine. The latter observation is explained by the absence of cytosine deaminase from *Tetrahymena.* Thymine and thymidine will serve as specific precursors for DNA and both thymidine phosphorylase and thymidine kinase activities have been found in extracts of *Tetrahymena* (69).

Two different enzymes capable of phosphorylating thymidine to dTMP are present in *Tetrahymena* (185). One, thymidine kinase, uses ATP as phosphate donor and is feedback inhibited by dTTP. The other catalyzes the transfer of a phosphate group from a nucleoside monophosphate e.g. AMP to thymidine. This latter enzyme has been named nucleoside phosphotransferase. The ratio between the two thymidine phosphorylating activities shows considerable variations with the growth conditions as well as during the cell cycle (183, 185). The physiological implications of the presence of these two activities are unclear.

Deoxyuridine is presumably metabolized by the same enzymes as thymidine, implying that it may either be phosphorylated to dUMP, or cleaved phosphorolytically to uracil and thereby serve as a pyrimidine source. Deoxycytidine is deaminated by cytidine/deoxycytidine deaminase. However, it has not been established whether *Tetrahymena* has the capacity of phosphorylating deoxycytidine to dCMP.

H. Summary

From the foregoing disucssion it appears that the capacity of different microorganisms to metabolize preformed pyrimidines varies considerably, although the ability to convert uracil directly to UMP seems ubiquitous. Table 3.5 shows the distribution of pyrimidine salvage enzymes in the different genera discussed above. However, it should be emphasized that the absence or presence of a given enzyme has not been established with equal certainty in all cases. In several instances it has only been deduced from physiological data available. Finally it must be stressed that the summary given in Table 3.5 does not imply that all species within a given genus necessarily have the same capacities for pyrimidine salvage. As illustrated by the pseudomonads, significant variations may be observed within the same genus.

IV. ENZYMES INVOLVED IN PYRIMIDINE SALVAGE

A. Uracil phosphoribosyltransferase (EC 2.4.2.9)

$$\text{uracil} + \text{PRPP} \xrightarrow{Mg^{++}} \text{UMP} + \text{pyrophosphate}$$

Uracil phosphoribosyltransferase has been purified from *E. coli* and from yeast. The *E. coli* enzyme (MW 75 000) consists of three subunits of similar size (MW 23 500). Substrate saturation curves with both uracil and PRPP are hyperbolic (Rasmussen, U. 1980. Thesis, University of Copenhagen). The enzyme is inhibited by uridine nucleotides and activated by GTP, the activation by GTP being dependent on the PRPP concentration. The activation by GTP is abolished by addition of ppGpp in concentrations equimolar with GTP (54, 118). The physiological implications of the effects of GTP and ppGpp on the enzyme have been discussed previously (p. 100). The yeast enzyme (MW 80 000) seems to contain two nonidentical subunits. It displays Michaelis–Menten kinetics towards both substrates. Several pyrimidine nucleotides inhibit the enzyme, the most effective inhibitor being UMP. It shows 50% inhibition at 0.3 mM. No positive effector for the enzyme was reported (122).

Crude preparations of uracil phosphoribosyltransferases from *Neisseria* and *Mycoplasma* resemble the *E. coli* enzyme in that they are activated by GTP (87, 114). In contrast, GTP has no effect on partially purified enzymes from *B. subtilis* (Neuhard, unpublished) and *Tetrahymena pyriformis* (138).

Uracil phosphoribosyltransferases are specific for uracil. They show no activity with cytosine, orotate, or thymine, although they are active with 5-fluorouracil, 5-azauracil, 2-thiouracil, and 6-azauracil (29, 40, 44, 65, 107, 121, 138).

Pyrimidine starvation of *E. coli* and *S. typhimurium* results in a 6–10 fold increase in the internal levels of the enzyme, indicating that the synthesis of uracil phosphoribosyltransferase is regulated by a pyrimidine nucleotide (Andersen, L. 1979. Thesis, University of Copenhagen).

TABLE 3.5
Pyrimidine salvage enzymes in certain microbial genera.

					Pseudomonas							
Enzyme	*Escherichia Salmonella*	*Bacillus*	*Lactobacillus*	*Diplococcus*	*P. acidovorans*	Other species	*Neisseria*	*Mycoplasma*	*Saccharomyces*	*Neurospora*	*Physarum*	*Tetrahymena*
Uracil phosphoribosyl transferase	+	+	+	+	+	+	+	+	+	+	+	+
Cytosine deaminase	+	(−)	?	+	+	+	−	−	+	?	−	−
Cytidine/deoxycytidine deaminase	+	+	(−)	+	−	?	−	(+)	+	+	+	+
Uridine phosphorylase	+	+	−	−	−	−	−	+	−	−	−	+
Pyrimidine nucleoside hydrolase	−	−	+	−	−	+	−	?	+	+	+	−
Thymidine phosphorylase	+	+	−	−	−	?	−	+	−	−	−	+
Thymidine kinase	+	+	+	+	+	?	−	+	−	−	+	+
Uridine/cytidine kinase	+	+	?	+	−	?	−	+	+	+	+	+
Deoxycytidine kinase	−	+	+	−	−	?	−	?	?	?	+	?
Trans-N-deoxyribosylase	−	−	+	−	−	?	−	−	−	−	−	?
Thymidine-thymine oxidation	−	−	−	−	−	?	−	−	?	+	−	?
Ribonucleotide-glycosylase	+	?	?	?	?	+	?	?	?	?	?	?

+ present; − absent.
Parenthesis indicate conflicting data are available.
? indicates not known.

B. Cytosine deaminase (EC 3.5.4.1)

$$\text{cytosine} + H_2O \longrightarrow \text{uracil} + NH_3$$

Cytosine deaminase is the only known enzyme with cytosine as substrate. A survey of the distribution of this enzyme activity in a variety of bacteria has been published (154). Since this survey was based on enzyme assays on cell-free extracts only positive results can be regarded as unequivocal. On the other hand, physiological studies combined with a failure to detect cytosine deaminase enzymatically have established the absence of the enzyme from certain microorganisms such as *B. subtilis*, *Neisseria meningitidis*, *Mycoplasma mycoides*, *Physarum polycephalum*, and *Tetrahymena* (Table 3.5).

The enzyme has been purified from *Serratia marscescens* (153), *Pseudomonas aureofaciens* (154), *E. coli* (Andersen, L. 1979. Thesis, University of Copenhagen), *S. typhimurium* (Schaumburg. 1970. Thesis, University of Copenhagen), and yeast (80). The bacterial enzymes are all large multimeric proteins with molecular weights between 300 000 and 650 000, while the yeast enzyme is a monomer with a molecular weight of about 33 000 (182).

Substrate specificity studies of the enzymes have shown that cytosine deaminases from *P. aureofaciens* and yeast can deaminate 5-methylcytosine to thymine, while the enzymes from the enteric bacteria cannot. Isocytosine is a substrate for the cytosine deaminases from *E. coli* and *S. typhimurium* (108; Schaumburg, 1970. Thesis, University of Copenhagen). This explains why these bacteria can use isocytosine as a total source of pyrimidines. In yeast isocytosine is not a substrate but appears to inhibit the enzyme (100).

The yeast enzyme is inhibited by a number of nucleosides and nucleotides (80), but the physiological implications of these effects are unknown.

In *E. coli* both pyrimidine starvation as well as partial nitrogen starvation induce cytosine deaminase synthesis several fold, whereas addition of purines causes severe repression (Andersen, L. 1979. Thesis, University of Copenhagen). The nature of the regulatory system responsible for this control has not been established, but it appears that the same system controls the synthesis of compounds involved in cytosine uptake (p. 282).

C. Cytidine deaminase (EC 3.5.4.5)

$$\text{cytidine deoxycytidine} + H_2O \longrightarrow \text{uridine deoxyuridine} + NH_3$$

Cytidine deaminase is found in most microorganisms (152). The only two microorganisms which have been established to be devoid of this activity are *Pseudomonas acidovorans* and *Neisseria meningitidis* (Table 3.5). The enzyme has been partially purified from *E. coli* (76, 177), *S. typhimurium* (Schaumburg, J. 1970. Thesis, University of Copenhagen), and yeast (79). All three enzymes deaminate both cytidine and deoxycytidine, the latter being the preferred substrate. In addition a large number of cytidine and deoxycytidine analogs including 5-methylcytidine, 5-methyldeoxycytosine, 5-fluorocytidine, 5-fluorodeoxycytidine, and arabinofuranosylcytidine are substrates for the enzymes from enteric bacteria (43, 53, 99, 108, 168). 3,4,5,6-tetrahydrouridine is a very potent competitive inhibitor of the enzyme ($K_i = 2.4 \times 10^{-7}$M) and may be regarded as a transition state analog (42, 53). Tetrahydrouridine also inhibits cytidine deaminase *in vivo*, though the effect is only temporary (168).

The regulation of cytidine deaminase synthesis in *E. coli* and *S. typhimurium* is quite complex and is described in detail elsewhere in this book (Chapter 5). Therefore it will only be summarized at this point. The expression of the cytidine deaminase (*cdd*) gene is controlled by two regulatory proteins. One is a repressor coded for by the *cytR* gene, and the other is the cyclic AMP-binding protein. In the presence of cyclic AMP the latter acts as a positive controlling element. Cytidine is the inducer in *E. coli*, whereas both cytidine and uridine induce enzyme synthesis in *S. typhimurium*.

The rate of cytidine deaminase synthesis is *not* increased in *B. subtilis* by growth in the presence of high concentrations of cytidine or uridine (Neuhard, unpublished).

D. Enzymes cleaving the glycosylic bond of pyrimidine nucleosides

The capacity of different microorganisms to cleave the glycosylic bond of pyrimidine nucleosides varies considerably (Table 3.5). The cleavage may be phosphorolytic catalyzed by nucleoside phosphorylases, hydrolytic catalyzed by nucleosidases i.e. nucleoside hydrolases, or consist in a transfer of the pentose moiety from a deoxyribonucleoside to a nucleobase. The latter reactions are catalyzed by *trans*-N-deoxyribosylases.

Microorganisms containing pyrimidine nucleoside phosphorylases are devoid of pyrimidine nucleoside hydrolase and *trans*-N-deoxyribosylase activities. Phosphorolysis of uridine and thymidine may be catalyzed either by two enzymes, i.e. uridine phosphorylase and thymidine phosphorylase as in enteric bacteria, or by a single enzyme, pyrimidine nucleoside phosphorylase as in certain bacteria such as *B. subtilis* and *Haemophilus influenzae*. Enzymes capable of cleaving cytidine and deoxycytidine phosphorolytically have never been found.

Pyrimidine nucleoside hydrolases are specific for *ribo*nucleosides. They occur either alone as in many pseudomonads, or together with *trans*-N-deoxyribosylases as in *Lactobacillus*. This implies that organisms containing only pyrimidine nucleoside hydrolases cannot cleave thymidine or deoxyuridine. In certain fungi this has been circumvented by the evolution of enzymes capable of oxidizing thymidine and deoxyuridine to the corresponding ribonucleosides which can be hydrolyzed (p. 115).

Certain bacteria such as *Diplococcus pneumoniae*, *Pseudomonas acidovorans*, and *Neisseria meningitidis* are totally deficient in pyrimidine nucleoside cleaving activities (cf. Table 3.5).

A more detailed discussion on the distribution of nucleoside cleaving enzymes in microorganisms, and their role in nucleoside catabolism, may be found in Chapter 5.

1. *Uridine phosphorylase* (EC 2.4.2.3)

$$\text{uridine} + \text{orthophosphate} \longrightarrow \text{uracil} + \text{ribose 1-phosphate}$$

Studies on a homogenous preparation of uridine phosphorylase from *E. coli* (103) showed that the enzyme consists of eight identical subunits of MW 22 000. The enzyme shows no activity towards cytidine as substrate, but will phosphorolyze both thymidine and deoxyuridine with 2% and 6% of the rate with uridine, respectively. Some physiological implications of this inherent weak thymidine cleaving activity of uridine phosphorylase have been discussed elsewhere (pp. 104, 182).

Uridine phosphorylase activity has been detected in a variety of bacteria by enzyme assays on crude extracts (77, 150).

In *E. coli* and *S. typhimurium* the intracellular levels of uridine phosphorylase are regulated. Although the uridine phosphorylase (*udp*) and *cdd* genes are not closely linked on the chromosomes, they belong to the same regulon (cf. Chapter 5). Thus the expression of the *udp* gene is regulated as described above for the *cdd* gene (p. 121).

2. *Thymidine phosphorylase* (EC 2.4.2.4)

$$\text{thymidine} + \text{orthophosphate} \longrightarrow \text{thymine} + \text{deoxyribose 1-phosphate}$$

E. coli and *S. typhimurium* mutants constitutive for the synthesis of thymidine phosphorylase (*see below*) have been used as sources for the purification of thymidine phosphorylase (72, 158). The highly purified enzymes from these organisms are very similar. They are composed of two identical subunits of MW around 45 000 and are specific for thymidine and deoxyuridine as substrates. They show no activity towards deoxycytidine, uridine, and purine nucleosides, though the enzymes will catalyze the phosphorolytic cleavage of certain 5-substituted deoxyuridine analogs such as FUdR (143). A large number of thymidine analogs carrying modifications in both the pentose and the pyrimidine moiety have been tested as substrates and inhibitors of the *E. coli* thymidine phosphorylase. From these studies the dimensions of the hydrophobic bonding region of the active site emerged, which allow the design of extremely potent and irreversible active-site-directed inhibitors of the enzyme (7, 8).

Both enzymes follow an ordered sequential reaction mechanism and both enzymes are inhibited by uridine, deoxyadenosine, and ribose 1-phosphate (24, 157).

The presence of thymidine phosphorylase in different microorganisms has been established either by the ability of thymine to satisfy the growth requirement of mutants lacking thymidylate synthase (Chapter 4), or by enzyme assays on crude cell-free extracts (77). The enzyme appears to be widely distributed although its absence from certain fungi and some bacteria has been established (Table 3.5).

The structural gene for thymidine phosphorylase, *deoA*, is part of the *deo*-operon which in addition includes the *deoB*,*deoC*, and *deoD* genes (the latter genes code for phosphopentomutase, phosphodeoxyriboaldolase, and purine nucleoside phosphorylase, respectively). The expression of the *deo*-operon is controlled by three regulatory proteins, two of which are repressors coded for by the *deoR* and *cytR* genes (64, 120). In addition the cyclic AMP-CRP complex is required for that part of the *deoA* expression which is controlled by the *cytR* repressor. The effector (inducer) for the *deoR* coded repressor is deoxyribose 5-phosphate, which *in vivo* may be derived from deoxyribonucleosides. Cytidine is the effector for the *cytR* coded repressor in *E. coli*, while both cytidine and uridine are effectors in *S. typhimurium*. A detailed account of the regulation of the *deo*-operon is presented in Chapter 5.

3. Pyrimidine nucleoside phosphorylase

Certain organisms such as *Bacillus stearothermophilus*, *Bacillus subtilis*, and *Haemophilus influenzae* contain pyrimidine nucleoside phosphorylases which will cleave a variety of pyrimidine nucleosides regardless of the nature of the pentose moieties. Thus uridine, deoxyuridine, uracil arabinoside, and thymidine, but not cytidine or deoxycytidine are substrates for the enzymes. The ratios between the rate of thymidine and uridine cleavage catalyzed by the purified enzymes from *B. stearothermophilus* and *H. influenzae* are 1.39 and 0.27 respectively (155, 160).

The synthesis of pyrimidine nucleoside phosphorylase is inducible in *Bacillus*. In *B. subtilis* thymidine as well as purine deoxyribonucleosides and acetaldehyde are exogenous inducers; ribonucleosides are inactive (149, 155; Neuhard, unpublished). In *B. stearothermophilus* both thymidine and uridine act as inducers (155).

4. Pyrimidine ribonucleoside hydrolases (EC 3.2)

$$\text{pyrimidine ribonucleoside} + H_2O \longrightarrow \text{pyrimidine} + \text{ribose}$$

In contrast to the reactions catalyzed by nucleoside phosphorylases, the hydrolytic cleavage of nucleosides are irreversible. Pyrimidine nucleoside hydrolases have so far been found in *Pseudomonas* (167), *Lactobacillus* (101, 166), fungi (110, 162, 163), and protozoans (48, 98). Although all the enzymes are strictly specific towards the β-D-ribofuranosyl glycosylic linkages found in ribonucleosides, they may be classified into two groups according to their specificities towards the base moieties of the substrates. The first group consists of enzymes from fungi such as *Saccharomyces cerevisiae* and *Neurospora crassa*. They are highly specific for uridine and ribosylthymine; neither cytidine nor purine ribonucleosides are cleaved by these enzymes. Hence they are frequently named uridine nucleosidases. The second group includes hydrolases with broader specificities towards the base moieties. They will use both pyrimidine and purine ribonucleosides as substrates.

Uridine nucleosidase from yeast is the only representative of the first group which has been highly purified (109). It is composed of two apparently identical subunits with a molecular weight of 16 500, and seems to contain approximately 1 g-atom of copper per mole of enzyme. The enzyme shows Michaelis–Menten kinetics with both uridine and ribosylthymine as substrates, the K_m-values being 0.9 and 1.7 mM respectively. The enzyme is inhibited by ribose 5-phosphate and other pentose phosphate pathway intermediates, as well as by various metal ions and chelating agents (110, 111).

Nucleoside hydrolases of the second group have been purified from *Lactobacillus delbrueckii* (166), *Pseudomonas fluorescens* (167) as well as from the protozoans *Crithidia fasciculata* (MW 150 000) (48), and *Leishmania donovani* (MW 180 000) (98). They can all use cytidine, uridine, ribosylthymine, and certain purine ribonucleosides as substrates.

5. Trans-*N-deoxyribosylases (nucleoside deoxyribosyltransferases)*

$$\text{deoxyribonucleoside}_1 + \text{nucleobase}_2 \longrightarrow \text{deoxyribonucleoside}_2 + \text{nucleobase}_1$$

Trans-N-deoxyribosylases catalyze the reversible, phosphate *in*dependent transfer of deoxyribosyl groups from a purine or pyrimidine deoxyribonucleoside to a purine or pyrimidine base. They have only been found in those species of the *Lactobacillus* genus which require deoxyribonucleosides for growth (22). Affinity chromatography allowed the separation and purification of two distinct *trans*-N-deoxyribosylases from extracts of *L. helveticus*. One, i.e. *trans*-N-deoxyribosylase I, catalyzes exclusively transfers between purines, while the other, *trans*-N-deoxyribosylase II, catalyzes the transfer to and from purine and pyrimidine bases (33, 34, 73, 74). Kinetic studies of the two *L. helveticus* enzymes established that the transfers in both cases proceed via a ping-pong mechanism (45, 46). Whether other *Lactobacillus* species also contain two *trans*-N-deoxyribosylases remains to be determined.

Starvation for deoxyribonucleosides causes derepression of the synthesis of *trans*-N-deoxyribosylase in all *Lactobacillus* species which contain the enzyme. However, the nature of the low molecular weight effector for this control system has not been established (22).

E. Hydroxylases (dioxygenases)

A unique pathway for converting thymidine to uracil and ribose is found in certain fungi (p. 114). It involves four hydroxylation reactions catalyzed by two hydroxylases, i.e. pyrimidine deoxyribonucleoside 2′-hydroxylase and thymine 7-hydroxylase, both of which appear to be 2-ketoglutarate dioxygenases catalyzing reactions of the following general type:

$$\text{Substrate} + O_2 + \begin{array}{l}\text{COOH}\\ |\\ \text{CH}_2\\ |\\ \text{CH}_2\\ |\\ \text{COCOOH}\end{array} \xrightarrow[\text{Fe}^{2+}]{\text{ascorbate}} \text{substrate} - O + \text{CO}_2 + \begin{array}{l}\text{COOH}\\ |\\ \text{CH}_2\\ |\\ \text{CH}_2\\ |\\ \text{CO}O\text{H}\end{array}$$

As shown by the equation one atom of molecular oxygen is incorporated into the substrate and one into succinate (75). The two hydroxylases have so far only been found in *Neurospora crassa* (66), *Rhodotorula glutinis* (180), and *Aspergillus nidulans* (161).

1. *Pyrimidine deoxyribonucleoside 2′-hydroxylase* (EC 1.14.11.3)

$$\text{thymidine} + \text{2-ketoglutarate} + \text{O}_2 \xrightarrow[\text{Fe}^{2+}]{\text{ascorbate}} \text{ribosylthymine} + \\ + \text{succinate} + \text{CO}_2$$

With partially purified preparations of the enzyme from *N. crassa* the stoichiometry of the reaction was established (9, 10, 106). No other 2-ketodicarboxylic acid can replace 2-ketoglutarate as cofactor in the reaction (10). Of the deoxyribonucleosides tested as substrates, activity was observed with deoxyuridine, thymidine, 5- hydroxymethyldeoxyuridine, 5-bromodeoxyuridine, and 6-azathymidine. Deoxycytidine and purine deoxyribonucleosides are not hydroxylated by the enzyme (10, 163). The requirement for ferrous ions is absolute and specific. Catalase stimulates the reaction about 3-fold, but the mechanism underlying this effect is not known (10, 106).

The amount of 2′-hydroxylase contained by *Neurospora* varies according to the growth conditions. A five-fold lower level of the enzyme activity was found in cultures grown with aeration in media lacking ammonia than in media supplied with ammonium ions. However, this effect of ammonium ions on the 2′-hydroxylase content was not observed when cultures were grown under non-aerated conditions (58).

2. Thymine 7-hydroxylase (EC 1.14.11.6)

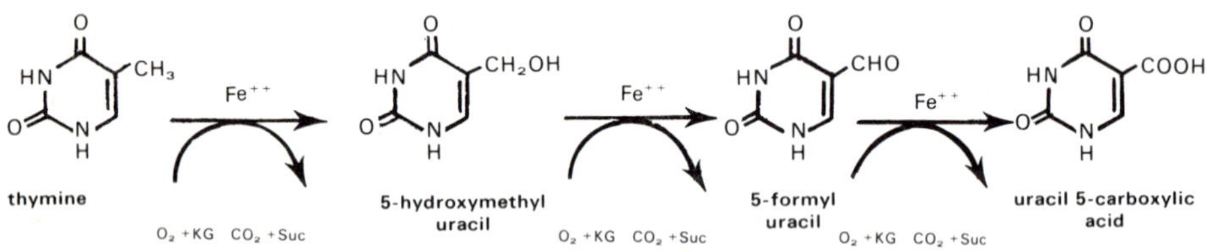

The oxidation of thymine to uracil-5-carboxylic acid through the intermediate formation of 5-hydroxymethyluracil and 5-formyluracil is catalyzed by one single protein, thymine 7-hydroxylase (MW 40 000 (10)), which has been purified 1300-fold from *N. crassa* (105).

All three dioxygenase reactions require oxygen, 2-ketoglutarate, ferrous ions, and a reducing agent such as ascorbate, yielding as products the oxidized substrates, succinate, and CO_2 in a 1 : 1 : 1 molar ratio (106). The requirement for ferrous ions cannot be met by other metal ions, whereas the ascorbate requirement may be partially satisfied by glutathione or dithiothreitol (66). 2-ketoglutarate cannot be replaced by other dicarboxylic acids or by reduced pyridine nucleotides (66).

The substrate specificity of thymine 7-hydroxylase is unusual since an alkane, an alcohol, and an aldehyde are substrates. A large number of pyrimidines chemically related to thymine have been tested as substrates for the enzyme, and it was found that 1-methyluracil, 1-ethyluracil, and 6-azathymine, but not 5-methylcytosine, would serve as substrates (11).

The requirement for ascorbate in the dioxygenase reaction is not understood. However, studies on a partially purified preparation of thymine 7-hydroxylase from *Rh. glutinis* have shown that ATP and certain other nucleoside di- and triphosphates may substitute for ascorbate. Hydrolysis of ATP is not required for stimulation (181).

F. Uracil-5-carboxylic acid decarboxylase

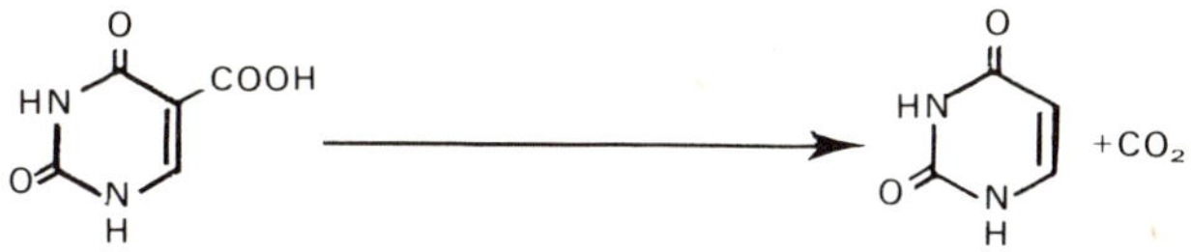

uracil 5-carboxylic acid — uracil

The final step in the conversion of thymidine to uracil in *Neurospora* consists of the decarboxylation of uracil-5-carboxylic acid to uracil catalyzed by uracil-5-carboxylic acid decarboxylase. The enzyme has been identified in *N. crassa*, although it has not been purified to any appreciable degree. With crude cell-free extracts the stoichiometry of the reaction was established, and it was shown that orotic acid cannot substitute for uracil-5-carboxylic acid as substrate for the enzyme. No cofactor was required for the reaction following Sephadex-G-25 chromatography of the crude cell-free extract (133).

G. Pyrimidine ribonucleoside kinase (EC 2.7.1.48)

$$\text{uridine/cytidine} + \text{GTP/ATP} \xrightarrow{\text{Mg}^{2+}} \text{UMP/CMP} + \text{GDP/ADP}$$

Pyrimidine ribonucleoside kinases have been purified from *E. coli* (MW 90 000) (172), *Bacillus stearothermophilus* (MW 109 000) (130, 132), and *Tetrahymena pyriformis* (MW 195 000) (137). In all three cases it was shown that one single enzyme is responsible for the phosphorylation of both uridine and cytidine with the relative activity towards the two substrates of 1 : 0.85. *B. subtilis* mutants selected for lack of cytidine kinase activity invariably also lack uridine kinase, and vice versa, indicating that uridine and cytidine kinase in this organism is coded for by one single gene (Neuhard, unpublished).

The purified enzymes are specific for uridine and cytidine as phosphate acceptors, deoxyribonucleosides and purine ribonucleosides are inactive as substrates. The *E. coli* enzyme will phosphorylate a number of pyrimidine ribonucleoside analogs, including 5-fluorouridine, 5-fluorocytidine, 6-azauridine, and 5-azacytidine (172).

The specificity towards the phosphate donor varies with the source of the enzyme. Thus the enzymes from *E. coli*, *Salmonella typhimurium*, and *B. subtilis* use GTP as the preferred donor, ATP giving less than 10% activity (172; Neuhard unpublished), while the *B. stearothermophilus* (130), *Mycoplasma mycoides* (114), and *Tetrahymena* (137) enzymes show the greatest activity with ATP as phosphate donor.

Pyrimidine ribonucleoside kinases are feedback inhibited by CTP and UTP. In addition, it was shown that the intracellular enzyme activity in *E. coli* and *S. typhimurium* is controlled via regulation of enzyme synthesis. Thus, cultures grown under conditions of partial pyrimidine starvation contain 10-fold increased levels of pyrimidine ribonucleoside kinase (172) suggesting some kind of pyrimidine control of the expression of the *udk* gene. Such a control system may explain the enhancement of uridine kinase activity observed in *E. coli* when grown in the presence of low doses of 5-azauracil or 5-azacytidine (39).

H. Thymidine kinase (EC 2.7.1.21)

$$\text{thymidine} + \text{ATP} \xrightarrow{\text{Mg}^{2+}} \text{dTMP} + \text{ADP}$$

Thymidine kinase is essential for the utilization of exogenous thymine or thymidine for thymidine nucleotide synthesis. A number of bacteria and fungi have been shown to lack thymidine kinase activity (Table 3.5 and p. 188). Since thymidine kinases can use FUdR as substrate, it is also essential in converting this analog to the potent thymidylate synthase inhibitor FdUMP.

Thymidine kinase has been purified to homogeneity from *E. coli* (37, 128, 148) and *Bacillus stearothermophilus* (97). The *E. coli* enzyme uses thymidine, deoxyuridine, and 5-halogenated deoxyuridine analogs as substrates with about equal effectivity, while ribosylthymine and thymine arabinoside are not phosphorylated by the enzyme. In attempts to design effective active-site-directed irreversible inhibitors a large number of thymidine analogs have been tested for their abilities to function as substrates or inhibitors for the enzyme (7, 147, 148). The specificity towards the phosphate donor is rather broad with ATP and dGTP as the most active (129).

E. coli thymidine kinase is an allosteric enzyme. It is feedback inhibited by dTTP and activated by deoxyribonucleoside di- and triphosphates, with dCTP being the most active. Kinetic studies show that inhibition by dTTP is competitive with thymidine, not with ATP or the activators, whereas activation by dCDP causes a decrease in K_m for thymidine as well as an increase in V_{max}. In the absence of activators the saturation curve for ATP is sigmoidal, while it is hyperbolic in their presence (129).

In the absence of effectors the enzyme is a monomer (MW 42 000) (81), which may exist in an active or an inactive form depending on the temperature. By lowering the temperature the K_m for the substrate decreases and the sigmoidicity of the ATP saturation curve weakens. This temperature effect is reversible, and it appears that the monomer exists mostly in the less active form even at moderate temperatures (82).

Both the activating and the inhibitory effectors cause dimerization of the enzyme concomitantly with a decrease in temperature sensitivity. This suggests that the enzyme can exhibit either an active or an inactive conformation depending on the nature of the effector present (dTTP or dCTP). Sedimentation experiments have indicated two alternative conformations of the dimer, since it was consistently observed that the sedimentation coefficient of the enzyme in the presence of dTTP is 5.9–6.0 S, whereas it is 5.3–5.6 S in the presence of an activating deoxyribonucleotide. The sedimentation coefficient of the monomer is 3.4–3.5 S (81). A model accounting for molecular states of *E. coli* thymidine kinase and their interconversions has been proposed (82).

Thymidine kinase from the obligate thermophile *B. stearothermophilus* uses thymidine and deoxyuridine as substrates with equal efficiency, and has a broad specificity towards the phosphate donor, the most active being ATP and dATP. It shows Michaelis–Menten kinetics towards both thymidine and ATP, and no evidence for any activating nucleotide was obtained. As with the *E. coli* enzyme, dTTP inhibits the *B. stearothermophilus* enzyme competitively with thymidine. A remarkable property of the enzyme is its extreme thermostability. No loss of activity was detected after an exposure of 60 min at 75°C. Even 30 min at 100°C only resulted in a 40% loss of activity (97).

Isoelectric focusing of plasmodial extracts of *Physarum polycephalum* has demonstrated the presence of multiple enzyme variants of thymidine kinase, which appear sequentially during the nuclear division cycle. Evidence has been presented that variants with high isoelectric points arise by dephosphorylation of a primary translation product with low isoelectric point (60, 61).

I. Deoxycytidine kinase (EC 2.7.1.74)

$$\text{deoxycytidine} + \text{(d)NTP} \xrightarrow{Mg^{++}} \text{dCMP} + \text{(d)NDP}$$

So far, deoxycytidine kinase activity has only been demonstrated with certainty in *Bacillus* (173, 145) and in *Lactobacillus* (50, 140). Although some evidence suggests that it may also be present in certain pneumococci, its distribution throughout the microbial world is quite restricted (Table 3.5).

Deoxycytidine kinases have been purified from *B. subtilis* (116) and *L. acidophilus* (47), and they were shown to utilize both deoxycytidine *and* deoxyadenosine as phosphate acceptors. Since the general properties of these two enzymes are discussed elsewhere in this book they will not be presented here.

K. Nucleoside phosphotransferases (EC 2.7.1.77)

$$\text{nucleoside}' + \text{nucleoside monophosphate} \longrightarrow \text{nucleoside}' \text{ monophosphate} + \text{nucleoside}$$

Nucleoside phosphotransferases are enzymes which catalyze the transfer of phosphate groups between nucleoside monophosphates and a hydroxyl group of the pentose moieties of nucleosides. In general these enzymes show broad specificities towards both the phosphate acceptor and donor. Both 3′- and 5′-mononucleotides, and to a varying degree also nucleoside di- and triphosphates may act as donors, although nucleoside monophosphates are generally more efficient. Cell-free extracts of a large number of bacteria have been screened for the presence of nucleoside phosphotransferases, and many of them showed activity (115). From these studies it appeared that genera having nucleoside phosphotransferases could be divided into two groups according to the product of the transfer: one, in which the products are mainly nucleoside-5′-monophosphates and another, in which nucleoside 3′- or 2′-monophosphates are the predominant products. Genera such as *Pseudomonas*, *Serratia*, and *Staphylococcus* belong to the first group, whereas *Escherichia*, *Proteus*, and *Salmonella* belong to the second. Nucleoside phosphotransferase activity seems to be absent from *Bacillus*.

The enzyme has been highly purified from *Erwinia herbicola* (MW 70 000) belonging to the first group (36), and from *E. coli* (MW 29 000) which belongs to the second group of nucleoside phosphotransferases (31).

(1) The *Erwinia* enzyme uses pyrimidine and purine nucleosides as phosphate acceptors, and both 5′- and 3′-ribo- and deoxyribonucleotides will serve as phosphate donors, although the 5′-monophosphates are the most active (36).
(2) The *E. coli* enzyme catalyzes the transfer of phosphate groups to the 2′ or 3′ hydroxyl groups of a nucleoside *and* its 5′-mono, di-, and triphosphate derivatives. Since nucleo*tides* are in most cases better acceptors than nucleo*sides*, the designation nucleo*tide* phosphotransferase has been suggested for the *E. coli* enzyme. Among the most effective phosphate donors are 2′(3′)-GMP, 3′-dGMP, 2′-AMP, and 3′-dAMP (30).
(3) *Tetrahymena pyriformis* contains a nucleoside phosphotransferase, which catalyzes the transfer of phosphate from 5′-AMP to the 5′hydroxyl group of thymidine (5). Since *T. pyriformis* also contains a thymidine kinase, the nucleoside phosphotransferase represents an additional capacity for

thymidine phosphorylation in this organism (185). While the thymidine kinase is inhibited by dTTP this is not the case for the nucleoside phosphotransferase. In contrast the nucleoside phosphotransferase is inhibited by ATP, which has no such effect on the thymidine kinase (184).

The possible physiological function of nucleoside phosphotransferases has only been speculatively explored (30, 185).

L. CMP/dCMP Kinase (EC 2.7.4.14?)

$$\text{CMP/dCMP} + \text{ATP} \xrightarrow{\text{Mg}^{2+}} \text{CDP/dCDP} + \text{ADP}$$

CMP and dCMP are not intermediates in the *de novo* synthesis of CTP and dCTP. Thus enzymes capable of phosphorylating CMP and dCMP to the corresponding diphosphates may be regarded as pyrimidine salvage enzymes. In contrast, UMP kinase is a *de novo* enzyme, which is essential for growth (78). The physiological role of CMP/dCMP kinase is most likely to rephosphorylate CMP arising from RNA turnover and as a product of phospholipid synthesis (35, 142). Accordingly, a *S. typhimurium cmk* mutant, which lacks CMP kinase activity, accumulates CMP intracellularly and excretes large amounts of pyrimidines into the medium (20). One may expect that CMP kinase activity, like UMP kinase activity, is ubiquitous among microorganisms.

It has been shown that the CMP and dCMP kinase activities in cell-free extracts of *E. coli* comigrate upon DEAE-cellulose column chromatography in a peak distinct from other nucleoside monophosphate kinase activities (70). *S. typhimurium* mutants lacking CMP kinase activity are also deficient in dCMP kinase activity, while they retain normal UMP kinase activity (20). Thus in these two enteric bacteria CMP and dCMP phosphorylation is catalyzed by one single protein distinct from UMP kinase.

A CMP–UMP kinase has been purified to homogeneity from *B. subtilis* (MW 25 000) (174). It appears to be the only enzyme with UMP kinase activity in *B. subtilis*. K_m values for CMP and UMP were 0.04 mM and 2.5 mM respectively, and the ratio between the maximal velocities with CMP and UMP as substrates was 6 : 1. With UMP as substrate, but not with CMP, the enzyme is feedback inhibited by CTP and CDP. The purified enzyme showed no activity with dCMP or dUMP as substrates. A dCMP kinase was subsequently purified to homogeneity from *B. subtilis* using Blue-Sepharose affinity chromatography (Neuhard, unpublished). The purified protein showed activity with both dCMP, CMP and UMP as substrates, but not with dUMP. The K_m values for the three substrates were 0.3 mM, 0.05 mM, and about 3.6 mM respectively, and the relative V_{max} values were 0.4 : 1 : 0.1. These results strongly suggest that the phosphorylation of dCMP, CMP, and UMP is catalyzed by one single protein in *B. subtilis*, and that the enzyme under certain conditions may selectively lose its dCMP kinase activity (174).

In *Tetrahymena pyriformis* CMP, dCMP, and UMP phosphorylation is likewise catalyzed by a single protein, which has been extensively purified (2). The specificity towards the phosphate acceptor in order of decreasing activity is CMP, UMP, dCMP. No activity with dUMP was detected.

Thus in *B. subtilis* and *T. pyriformis* as well as in several mammalian systems (131) phosphorylations of CMP and dCMP are alternate activities of an enzyme catalyzing an essential biosynthetic reaction, i.e. phosphorylation of UMP to UDP. With *B. subtilis* this may be rationalized, since the turnover of mRNA in *B. subtilis* is largely phosphorolytic, i.e. generating nucleoside diphosphates (49) rather than hydrolytic as in *E. coli* (35). Thus the amount of endogenously formed CMP, which has to be rephosphorylated is considerably less in *B. subtilis* than in enteric bacteria (174).

M. Ribonucleoside monophosphate glycosylases

$$\text{ribonucleoside 5}'\text{-monophosphate} + H_2O \longrightarrow \text{base} + \text{ribose-5-phosphate}$$

When endogenously generated ribonucleoside 5′-monophosphates are present in excess of what can be rephosphorylated and used for nucleic acid synthesis, they are degraded, and the products excreted into the growth medium. This degradation may involve dephosphorylation followed by cleavage of the glycosylic bond catalyzed by nucleoside hydrolases or phosphorylases. However, another class of enzymes, the ribonucleoside monophosphate glycosylases have been identified, which catalyze the hydrolysis of the glycosylic bond of ribonucleoside 5′-monophosphate without prior dephosphorylation. The products of these reactions are the free nucleobases and ribose-5-phosphate.

Ribonucleotide glycosylases, which will cleave UMP and CMP have so far been identified in *Pseudomonas oleovorans* (151), and in *E. coli* and *S. typhimurium* (Neuhard, unpublished). A crude enzyme preparation from *P. oleovorans* was shown to catalyze the hydrolysis of all four naturally occurring ribonucleoside-5′-monophosphates to the corresponding nucleobases and ribose-5-phosphate. However, it was not established whether the individual substrates were hydrolyzed by specific glycosylases.

A ribonucleotide glycosylase has been purified about 200-fold from *S. typhimurium* (MW 60 000) (Eisenhardt, A. 1971. Thesis, University of Copenhagen). The enzyme catalyzes the hydrolysis of GMP, CMP, and UMP with the following relative activities: 1 : 0.7 : 0.04. AMP is a very poor substrate, and deoxyribonucleoside-5′-monophosphates are not cleaved at all. The K_m

value for CMP was found to be 0.07 mM at the optimal pH of 6.6. ATP at 10^{-4} M activates the reaction 3-fold with all substrates, whereas Mg^{2+} ions completely prevent this activation. It may be recalled that *E. coli* and certain other bacteria have been shown to contain a specific, allosterically regulated AMP glycosylase (p. 38).

Although the ribonucleotide glycosylases of *E. coli* and *S. typhimurium* have been shown to function *in vivo* under certain conditions (p. 104) their physiological role cannot be ascertained until mutants defective in this enzyme activity have been characterized.

V. PHYSIOLOGICAL ASPECTS OF PYRIMIDINE SALVAGE

A. Utilization of exogenous pyrimidine compounds

The pyrimidine salvage pathways may be regarded as serving two different functions. One is to scavenge pyrimidine compound for nucleic acid synthesis, and the other is to make the pentose moieties of nucleosides and nucleotides as well as the 4-amino group of cytosine compounds available as sources of carbon, energy, and nitrogen. This is accomplished by developing two types of salvage enzymes: (a) the anabolic enzymes, comprising uracil phosphoribosyltransferases, nucleoside kinases, nucleotide kinases, *trans*-N-deoxyribosylases, and cytosine deaminases; and (b) the catabolic enzymes which include cytidine deaminases, nucleoside phosphorylases, nucleoside hydrolases, nucleotidases, and ribonucleotide glycosylases. The anabolic function of the first type of enzymes is emphasized by the way they are regulated. Thus the synthesis of uracil phosphoribosyltransferase, cytosine deaminase, and uridine kinase are derepressed during pyrimidine starvation, and the synthesis of *trans*-N-deoxyribosylase is derepressed by deoxyribonucleoside starvation. In addition the activities of uridine kinase, thymidine kinase, and deoxycytidine kinase are feedback inhibited by the end products of the pathways. In contrast the synthesis of many of the catabolic enzymes is inducible (cf. Chapter 5).

Many microorganisms possess two alternative routes for metabolizing nucleosides: cytidine may be deaminated to uridine, and it may be phosphorylated to CMP; uridine and thymidine may be cleaved to the free bases, and they may be phosphorylated to UMP and dTMP, respectively. This is illustrated in Fig. 3.3(b), which shows that about 50% of the uridine offered to a growing culture of *E. coli* is phosphorylated to UMP, the rest being cleaved to uracil and ribose 1-phosphate. In contrast (Fig. 3.3(c)), if uridine is fed slowly to the cells in the form of UMP, it is used exclusively via uridine kinase. Similarly, it has been shown that the incorporation of exogenous thymidine into DNA ceases after a short time due to the rapid breakdown of thymidine to thymine, whereas exogenous dTMP is incorporated over an extended period of time (28). This indicates that when pyrimidine nucleosides are present in low concentrations they escape the catabolic enzymes, and are preferentially phosphorylated directly to the corresponding nucleoside 5′-monophosphate. In this context it must be emphasized that the metabolism of the nucleosides by either pathway requires a functional nucleoside transport system (cf. Chapter 6).

B. Interrelation between pyrimidine *de novo* and salvage pathways

When wild type strains of *E. coli* are grown in media containing 0.05 mM uracil, 80–90% of the pyrimidine residues of the nucleic acids are derived from exogenous uracil (38, 119). Thus the cells must have means of preventing *de novo* synthesis when preformed pyrimidines are available exogenously.

The regulation of the *de novo* synthesis of UMP in different microorganisms has been the subject of numerous studies (for reviews see references 112, 127). In general it appears that the flow through the pathway is regulated by negative feedback control either via feedback inhibition of the first enzymes of the pathway, i.e. carbamoylphosphate synthase and aspartate transcarbamylase, or via repression of the synthesis of the six enzymes of the pathway. It may therefore be proposed that increased endogenous concentrations of pyrimidine nucleotides, derived from exogenous precursors, cause increased feedback inhibition and repression of synthesis of the *de novo* enzymes. In accordance with this proposition it has been found that wild type strains of *E. coli*, *S. typhimurium*, and *B. subtilis*, growing exponentially in the presence of uracil in the medium, contain lower levels of the six *de novo* enzymes than when they are growing in the absence of uracil (23, 93, 135, 139, 159). In *E. coli* and *S. typhimurium* it was further established that the endogenous pyrimidine nucleotide pools are 30–50% higher when the cells are grown in the presence of exogenous uracil than when grown without (6, 93, 159).

The effects of exogenous uracil on the *de novo* synthesis of pyrimidine nucleotides in *E. coli* have been studied in greater detail by Finch and coworkers (6, 38). In these studies the levels of certain of the pyrimidine *de novo* enzymes, as well as the endogenous nucleotide pools and the pools of several intermediates of the pathways were measured as a function of time following the addition of uracil. The results indicate that although the swelling of the pyrimidine nucleotide pools observed is in accordance with an increased feedback inhibition of the *de novo* pathway, other factors, such as changes in the pools of PRPP and of aspartate, may contribute to the decreased flow through the pathway in the earlier stages, where repression and growth have not yet diluted out the enzymes of the *de novo* pathway.

It should be recalled that the anabolic salvage enzymes, uracil phosphoribosyltransferase and uridine/cytidine kinase, are also regulated enzymes (pp. 118, 128). Like the *de novo* enzymes their syntheses are derepressed during pyrimidine starvation, and their activities are feedback inhibited by pyrimidine nucleotides. This implies that the concentrations of the pyrimidine nucleotide effectors required to inhibit these enzymes must be higher than required for inhibition of the *de novo* pathway or that different effectors are involved in the regulation of the two pathways.

C. Endogenous turnover of pyrimidine compounds

Large quantities of ribonucleotides are continuously formed from RNA as a result of messenger RNA turnover and of the processing of transfer RNA and ribosomal RNA precursors. In addition, significant amounts of ribonucleoside 5′-mono- and diphosphates are generated during the synthesis of polysaccharides and phospholipids from nucleotide activated precursors. Since microorganisms do not normally excrete nucleic acid precursors during balanced growth, they apparently reutilize the ribonucleotides. Accordingly, a *S. typhimurium* mutant lacking CMP kinase (*cmk* mutant) and hence unable to reutilize 5′-CMP by rephosphorylation, was shown to contain a 15-fold increased endogenous CMP pool and to excrete large amounts of pyrimidines into the medium. This shows that the cells normally have the capacity to dispose of ribonucleotides, when they accumulate inside the cells (20).

Degradation of ribosomal RNA is promoted when *E. coli* cells are deprived of a source of carbon. The acid soluble pyrimidine-containing degradation products formed under these non-growing conditions were shown to consist of 75% nucleobases, the residual 25% being made up of about equal amounts of nucleoside diphosphates, nucleoside 5′- and 3′-monophosphates, and nucleosides. The majority of the products, including the phosphorylated compounds, was excreted into the medium (89). Even though the cytoplasmic membrane is impermeable to exogenous nucleotides (104), it seems to allow them to exit from the starved cells.

A number of *E. coli* mutants defective in enzymes believed to be involved in the catabolism of RNA and nucleotides were subjected to starvation for carbon source, and the acid-soluble RNA degradation products were identified (41, 89). The following results were obtained: (a) Mutants defective in polynucleotide phosphorylase did not excrete nucleoside diphosphates. (b) Mutants lacking a functional RNase I did not excrete nucleoside 3′-monophosphates. Since RNase I is a periplasmic enzyme it implies that endogenous oligonucleotides can in some way gain access to the periplasmic enzymes. (c) Mutants defective in the periplasmic 5′-nucleotidase accumulated larger amounts of nucleoside 5′-monophosphates and diphosphates than the wild type does, although nucleobases still constituted a major fraction of the excreted breakdown products. This indicates that 5′-nucleotidase is involved in the degradation of endogenous nucleoside 5′-monophosphates, although it cannot account for the total catabolism observed. It seems possible that ribonucleotide glycosylases (p. 132) which catalyze the hydrolysis of ribonucleoside 5′-monophosphates to nucleobases and ribose 5-phosphate, might contribute significantly under these conditions. This would be in accordance with the finding that nucleobases are the predominant products of ribonucleotide degradation in this mutant.

Significant amounts of ribonucleotides are also degraded under conditions where they do not accumulate endogenously, that is during exponential growth. However, the degradation products (nucleobases and nucleosides) are under these conditions immediately converted back to ribonucleoside 5′-monophosphates via the salvage pathways, and therefore do not accumulate in the growth medium. Thus, strains of *E. coli* unable to convert uracil to UMP, due to mutations in the uracil phosphoribosyltransferase gene, accumulate uracil in the growth medium (136); and *E. coli* mutants deficient in cytosine deaminase activity excrete cytosine (Neuhard, unpublished). Since cytidine is not converted to cytosine by *E. coli*, it follows that the excreted cytosine is formed by cleavage of the glycosylic bond of CMP, presumably catalyzed by a ribonucleotide glycosylase (p. 104).

E. coli mutants unable to metabolize cytidine due to mutational inactivation of both cytidine deaminase and uridine/cytidine kinase contain induced levels of uridine phosphorylase when grown in the absence of inducer (63). Since cytidine is the inducer of uridine phosphorylase in *E. coli* (p. 122), it follows that the mutants contain an elevated endogenous pool of cytidine. However, the accumulation of cytidine is probably not very extensive because pyrimidines were not excreted into the growth medium by the mutants (63). The endogenous induction of uridine phosphorylase in the double mutant is entirely dependent on the presence of a functional cytidine transport system. The implication of this observation is discussed in Chapter 6.

Saccharomyces cerevisiae mutants defective in uracil phosphoribosyltransferase (*ups* mutants) excrete uracil into the medium. This loss of uracil is prevented by the introduction of an additional mutation (*urid-rh*) which inactivates uridine hydrolase. Such a *urid-rh,ups* double mutant excretes uridine instead of uracil. Also mutants defective in uracil permease excrete small amounts of uracil (57). Thus, in yeast UMP is continuously degraded to uridine, which is further hydrolyzed to uracil. In a wild type strain the uracil thus formed is converted back to UMP via a pathway, which involves uracil permease and uracil phosphoribosyltransferase.

The physiological significance of the turnover of ribonucleoside 5′-monophosphates through the intermediate formation of nucleobases and nucleosides can only be speculated about. The flow through the ribonucleoside 5′-monophosphate pools is dependent on several factors such as the rate of *de novo* synthesis, the rate of mRNA synthesis, and the rates by which polysaccharides and phospholipids are synthesized. Thus, one may expect considerable fluctuations in these pools under different physiological conditions. It could therefore be advantageous for a cell to contain enzymes designed to degrade excess 5′-ribonucleotides which may tend to accumulate, when growth is not balanced, i.e. during nutritional shifts. The degradation of ribonucleoside 5′-monophosphates observed during balanced growth may simply be a consequence of the presence of such catabolic enzymes. The effectivity with which the breakdown products are recycled back to the ribonucleotide pools would then determine the end products of these combined processes. Since phosphoribosyltransferases and ribonucleoside kinases are regulated enzymes these reactions may be controlled according to the needs for ribonucleotides.

The endogenous degradation of stable RNA observed under certain conditions will also result in the production of small amounts of modified ribonucleotides such as ribosyl TMP, 5-methylCMP, dihydroUMP, and pseudoUMP. Since the synthesis of these nucleotides occurs at the polynucleotide level, the cell must be able to dispose of such modified nucleotides. Most of the nucleotides are probably dephosphorylated to the corresponding nucleosides. 5-methylcytidine is a substrate of cytidine deaminase (p. 121) which converts it to ribosylthymine, and ribosylthymine may be cleaved by nucleoside phosphorylases or hydrolases.

An enzyme, pseudoUMP synthetase, which may account for the salvage of the uracil moiety of pseudoUMP has been found in *E. coli* (27), *Tetrahymena pyriformis* (68), and *Agrobacterium tumefaciens* (165). It catalyzes the following reversible reaction:

$$\text{pseudoUMP} + H_2O \rightleftharpoons \text{uracil} + \text{ribose 5-phosphate}$$

In *E. coli* the synthesis of pseudoUMP synthetase is repressed when glucose is present in the medium. Addition of cyclic AMP to a glucose containing medium or replacement of glucose with glycerol enhances enzyme synthesis (164). This indicates that the physiological function of the enzyme is that of cleaving pseudoUMP to uracil and ribose 5-phosphate. *E. coli* also contains a pseudouridine kinase which catalyzes the ATP and Mg^{2+} dependent phosphorylation of pseudouridine to pseudoUMP. This enzyme appears to be distinct from uridine and thymidine kinases. Thus, any pseudouridine which may be formed from pseudoUMP by dephosphorylation may still be converted to uracil and ribose 5-phosphate by the sequential action of the kinase and the synthetase.

REFERENCES

1. Alphen, van, W., Selm, van, N., and Lugtenberg, B. (1978). Pores in the outer membrane of *Escherichia coli* K 12. Involvement of proteins b and e in the functioning of pores for nucleotides. *Molec. Gen. Genet.* **159**, 75–83.
2. Anderson, E. P. (1978). UMP–CMP kinase from *Tetrahymena pyriformis*. *In* "Methods of Enzymology" (P. A. Hoffee, and M. E. Jones, *Eds.*), Vol. **51**, pp. 331–337. Academic Press, New York.
3. Anraku, Y. (1964). A new cyclic phosphodiesterase having a 3′-nucleotidase activity from *Escherichia coli* B. I. Purification and some properties of the mutant. *J. Biol. Chem.* **239**, 3412–3419.
4. Anraku, Y. (1964). A new cyclic phosphodiesterase having a 3′-nucleotidase activity from *Escherichia coli* B. II. Further studies on substrate specificity and mode of action of the enzyme. *J. Biol. Chem.* **239**, 3420–3424.
5. Arima, T., Masaka, M., Shiosaka, T., Okuda, H., and Fujii, S. (1971). Alternative methods of thymidine phosphorylation in different organisms. *Biochim. Biophys. Acta* **246**, 184–193.
6. Bagnara, A. S., and Finch, L. R. (1974). The effects of bases and nucleosides on the intracellular contents of nucleotides and 5-phosphoribosyl 1-pyrophosphate in *Escherichia coli*. *Eur. J. Biochem.* **41**, 421–430.
7. Baker, B. R. (1967). "Design of active-site-directed irreversible enzyme inhibitors", pp. 79–92. Wiley, New York.
8. Baker, B. R., and Hopkins, S. E. (1970). Irreversible enzyme inhibitors. CLXII. Thymidine phosphorylase. X. On the nature and dimension of the hydrophobic bonding region II. *J. Med. Chem.* **1**, 87–89.
9. Bankel, L., Holme, E., Lindstedt, G., and Lindstedt, S. (1972). Oxygenases involved in thymine and thymidine metabolism in *Neurospora crassa*. *FEBS Letts.* **21**, 135–138.
10. Bankel, L., Lindstedt, G., and Lindstedt, S. (1972). Thymidine 2′-hydroxylation in *Neurospora crassa*. *J. Biol. Chem.* **247**, 6128–6134.
11. Bankel, L., Lindstedt, G., and Lindstedt, S. (1977). Thymine 7-hydroxylase from *Neurospora crassa*. Substrate specificity studies. *Biochim. Biophys. Acta* **481**, 431–437.
12. Beacham, I. R., and Garrett, S. (1980). Isolation of *Escherichia coli* mutants (*cpdB*) deficient in periplasmic 2′:3′-cyclic phosphodiesterase and genetic mapping of the *cpdB* locus. *J. Gen. Microbiol.* **119**, 31–34.
13. Beacham, I. R., Haas, D., and Yagil, E. (1977). Mutants of *Escherichia coli* "cryptic" for certain periplasmic enzymes. Evidence for an alteration of the outer membrane. *J. Bacteriol.* **129**, 1034–1044.
14. Beacham, I. R., Kahana, R., Levy, L., and Yagil, E. (1973). Mutants of *Escherichia coli* K 12 "cryptic", or deficient in 5′-nucleotidase (uridine diphosphate-sugar hydrolase) and 3′-nucleotidase (cyclic phosphodiesterase) activity. *J. Bacteriol.* **116**, 957–964.
15. Beacham, I. R., and Pritchard, R. H. (1971). The role of nucleoside phosphorylases in the degradation of deoxyribonucleosides by thymine-requiring mutants of *E. coli*. *Molec. Gen. Genet.* **110**, 289–298.
16. Bean, B., and Tomasz, A. (1971). Inhibitory effects and metabolism of 5-fluoropyrimidine derivatives in *Pneumococcus*. *J. Bacteriol.* **106**, 412–420.
17. Beck, C. F., and Ingraham, J. L. (1971). Location on the chromosome of *Salmonella typhimurium* of genes governing pyrimidine metabolism. *Molec. Gen. Genet.* **111**, 303–316.

18. Beck, C. F., Ingraham, J. L., and Neuhard, J. (1972). Location on the chromosome of *Salmonella typhimurium* of genes governing pyrimidine metabolism. II. Uridine kinase, cytosine deaminase, and thymidine kinase. *Molec. Gen. Genet.* **115**, 208–215.
19. Beck, C. F., Ingraham, J. L., Neuhard, J., and Thomassen, E. (1972). Metabolism of pyrimidines and pyrimidine nucleosides by *Salmonella typhimurium. J. Bacteriol.* **110**, 219–228.
20. Beck, C. F., Neuhard, J., Thomassen, E., Ingraham, J. L., and Kleker, E. (1974). *Salmonella typhimurium* mutants defective in cytidine monophosphate kinase (*cmk*). *J. Bacteriol.* **120**, 1370–1379.
21. Beck, W. S., and Kashket, S. (1964). Metabolic abnormalities associated with unbalanced growth in *Lactobacillus leichmannii. Medicine* **43**, 195–206.
22. Beck, W. S., and Levin, M. (1963). Purification, kinetics and repression control of bacterial *trans*-N-deoxyribosylase. *J. Biol. Chem.* **238**, 702–709.
23. Beckwith, J. R., Pardee, A. B., Austrian, P., and Jacob, F. (1962). Coordination of the synthesis of the enzymes in the pyrimidine pathway of *E. coli. J. Mol. Biol.* **5**, 618–634.
24. Blank, J. G., and Hoffee, P. A. (1975). Purification and properties of thymidine phosphorylase from *Salmonella typhimurium. Arch. Biochem. Biophys.* **168**, 259–265.
25. Block, E. R., and Bennett, J. E. (1972). Pharmacological studies with 5-fluorocytosine. *Antimicrobial Agents and Chemotherapy* **1**, 476–482.
26. Bolton, E. T., and Reynard, A. M. (1954). Utilization of purine and pyrimidine compounds in nucleic acid synthesis by *Escherichia coli. Biochim. Biophys. Acta* **13**, 381–385.
27. Breitman, T. R. (1970). Pseudouridylate synthetase of *Escherichia coli*: correlation of its activity with utilization of pseudouridine for growth. *J. Bacteriol.* **103**, 263–264.
28. Breitman, T. R., Bradford, R. M., and Cannon, W. D. (1967). Use of exogenous deoxythymidylic acid to label the deoxyribonucleic acid of growing wild-type *Escherichia coli. J. Bacteriol.* **93**, 1471–1472.
29. Brockman, R. W., Davis, J. M., and Stutts, P. (1960). Metabolism of uracil and 5-fluorouracil by drug-sensitive and by drug-resistant bacteria. *Biochim. Biophys. Acta* **40**, 22–32.
30. Brunngraber, E. F., and Chargaff, E. (1973). A nucleotide phosphotransferase from *Escherichia coli.* Purification and properties. *Biochemistry* **12**, 3005–3012.
31. Brunngraber, E. F., and Chargaff, E. (1977). Nucleotide phosphotransferase of *Escherichia coli*: Purification by affinity chromatography. *Proc. Natl. Acad. Sci. U.S.A.* **74**, 3226–3229.
32. Buxton, R. S., Hammer-Jespersen, K., and Hansen, T. D. (1978). Insertion of bacteriophage lambda into the *deo* operon of *Escherichia coli* K 12 and isolation of plaque-forming λdeo^{+} transducing bacteriophages. *J. Bacteriol.* **136**, 668–681.
33. Cardinaud, R. (1978). Nucleoside deoxyribosyltransferase from *Lactobacillus helveticus. In* "Methods of Enzymology" (P. A. Hoffee, and M. E. Jones, *Eds.*), Vol. **51**, pp. 446–455. Academic Press, New York.
34. Cardinaud, R., and Holguin, R. (1979). Nucleoside deoxyribosyltransferase-II from *Lactobacillus helveticus.* Substrate specificity studies. Pyrimidine bases as acceptors. *Biochim. Biophys. Acta* **568**, 339–347.
35. Chaney, S. G., and Boyer, P. D. (1972). Incorporation of water oxygen into intracellular nucleotides and RNA. II. Predominantly hydrolytic RNA turnover in *Escherichia coli. J. Mol. Biol.* **64**, 581–591.

36. Chao, H. M. (1976). Nucleoside phosphotransferase from *Erwinia herbicola*, a new membrane-bound enzyme. *J. Biol. Chem.* **251**, 2330–2333.
37. Chen, M. S., and Prusoff, W. H. (1978). Thymidine kinase from *Escherichia coli*. *In* "Methods of Enzymology" (P. A. Hoffee, and M. E. Jones, *Eds*.), Vol. **51**, pp. 354–360. Academic Press, New York.
38. Christopherson, R. I., and Finch, L. R. (1978). Response of the pyrimidine pathway of *Escherichia coli* K 12 to exogenous adenine and uracil. *Eur. J. Biochem.* **90**, 347–358.
39. Cihak, A. (1975). Enhancement of uridine kinase activity in *Escherichia coli* B cultivated in the presence of metabolic inhibitors. *Collection Czechoslov. Chem. Commun.* **40**, 913–919.
40. Cihak, A., Skoda, J., and Sorm, F. (1964). Ribosylation and phosphoribosylation of 5-azauracil-2,4-^{14}C in a cell-free extract of *Escherichia coli. Collection Czechoslov. Chem. Commun.* **29**, 814–824.
41. Cohen, L., and Kaplan, R. (1977). Accumulation of nucleotides by starved *Escherichia coli* cells as a probe for the involvement of ribonucleases in ribonucleic acid degradation. *J. Bacteriol.* **129**, 651–657.
42. Cohen, R. M., and Wolfenden, R. (1971). Cytidine deaminase from *Escherichia coli*. Purification, properties, and inhibition by the potential transition state analog 3,4,5,6-tetrahydrouridine. *J. Biol. Chem.* **246**, 7561–7565.
43. Cohen, R. M., and Wolfenden, R. (1971). The equilibrium of hydrolytic deamination of cytidine and N^4-methyl-cytidine. *J. Biol. Chem.* **246**, 7566–7568.
44. Crawford, I., Kornberg, A., and Simms, E. S. (1957). Conversion of uracil and orotate to uridine 5′-phosphate by enzymes in lactobacilli. *J. Biol. Chem.* **266**, 1093–1101.
45. Danzin, C., and Cardinaud, R. (1974). Deoxyribosyl transfer catalysis with *trans*-N-deoxyribosylase. Kinetic studies of purine-to-purine *trans*-N-deoxyribosylase. *Eur. J. Biochem.* **48**, 255–262.
46. Danzin, C., and Cardinaud, R. (1976). Deoxyribosyl transfer catalysis with *trans*-N-deoxyribosylase. Kinetic study of purine (pyrimidine) to pyrimidine (purine) *trans*-N-deoxyribosylase. *Eur. J. Biochem.* **62**, 365–372.
47. Deibel, M. R., and Ives, D. H. (1978). Deoxynucleoside kinases from *Lactobacillus acidophilus* R-26. *In* "Methods of Enzymology (P. A. Hoffee, and M. E. Jones, *Eds*.), Vol. **51**, 346–354. Academic Press, New York.
48. Dewey, V. C., and Kidder, G. W. (1973). Partial purification and properties of a nucleoside hydrolase from *Chritidia. Arch. Biochem. Biophys.* **157**, 380–387.
49. Duffy, J. J., Chaney, S. G., and Boyer, P. D. (1972). Incorporation of water oxygen into intracellular nucleotides and RNA. I. Predominantly non-hydrolytic RNA turnover in *Bacillus subtilis. J. Mol. Biol.* **64**, 565–579.
50. Durham, J. P., and Ives, D. H. (1971). The metabolism of deoxyribonucleosides in *Lactobacillus acidophilus*: regulation of deoxyadenosine, deoxycytidine, deoxyguanosine, and deoxythymidine kinase activities by nucleotides. *Biochim. Biophys. Acta* **228**, 9–25.
51. Dvorak, H. F., Brockman, R. W., and Heppel, L. A. (1967). Purification and properties of two acid phosphatase fractions isolated from osmotic shock fluid of *Escherichia coli. Biochemistry* **6**, 1743–1751.
52. Edlin, G., and Neuhard, J. (1967). Regulation of nucleoside triphosphate pools in *Escherichia coli. J. Mol. Biol.* **24**, 225–230.
53. Evans, B. E., Mitchell, G. N., and Wolfenden, R. (1975). The action of bacterial cytidine deaminase on 5,6-dehydrocytidine. *Biochemistry* **14**, 621–624.
54. Fast, R., and Sköld, O. (1977). Biochemical mechanism of uracil uptake

regulation in *Escherichia coli* B. Allosteric effects on uracil phosphoribosyltransferase under stringent conditions. *J. Biol. Chem.* **252**, 7620–7624.
55. Fink, K., and Nygaard, P. (1978). Pyrimidine metabolism in microplasmodia of *Physarum polycephalum. Eur. J. Biochem.* **89**, 417–424.
56. Firshein, W., and Hasselbacher, P. (1970). Utilization of deoxyribonucleosides by virulent and avirulent *Pneumococci. Biochim. Biophys. Acta* **204**, 60–81.
57. Grenson, M. (1969). The utilization of exogenous pyrimidines and the recycling of uridine 5′-phosphate derivatives in *Saccharomyces cerevisiae*, as studied by means of mutants affected in pyrimidine uptake and metabolism. *Eur. J. Biochem.* **11**, 249–260.
58. Griswold, W. R., Madrid, V. O., Sharfer, P. M., Tappen, D. C., Pugh, C. S. G., and Abbot, M. T. (1976). Regulation of thymidine metabolism in *Neurospora crassa. J. Bacteriol.* **125**, 1040–1047.
59. Grivell, A. R., and Jackson, J. F. (1968). Thymidine kinase: evidence for its absence from *Neurospora crassa* and some other microorganisms, and the relevance of this to specific labelling of deoxyribonucleic acid. *J. Gen. Microbiol.* **54**, 307–317.
60. Gröbner, P. (1979). Thymidine kinase enzyme variants in *Physarum polycephalum. In vitro* interconversion of the enzyme variants. *J. Biochem.* **86**, 1595–1605.
61. Gröbner, P. (1979). Thymidine kinase enzyme variants in *Physarum polycephalum.* Kinetics and properties of the enzyme variants. *J. Biochem.* **86**, 1607–1614.
62. Gröbner, P., and Sachsenmaier, W. (1976). Thymidine kinase enzyme variants in *Physarum polycephalum*: change of pattern during the synchronous mitotic cycle. *FEBS Letts.* **71**, 181–184.
63. Hammer-Jespersen, K., and Munch-Petersen, A. (1973). Mutants of *Escherichia coli* unable to metabolize cytidine: Isolation and characterization. *Molec. Gen. Genet.* **126**, 177–186.
64. Hammer-Jespersen, K., and Munch-Petersen, A. (1975). Multiple regulation of nucleoside catabolizing enzymes: Regulation of the *deo* operon by the *cytR* and *deoR* gene products. *Molec. Gen. Genet.* **137**, 327–335.
65. Handschumacher, R. E. (1957). Studies of bacterial resistance to 6-azauracil and its riboside. *Biochim. Biophys. Acta* **23**, 428–430.
66. Hayaishi, O., Nozaki, M., and Abbott, M. T. (1975). Oxygenases: dioxygenases. *In* "The Enzymes" (P. D. Boyer, *Ed.*), Vol. **12**, pp. 119–189. Academic Press, New York.
67. Henner, D. J., and Hoch, J. A. (1980). The *Bacillus subtilis* chromosome. *Microbiol. Rev.* **44**, 57–82.
68. Henrikson, R. L., and Goldwasser, E. (1964). Studies on the biosynthesis of 5-ribosyluracil 5′-monophosphate in *Tetrahymena pyriformis. J. Biol. Chem.* **239**, 1177–1187.
69. Hill, D. L. (1972). The biochemistry and physiology of *Tetrahymena.* Academic Press, New York.
70. Hiraga, S., and Sugino, Y. (1966). Nucleoside monophosphokinases of *Escherichia coli* infected and uninfected with an RNA phage. *Biochim. Biophys. Acta* **114**, 416–418.
71. Hoeprich, P. D., Ingraham, J. L., Kleker, E., and Winship, M. J. (1974). Development of resistance to 5-fluorocytosine in *Candida parapsilosis* during therapy. *J. Infectious Diseases* **130**, 112–118.
72. Hoffee, P. A., and Blank, J. (1978). Thymidine phosphorylase from *Salmonella*

typhimurium. In "Methods in Enzymology" (P. A. Hoffee, and M. E. Jones, *Eds.*), Vol. **51**, pp. 437–442. Academic Press, New York.

73. Holguin, J., and Cardinaud, R. (1975). *Trans*-N-deoxyribosylase: Purification by affinity chromatography and characterization. *Eur. J. Biochem.* **54**, 505–514.
74. Holguin, J., Cardinaud, R., and Salemink, C. A. (1975). *Trans*-N-deoxyribosylase: Substrate specificity studies. Purine bases as acceptors. *Eur. J. Biochem.* **54**, 515–520.
75. Holme, E., Lindstedt, G., Lindstedt, S., and Toft, M. (1971). ^{18}O studies of the 2-ketoglutarate-dependent sequential oxygenation of thymine to 5-carboxyuracil. *J. Biol. Chem.* **246**, 3314–3319.
76. Hosono, H., and Kuno, S. (1973). The purification and properties of cytidine deaminase from *Escherichia coli. J. Biochem.* **74**, 797–803.
77. Imada, A., and Igarasi, S. (1967). Ribosyl and deoxyribosyl transfer by bacterial enzyme systems. *J. Bacteriol.* **94**, 1551–1559.
78. Ingraham, J. L., and Neuhard, J. (1972). Cold-sensitive mutants of *Salmonella typhimurium* defective in uridine monophosphate kinase (*pyrH*). *J. Biol. Chem.* **247**, 6259–6265).
79. Ipata, P. L., and Cercignani, G. (1978). Cytosine and cytidine deaminase from yeast. *In* "Methods of Enzymology" (P. A. Hoffee, and M. E. Jones, *Eds.*), Vol. **51**, pp. 394–404. Academic Press, New York.
80. Ipata, P. L., Marmocchi, F., Magni, G., Felicioli, R., and Polidoro, G. (1971). Baker's yeast cytosine deaminase. Some enzymic properties and allosteric inhibition by nucleosides and nucleotides. *Biochemistry* **10**, 4270–4276.
81. Iwatsuki, N., and Okazaki, R. (1967). Mechanism of regulation of deoxythymidine kinase of *Escherichia coli.* I. Effect of regulatory deoxynucleotides on the state of aggregation of the enzyme. *J. Mol. Biol.* **29**, 139–154.
82. Iwatsuki, N., and Okazaki, R. (1967). Mechanism of regulation of deoxythymidine kinase of *Escherichia coli.* II. Effect of temperature on the enzyme activity and kinetics. *J. Mol. Biol.* **29**, 155–165.
83. Janion, C. (1977). On the ability of *Salmonella typhimurium* cells to form deoxycytidine nucleotides. *Molec. Gen. Genet.* **153**, 179–183.
84. Jensen, K. F. (1979). Apparent involvement of purines in the control of expression of *Salmonella typhimurium pyr* genes: Analysis of a leaky *guaB* mutant resistant to pyrimidine analogs. *J. Bacteriol.* **138**, 731–738.
85. Jund, R., and Lacraute, F. (1970). Genetic and physiological aspects of resistance to 5-fluoropyrimidines in *Saccharomyces cerevisiae. J. Bacteriol.* **102**, 607–615.
86. Jyssum, S., and Bøvre, K. (1974). Search for thymidine phosphorylase, nucleoside deoxyribosyltransferase and thymidine kinase in *Moraxella*, *Acinetobacter*, and allied bacteria. *Acta Path. Microbiol. Scand. Section B.* **82**, 57–66.
87. Jyssum, S., and Jyssum, K. (1979). Metabolism of pyrimidine bases and nucleosides in *Neisseria meningitidis. J. Bacteriol.* **138**, 320–323.
88. Kalckar, H. M. (1951). Biosynthetic aspects of nucleosides and nucleic acids. *Publ. Staz. Zool. Napoli.* **23** Supplement 87–103.
89. Kaplan, R., Cohen, L., and Yagil, E. (1975). Acid-soluble degradation products of ribonucleic acid in *Escherichia coli* and the role of nucleosidases in their catabolism. *J. Bacteriol.* **124**, 1159–1164.
90. Karlström, O. (1968). Mutants of *Escherichia coli* defective in deoxyribonucleoside catabolism. *J. Bacteriol.* **95**, 1069–1077.
91. Karlström, O. (1970). Inability of *Escherichia coli* B to incorporate added deoxycytidine, deoxyadenosine, and deoxyguanosine in to DNA. *Eur. J. Biochem.* **17**, 68–71.
92. Karlström, O., and Larsson, A. (1967). Significance of ribonucleotide reduction

in the biosynthesis of deoxyribonucleotides in *Escherichia coli*. *Eur. J. Biochem.* **3**, 164–170.
93. Kelln, R. A., Kinahan, J. J., Folterman, K. F., and O'Donovan, G. A. (1975). Pyrimidine biosynthetic enzymes of *Salmonella typhimurium*, repressed specifically by growth in the presence of cytidine. *J. Bacteriol.* **124**, 764–774.
94. Kelln, R. A., and Warren, R. A. J. (1974). Pyrimidine metabolism in *Pseudomonas acidovorans*. *Can. J. Microbiol.* **20**, 427–433.
95. Kier, L. D., Weppelman, R., and Ames, B. N. (1977). Resolution and purification of three periplasmic phosphatases of *Salmonella typhimurium*. *J. Bacteriol.* **130**, 399–410.
96. Kier, L. D., Weppelman, R. M., and Ames, B. N. (1979). Regulation of nonspecific acid phosphatase in *Salmonella*: *phoN* and *phoP* genes. *J. Bacteriol.* **138**, 155–161.
97. Kobayashi, S., Hubbell, H. R., and Orengo, A. (1974). A homogenous, thermostable deoxythymidine kinase from *Bacillus stearothermophilus*. *Biochemistry* **13**, 4537–4543.
98. Koszalka, G. W., and Krenitsky, T. A. (1979). Nucleosidases from *Leishmania donovani*. Pyrimidine ribonucleosidase, purine ribonucleosidase, and a novel purine 2′-deoxyribonucleosidase. *J. Biol. Chem.* **254**, 8185–8193.
99. Krajewska, E., and Shugar, D. (1975). Pyrimidine nucleoside analogues as inducers of pyrimidine nucleoside catabolizing enzymes in *Salmonella typhimurium*. *Mol. Biol. Rep.* **2**, 295–301.
100. Kream, J., and Chargaff, E. (1952). On the cytosine deaminase of yeast. *J. Am. Chem. Soc.* **74**, 5157–5160.
101. Lampen, J. O., and Wang, T. P. (1952). The mechanism of action of *Lactobacillus pentosus* nucleosidase. *J. Biol. Chem.* **198**, 385–395.
102. Lazzarini, R. A., and Dahlberg, A. E. (1971). The control of ribonucleic acid synthesis during amino acid deprivation in *Escherichia coli*. *J. Biol. Chem.* **246**, 420–429.
103. Leer, J. C., Hammer-Jespersen, K., and Schwartz, M. (1977). Uridine phosphorylase from *Escherichia coli*. Physical and chemical characterization. *Eur. J. Biochem.* **75**, 217–224.
104. Lichtenstein, J., Barner, H. D., and Cohen, S. S. (1960). The metabolism of exogenously supplied nucleotides by *Escherichia coli*. *J. Biol. Chem.* **235**, 457–465.
105. Lin, C., Hsu, C., and Abbott, M. T. (1973). Catalysis of three sequential deoxygenase reactions by thymine 7-hydroxylase. *Arch. Biochem. Biophys.* **159**, 180–187.
106. Lin, C. K., Shaffer, P. M., Slaughter, R. S., McCroskey, R. P., and Abbott, M. T. (1972). Stochiometry of the pyrimidine deoxyribonucleoside 2′-hydroxylase reaction and of the conversions of 5-hydroxy-methyluracil to 5-formyluracil and of the latter to uracil 5-carboxylic acid. *Biochemistry* **11**, 2172–2176.
107. Lindsay, R. H., Tillery, C. R., and Yu, M. W. (1972). Conversion of the antithyroid drug 2-thiouracil to 2-thio-5′-UMP by UMP pyrophosphorylase. *Arch. Biochem. Biophys.* **148**, 466–474.
108. Lisy, V., and Skoda, J. (1966). Deamination of some cytosine and cytidine analogues by an *Escherichia coli* extract. *Collection Czechoslov. Chem. Commun.* **31**, 3020–3022.
109. Magni, G. (1978). Uridine nucleosidase from yeast. *In* "Methods of Enzymology" (P. A. Hoffee, and M. E. Jones, *Eds.*), Vol. **51**, pp. 290–296. Academic Press, New York.
110. Magni, G., Fioretti, E., Ipata, P. L., and Natalini, P. (1975). Baker's yeast uridine

nucleosidase. Purification, composition, and physical and enzymatic properties. *J. Biol. Chem.* **250**, 9–13.

111. Magni, G., Natalini, P., Ruggieri, S., and Vita, A. (1976). Baker's yeast uridine nucleosidase is a regulatory copper containing protein. *Biochem. Biophys. Res. Commun.* **69**, 724–730.
112. Makoff, A. J., and Radford, A. (1978). Genetics and biochemistry of carbamoyl phosphate biosynthesis and its utilization in the pyrimidine biosynthetic pathway. *Microbiol. Rev.* **42**, 307–328.
113. Mitchell, A., and Finch, L. R. (1977). Pathways of nucleotide biosynthesis in *Mycoplasma mycoides* subsp. *mycoides. J. Bacteriol.* **130**, 1047–1054.
114. Mitchell, A., and Finch, L. R. (1979). Enzymes of pyrimidine metabolism in *Mycoplasma mycoides* subsp. *mycoides. J. Bacteriol.* **137**, 1073–1080.
115. Mitsugi, K., Komagata, K., Takahashi, M., Iizaka, H., and Katagiri, H. (1964). Bacterial synthesis of nucleotides. Part II. Distribution of nucleoside phosphotransferases in bacteria. *Agric. Biol. Chem.* **28**, 586–600.
116. Møllgaard, H. (1980). Deoxyadenosine (deoxycytidine) kinase from *Bacillus subtilis*. Purification, characterization, and physiological function. *J. Biol. Chem.* **255**, 8216–8220.
117. Møllgaard, H., and Neuhard, J. (1978). Deoxycytidylate deaminase from *Bacillus subtilis*. Purification, characterization, and physiological function. *J. Biol. Chem.* **253**, 3536–3542.
118. Molloy, A., and Finch, L. R. (1969). Uridine 5′-monophosphate pyrophosphorylase activity from *Escherichia coli. FEBS Letts.* **5**, 211–213.
119. Mueller, K., and Bremer, H. (1968). Rate of synthesis of messenger ribonucleic acid in *Escherichia coli. J. Mol. Biol.* **38**, 329–353.
120. Munch-Petersen, A., Nygaard, P., Hammer-Jespersen, K., and Fiil, N. (1972). Mutants constitutive for nucleoside catabolizing enzymes in *Escherichia coli* K 12. Isolation, characterization and mapping. *Eur. J. Biochem.* **27**, 208–215.
121. Natalini, P., Fioretti, E., Ruggieri, S., Vita, A., and Magni, G. (1975). Presence of a specific uridine 5′-monophosphate pyrophosphorylase in baker's yeast. *Experientia* **31**, 1008–1009.
122. Natalini, P., Ruggieri, S., Santarelli, I., Vita, A., and Magni, G. (1979). Baker's yeast UMP: pyrophosphate phosphoribosyltransferase. Purification, enzymatic and kinetic properties. *J. Biol. Chem.* **254**, 1558–1563.
123. Neu, H. C. (1967). The 5′-nucleotidase of *Escherichia coli.* I. Purification and properties. *J. Biol. Chem.* **242**, 3896–3904.
124. Neu, H. C. (1968). The 5′-nucleotidases (uridine diphosphate sugar hydrolases) of the *Enterobacteriaceae. Biochemistry* **7**, 3766–3773.
125. Neuhard, J. (1968). Pyrimidine nucleotide metabolism and pathways of thymidine triphosphate biosynthesis in *Salmonella typhimurium. J. Bacteriol.* **96**, 1519–1527.
126. Neuhard, J., and Ingraham, J. (1968). Mutants of *Salmonella typhimurium* requiring cytidine for growth. *J. Bacteriol.* **95**, 2431–2433.
127. O'Donovan, G. A., and Neuhard, J. (1970). Pyrimidine metabolism in microorganism. *Bacteriol. Rev.* **34**, 278–343.
128. Okazaki, R., and Kornberg, A. (1964). Deoxythymidine kinase of *Escherichia coli.* I. Purification and some properties of the enzyme. *J. Biol. Chem.* **239**, 269–274.
129. Okazaki, R., and Kornberg, A. (1964). Deoxythymidine kinase of *Escherichia coli.* II. Kinetics and feedback inhibition. *J. Biol. Chem.* **239**, 275–284.

130. Orengo, A., and Kobayashi, S. (1978). Uridine-cytidine kinase from Novikoff ascites rat tumor and *Bacillus stearothermophilus*. *In* "Methods of Enzymology" (P. A. Hoffee, and M. E. Jones, *Eds.*), Vol. **51**, pp. 299–307. Academic Press, New York.
131. Orengo, A., and Maness, P. (1978). Pyrimidine nucleoside monophosphate kinase from rat liver and rat Novikoff ascites hepatomas. *In* "Methods of Enzymology (P. A. Hoffee, and M. E. Jones, *Eds.*), Vol. **51**, pp. 321–331. Academic Press, New York.
132. Orengo, A., and Saunders, G. F. (1972). Regulation of a thermostable pyrimidine ribonucleoside kinase by cytidine triphosphate. *Biochemistry* **11**, 1761–1767.
133. Palmatier, R. D., McCroskey, R. P., and Abbot, M. T. (1970). The enzymatic conversion of uracil 5-carboxylic acid to uracil and carbon dioxide. *J. Biol. Chem.* **245**, 6706–6710.
134. Palmer, L. M., Scazzocchio, C., and Cove, D. J. (1975). Pyrimidine biosynthesis in *Aspergillus nidulans*. Isolation and characterization of mutants resistant to fluoropyrimidines. *Molec. Gen. Genet.* **140**, 165–173.
135. Piérard, A., Glansdorff, N., Gigol, D., Crabeel, M., Halleux, P., and Thiry, L. (1976). Repression of *Escherichia coli* carbamoylphosphate synthase: Relationship with enzyme synthesis in the arginine and pyrimidine pathways. *J. Bacteriol.* **127**, 291–301.
136. Piérard, A., Glansdorff, N., and Yashphe, J. (1972). Mutations affecting uridine monophosphate pyrophosphorylase or the *argR* gene in *Escherichia coli*. *Molec. Gen. Genet.* **118**, 235–245.
137. Plunkett, W., and Moner, J. G. (1971). The role of uridine-cytidine kinase in the regulation of pyrimidine ribonucleotide synthesis in *Tetrahymena pyriformis*. *Biochim. Biophys. Acta* **250**, 92–102.
138. Plunkett, W., and Moner, J. G. (1978). UMP pyrophosphorylase of *Tetrahymena pyriformis*. Partial purification and properties. *Arch. Biochem. and Biophys.* **187**, 264–271.
139. Potvin, B. W., Kelleher, R. J., and Gooder, H. (1975). Pyrimidine biosynthetic pathway of *Bacillus subtilis*. *J. Bacteriol.* **123**, 604–615.
140. Powell, J. W., and Wachsmann, J. T. (1973). Evidence for four deoxynucleoside kinase activities in extracts of *Lactobacillus leichmannii*. *Applied Microbiol.* **25**, 869–872.
141. Pritchard, R. H., and Ahmad, S. I. (1971). Fluorouracil and the isolation of mutants lacking uridine phosphorylase in *Escherichia coli*: Location of the gene. *Molec. Gen. Genet.* **111**, 84–88.
142. Raetz, C. H. R., and Kennedy, E. R. (1973). Function of cytidine-diglyceride and deoxycytidine-diglyceride in the biosynthesis of membrane lipids in *Escherichia coli*. *J. Biol. Chem.* **248**, 1098–1105.
143. Razzell, W. E., and Casshyap, P. (1964). Substrate specificity and induction of thymidine phosphorylase in *Escherichia coli*. *J. Biol. Chem.* **239**, 1789–1793.
144. Reid, T. W., and Wilson, I. B. (1971). *E. coli* alkaline phosphatase. *In* "The Enzymes" (P. D. Boyer, *Ed.*), Vol. **IV**, pp. 373–415. Academic Press, New York.
145. Rima, B. K., and Takahashi, I. (1977). Metabolism of pyrimidine bases and nucleosides in *Bacillus subtilis*. *J. Bacteriol.* **129**, 574–579.
146. Rima, B. K., and Takahashi, I. (1978). Synthesis of thymidine nucleotides in *Bacillus subtilis*. *Can. J. Biochem.* **56**, 158–160.
147. Rohde, W. (1977). Irreversible inhibition of thymidine kinase from *Escherichia coli*. *FEBS Letts.* **82**, 118–120.

148. Rohde, W., and Lezins, A. G. (1971). The purification of thymidine kinase from *Escherichia coli* by affinity chromatography. *Hoppe-Seyler's Z. Physiol. Chem.* **352**, 1507–1516.
149. Rumyantseva, E. V., Sukhodolets, V. V., and Smirnov, Y. V. (1979). Isolation and characterization of mutants for genes of nucleoside catabolism in *Bacillus subtilis*. *Genetika* **15**, 594–604.
150. Sakai, T., Tochikina, T., and Ogata, K. (1965). Metabolism of nucleosides in bacteria. Part II. Studies on nucleoside-N-ribosyl group transfer reaction in bacteria (1). *Agric. Biol. Chem.* **29**, 742–750.
151. Sakai, T., Watanabe, T., and Chibata, I. (1971). Metabolism of pyrimidine nucleotides in bacteria. II. Studies on the regulation system of the degradation of nucleotides in *Pseudomonas oleovorans*. *J. Ferment. Technol.* **49**, 488–498.
152. Sakai, T., Yu, T., and Omata, S. (1976). Distribution of enzymes related to cytidine degradation in bacteria. *Agric. Biol. Chem.* **9**, 1893–1895.
153. Sakai, T., Yu, T., Tabe, H., and Omata, S. (1975). Purification of cytosine deaminase from *Serratia marscesens*. *Agric. Biol. Chem.* **39**, 1623–1629.
154. Sakai, T., Yu, T., Taniguchi, K., and Omata, S. (1975). Purification of cytosine deaminase from *Pseudomonas aureofaciens*. *Agric. Biol. Chem.* **39**, 2015–2020.
155. Saunders, P. P., Wilson, B. A. and Saunders, G. F. (1969). Purification and comparative properties of a pyrimidine nucleoside phosphorylase from *Bacillus stearothermophilus*. *J. Biol. Chem.* **244**, 3691–3697.
156. Schlesinger, M. J., and Olsen, R. (1968). Expression and localization of *Escherichia coli* alkaline phosphatase synthesized in *Salmonella typhimurium* cytoplasm. *J. Bacteriol.* **96**, 1601–1605.
157. Schwartz, M. (1971). Thymidine phosphorylase from *Escherichia coli*. Properties and kinetics. *Eur. J. Biochem.* **21**, 191–198.
158. Schwartz, M. (1978). Thymidine phosphorylase from *Escherichia coli*. *In* "Methods of Enzymology" (P. A. Hoffee, and M. E. Jones, *Eds.*), Vol. **51**, pp. 442–445. Academic Press, New York.
159. Schwartz, M., and Neuhard, J. (1975). Control of expression of the *pyr* genes in *Salmonella typhimurium*: Effects of variations in uridine and cytidine nucleotide pools. *J. Bacteriol.* **121**, 814–822.
160. Scocca, J. J. (1971). Purification and substrate specificity of pyrimidine nucleoside phosphorylase from *Haemophilus influenzae*. *J. Biol. Chem.* **246**, 6606–6610.
161. Shaffer, P. M., Chabo, J. L., Dorman, D. C., Driscoll, B. A., Lott, A. M., Oesterle, E. B., and Taggart, M. M. (1979). Nucleoside and pyrimidine α-ketoglutarate dioxygenase activities in *Aspergillus nidulans*. Abst. Int. Congress of Biochem., Toronto.
162. Shaffer, P. M., Hsu, C-A., and Abbott, M. T. (1975). Metabolism of pyrimidine deoxyribonucleosides in *Neurospora crassa*. *J. Bacteriol.* **121**, 648–655.
163. Shaffer, P. M., McCroskey, R. P., and Abbott, M. T. (1972). Substrate specificity of the hydroxylase reaction in which thymidine is converted to thymine ribonucleoside. *Biochim. Biophys. Acta* **258**, 387–394.
164. Solomon, L. R., and Breitman, T. R. (1971). Pseudouridylate synthetase of *Escherichia coli*: a catabolite-repressible enzyme. *J. Bacteriol.* **107**, 535–542.
165. Suzuki, T., and Hochster, R. M. (1966). On the biosynthesis of pseudouridine and of pseudouridylic acid in *Agrobacterium tumefaciens*. *Can. J. Biochem.* **44**, 259–272.

166. Takagi, Y., and Horecker, B. L. (1957). Purification and properties of a bacterial riboside hydrolase. *J. Biol. Chem.* **225**, 77–86.
167. Terada, M., Tatibana, M., and Hayaishi, O. (1967). Purification and properties of nucleoside hydrolase from *Pseudomonas fluorescens*. *J. Biol. Chem.* **242**, 5578–5585.
168. Trimble, R. B., and Maley, F. (1971). Metabolism of 4-N-hydroxy-cytidine in *Escherichia coli*. *J. Bacteriol.* **108**, 145–153.
169. Uerkvitz, W. (1971). Trans-N-deoxyribosylase from *Lactobacillus helveticus*. Crystallization and properties. *Eur. J. Biochem.* **23**, 387–395.
170. Uerkvitz, W., and Beck, C. F. (1981). Periplasmic phosphatases in *Salmonella typhimurium* LT2. A biochemical, physiological, and partial genetic analysis of three nucleoside monophosphate dephosphorylating enzymes. *J. Biol. Chem.* **256**, 382–389.
171. Uerkvitz, W., Karlström, O., and Munch-Petersen, A. (1973). A deoxyuridine phosphatase detected in mutants of *Escherichia coli* lacking alkaline phosphatase and 5′-nucleotidase. *Molec. Gen. Genet.* **121**, 337–346.
172. Valentin-Hansen, P. (1978). Uridine-cytidine kinase from *Escherichia coli*. *In* "Methods of Enzymology" (P. A. Hoffee, and M. E. Jones, *Eds.*), Vol. **51**, pp. 308–314. Academic Press, New York.
173. Wachsman, J. T., and Morgan, D. D. (1973). Demonstration of a deoxycytidine kinase activity in extracts of *Bacillus megaterium* KM. *Appl. Microbiol.* **25**, 506–508.
174. Waleh, N. S., and Ingraham, J. L. (1976). Pyrimidine ribonucleoside monophosphokinase and the mode of RNA turnover in *Bacillus subtilis*. *Arch. Microbiol.* **110**, 49–54.
175. Wang, T. P., and Lampen, J. O. (1951). The cleavage of adenosine, cytidine, and xanthosine by *Lactobacillus pentosus*. *J. Biol. Chem.* **192**, 339–347.
176. Wehr, C. T. (1973). Isolation and properties of a ribonuclease-deficient mutant of *Salmonella typhimurium*. *J. Bacteriol.* **114**, 96–102.
177. Wentworth, D. F., and Wolfenden, R. (1978). Cytidine deaminases (from *Escherichia coli* and human liver). *In* "Methods of Enzymology" (P. A. Hoffee, and M. E. Jones, *Eds.*), Vol. **51**, pp. 401–407. Academic Press, New York.
178. Weppelman, R., Kier, L. D., and Ames, B. N. (1977). Properties of two phosphatases and a cyclic phosphodiesterase of *Salmonella typhimurium*. *J. Bacteriol.* **130**, 411–419.
179. Williams, L. G., and Mitchell, H. K. (1969). Mutants affecting thymidine metabolism in *Neurospora crassa*. *J. Bacteriol.* **100**, 383–389.
180. Wondrack, L. M., Hsu, C. A., and Abbot, M. T. (1978). Thymine 7-hydroxylase and pyrimidine deoxyribonucleoside 2′-hydroxylase activities in *Rhodotorula glutinis*. *J. Biol. Chem.* **253**, 6511–6515.
181. Wondrack, L. M., Warn, B. J., Seawert, M. D., and Abbot, M. T. (1979). Substitution of nucleoside triphosphates for ascorbate in the thymine 7-hydroxylase reaction of *Rhodotorula glutinis*. *J. Biol. Chem.* **254**, 26–29.
182. Yergatian, S., Lee, J. B., Geisow, M. J., and Ireson, J. C. (1977). Cytosine deaminase: Structural modification studies. *Experientia* **33**, 1570–1571.
183. Young, P. G., Corff, S., and Yuyama, S. (1978). Activity of thymidine kinase during the cell cycle in *Tetrahymena pyriformis*. *Cytobios.* **20**, 191–198.
184. Yuyama, S., Corff, S., and Young, P. G. (1978). Identification of true thymidine kinase in *Tetrahymena pyriformis ST*. *Arch. Biochem. Biophys.* **188**, 64–69.

185. Yuyama, S., Corff, S. C., and Young, P. G. (1979). Levels of true thymidine kinase and nucleoside phosphotransferase in two strains of *Tetrahymena pyriformis* under different growth conditions. *Comp. Biochem. Physiol.* **62B**, 325–328.

186. Øvrebø, S., and Kleppe, K. (1973). Pyrimidine metabolism in *Acinetobacter calcoaceticus*. *J. Bacteriol.* **116**, 331–336.

4

Biosynthesis of deoxythymidine triphosphate

HENRIK MØLLGAARD and JAN NEUHARD

I.	Introduction	149
II.	Endogenous synthesis of dTTP	150
	A. *Escherichia coli* and *Salmonella typhimurium*	150
	B. *Bacillus*	169
	C. *Lactobacillus*	174
	D. General discussion	177
III.	Synthesis of dTTP from exogenous thymine and thymidine	179
	A. *Escherichia coli* and *Salmonella typhimurium*	179
	B. Other microorganisms	187

I. INTRODUCTION

Deoxythymidine triphosphate (dTTP) may be synthesized via two different routes in microorganisms. One, the endogenous pathway, is initiated by the reduction of pyrimidine ribonucleotides to the corresponding deoxyribonucleotides. Subsequently, the pyrimidine deoxyribonucleotides are converted to dUMP, which is methylated in the 5-position to dTMP. Finally dTMP is converted to dTTP by two consecutive phosphorylations. With very few exceptions, such as certain lactobacilli and certain members of the genus *Mycoplasma*, all microorganisms contain an endogenous pathway for dTTP synthesis.

Preformed thymine and thymidine are utilized for dTTP synthesis via the second route, the exogenous pathway. In this thymine reacts with deoxyribose 1-phosphate to form thymidine, which subsequently is phosphorylated to dTMP. The synthesis of dTTP from exogenous and endogenous sources is tightly coupled and is therefore treated together in this chapter. The close connection between the two pathways stems from the fact, that the deoxyribose 1-phosphate required for the utilization of preformed thymine,

METABOLISM OF NUCLEOTIDES, NUCLEOSIDES
AND NUCLEOBASES IN MICROORGANISMS
ISBN: 0-12-510580-0

originates from dUMP. Thus, an understanding of the biosynthesis of dUMP and its regulation is imperative for the understanding of dTTP synthesis.

In the present discussion of microbial dTTP biosynthesis we wish to emphasize the progress which has occurred in this field over the last 10 years. Therefore we are preferentially referring to literature published after 1970. Readers interested in earlier literature are referred to a previous review (145).

II. ENDOGENOUS SYNTHESIS OF dTTP

A. *Escherichia coli* and *Salmonella typhimurium*

The pathway by which dTTP is synthesized from UMP in *Escherichia coli* and *Salmonella typhimurium* is shown in Fig. 4.1, which also shows that some of the intermediates may be supplied from exogenous sources via the salvage pathway (cf. Chapter 3). With one exception, i.e. the possible dephosphorylation of dUDP to dUMP, all the enzymes involved in the pathway have been characterized, and mutants defective in each of these enzymes have been isolated. Studies of the mutants have established the physiological significance of each of the enzymes in the overall conversion.

The intracellular concentration of the intermediates of the entire pathway can be determined accurately by simple chromatographic techniques. This

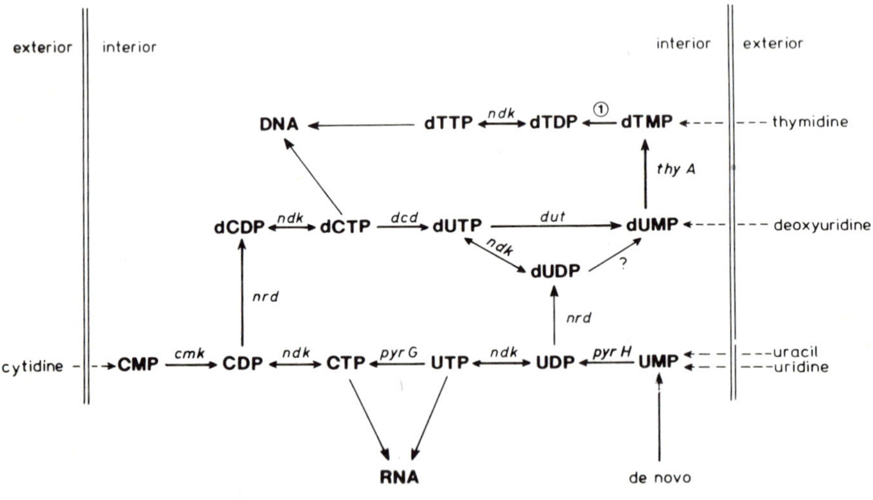

FIG. 4.1. Pyrimidine nucleotide biosynthesis in *E. coli* and *S. typhimurium*. The enzymes involved are identified by their gene symbols as follows: *cmk*, CMP(dCMP) kinase; *dcd*, dCTP deaminase; *dut*, dUTPase; *ndk*, nucleoside diphosphokinase; *nrd*, ribonucleoside diphosphate reductase; *pyrG*, CTP synthetase; *pyrH*, UMP kinase; *thyA*, thymidylate synthase. dTMP kinase is identified by 1.

has facilitated studies of the *in vivo* regulation of dTTP synthesis under a variety of physiological conditions.

In this section we will discuss those steps of the endogenous pathway in Fig. 4.1, which are specifically involved in deoxyribonucleotide interconversion, i.e. the reaction sequences which are initiated with the reduction of CDP and UDP. Subsequently we will discuss what is known about the *in vivo* functioning of the pathway, and how it appears to be regulated in the growing cell.

1. Ribonucleoside diphosphate reductase

Ribonucleoside diphosphate reductase (ribonucleotide reductase (EC 1.17.4.1) catalyzes the reduction of ribonucleoside diphosphates to 2′-deoxyribonucleoside diphosphates (review, 186).

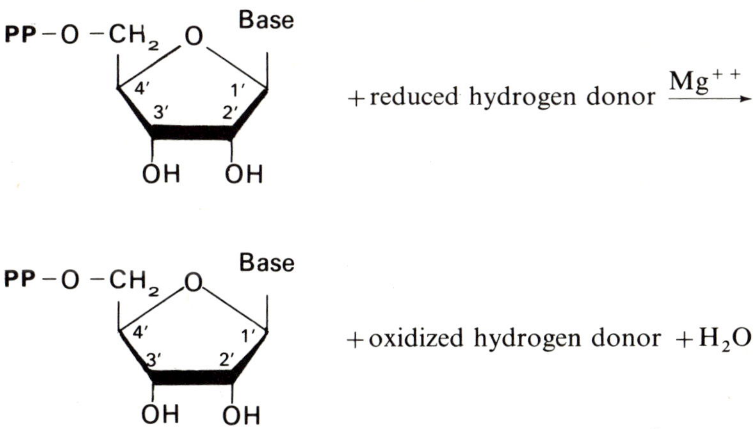

The base can be either uracil, cytosine, adenine, or guanine. The physiological hydrogen donor is not known with certainty, but reduced dithiols such as dithiothreitol or thioredoxin will function *in vitro*. The enzyme has an absolute requirement for magnesium ions and is allosterically regulated.

Molecular structure. Ribonucleoside diphosphate reductase of *E. coli* consists of two nonidentical subunits, protein B1 and B2, both of which are required for activity. Protein B1 (MW 160 000) is composed of two polypeptide chains of approximately equal size, but different primary structure (183). Each peptide chain contains two thiol groups, which participate in the catalytic process (184). In addition protein B1 has two substrate binding sites in the vicinity of the active thiol groups, and four effector binding sites located elsewhere in the molecule (43, 44).

Protein B2 (MW 78 000) consists of two apparently identical polypeptides, each containing one iron atom and one free radical (6, 50, 183). The radical is essential for catalytic activity and is located on the benzene ring of a tyrosine residue in the peptide chain (179, 180). The radical is destroyed by hydroxyurea or hydroxylamine (50) rendering the enzyme inactive. The inactivated enzyme can be reactivated only by removal of iron followed by reconstitution. The iron atoms are probably required for the generation and stabilization of the radical (6, 152).

In the presence of magnesium ions, proteins B1 and B2 form a 1:1 complex, which is catalytically active when positive effectors are added (23, 183). A model for the structure of the active enzyme is shown in Fig. 4.2. As indicated on the figure the active site of the enzyme is thought to be formed by both the B1 and the B2 subunit in such a way that the active thiols of protein B1 are in close contact with the free radicals of protein B2 (185). In the presence of magnesium ions and the negative effector dATP, a catalytically inactive complex is formed consisting of two molecules B1 and two molecules B2.

Kinetics and regulation. Kinetic studies suggest that the enzyme acts by a ping pong mechanism, i.e. reduction of the B1 protein followed by reduction of the ribonucleoside diphosphate substrate (184). The hydrogen atom substitutes the hydroxyl of the pentose moiety with retention of the steric configuration at position 2′,

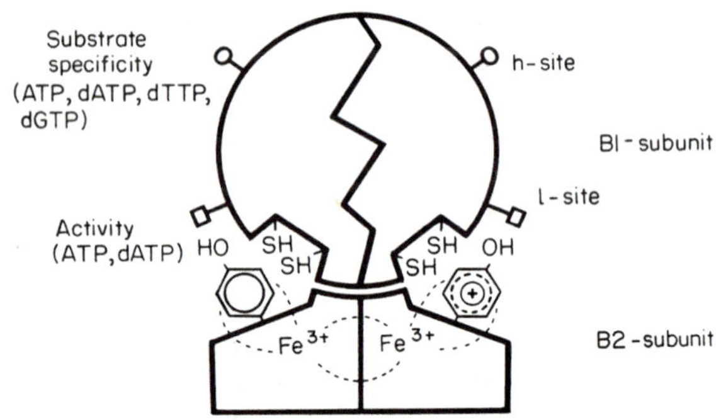

FIG. 4.2. Model of ribonucleoside diphosphate reductase from *E. coli* (186).

The allosteric control of ribonucleotide reductase is linked to the four effector sites on protein B1. Two of these, the *l*-sites, have low affinities for dATP, whereas the other two, the *h*-sites, have 10-fold higher affinities for dATP. The *l*-sites determine the general activity of the enzyme and can bind ATP and dATP resulting in an active and inactive enzyme, respectively. The *h*-sites determine the substrate specificity of the enzyme through binding of ATP, dATP, dTTP, or dGTP (43, 44). In Table 4.1 the more important features of this complex regulation are summarized. Thus, with ATP bound in

TABLE 4.1

Regulation of ribonucleotide reductase by nucleoside triphosphates.

Effector in activity sites	Effector in specificity sites	Preferred substrate(s)
l-sites	h-sites	
ATP	ATP	CDP, UDP
ATP	dATP	CDP, UDP
ATP	dTTP	GDP
ATP	dGTP	ADP
dATP	—	no activity

the activity sites, the affinity of the enzyme towards the different substrates is determined by the effector bound in the substrate specificity sites. The maximal velocity obtained with a given substrate is also influenced by the effectors. The effect of dATP is more complex since it can bind to both classes of effector sites: at low concentrations ($\sim 1\ \mu M$) dATP stimulates the reduction of pyrimidine nucleotides, whereas at higher concentrations ($\sim 100\ \mu M$) it is a general inhibitor of the enzyme.

The effector concentrations required *in vitro* to obtain maximal regulatory responses are in the range of 10^{-6} to 10^{-4} M for deoxyribonucleoside triphosphates and 10^{-3} M for ATP (24). This is comparable to the intracellular concentrations of these compounds in exponentially growing cells of *E. coli*.

The results discussed above have all been obtained with highly purified, soluble ribonucleotide reductase. However, the enzyme may turn out to have different properties *in vivo* because of its apparent attachment to the cytoplasmic membrane. Thus, the specific activity of the soluble ribonucleotide reductase can only account for the synthesis of about 10% of the deoxyribonucleotides required for DNA synthesis *in vivo*, whereas the specific activity found in ether-treated cells (198), and in carefully prepared membrane fractions is about 10-fold higher (119, 193). In the latter preparation altered regulatory properties of the enzyme were also observed (193). It has been suggested that the ribonucleotide reductase is part of a membrane associated, replicating complex (40, 42, 122), since it cosediments with DNA polymerase, thioredoxin, and thymidylate synthase in a membrane fraction (119).

Very little work has been done on ribonucleotide reduction in *S. typhimurium*, but experiments with ether-treated cells suggest that the overall pattern of regulation is similar to that of *E. coli* (Fig. 4.3).

Genetics. Mutants totally lacking ribonucleotide reductase activity cannot be

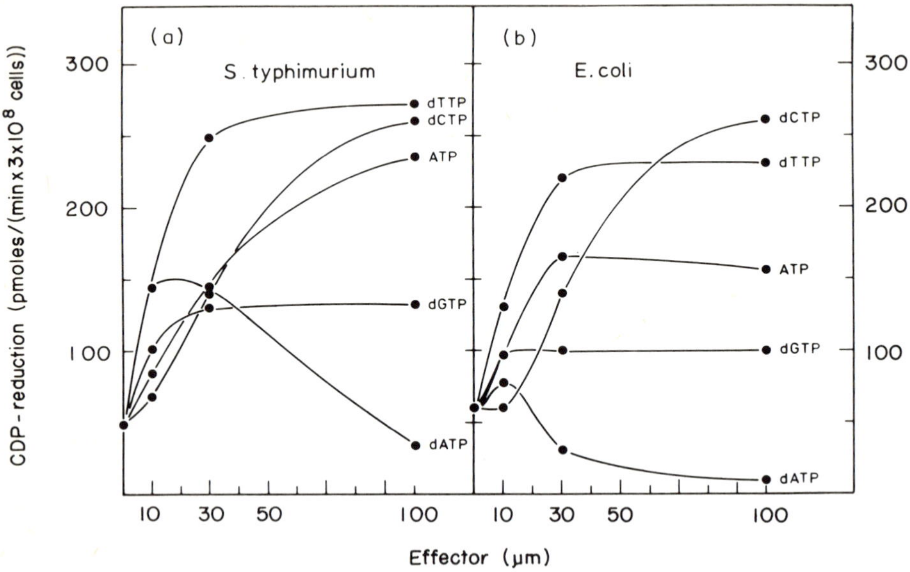

FIG. 4.3. Effect of nucleoside triphosphates on ribonucleotide reductase activity in ether-treated cells (13, 198) of thymine-requiring strains of *E. coli* and *S. typhimurium*. The cells were derepressed for ribonucleotide reductase synthesis by 4 hrs of thymine starvation prior to ether-treatment (Neuhard and Stauning, unpublished results).

isolated in *E. coli* and *S. typhimurium*, since these organisms are devoid of deoxyadenosine, deoxyguanosine, and deoxycytidine kinase activities and thus cannot be supplied with deoxyribonucleotides from the medium.* A mutant of *E. coli* with a modified protein B2 (*nrdB*), as well as conditional mutants containing a temperature sensitive protein B1 (*nrdA*) have been isolated (65, 67, 68). The *nrdA* and *nrdB* genes are located very close to each other on the *E. coli* chromosome at around 48 min (*see* Appendix p. 307). A defective λ phage carrying both *nrd*-genes has been isolated (51) and from that the *nrd*-genes have been transferred to the multicopy plasmid pBR 322 (154). A more exact location of the *nrd*-genes relative to the *ubiG* and *nalA* genes was established by restriction mapping of this hybrid plasmid (154).

Genetic regulation. For many years it has been known that thymine starvation in *E. coli* causes derepression of ribonucleotide reductase synthesis. Under these conditions the B1 and B2 subunits are coordinately derepressed (64). In a more detailed study Filpula and Fuchs have shown a negative correlation between the specific activity of ribonucleotide reductase and the

* Evidence for the existence of deoxycytidine kinase activity in certain strains of *E. coli* C has been presented (146).

DNA/mass ratio in *E. coli.* It appears that any condition, which tends to decrease the DNA/mass ratio, results in elevated levels of the enzyme. These conditions include thymine limitation, introduction of a *rep* mutation, increase in growth rate, due to enrichment of the growth medium, and inhibition of DNA synthesis either in temperature sensitive *dna* mutants or by inhibitors (56, 57). Based on their results they proposed a model, in which the signal that causes initiation of new DNA replication forks also triggers the synthesis of ribonucleotide reductase and possibly of other proteins needed for DNA replication.

2. Hydrogen donor in ribonucleotide reduction

The physiological hydrogen donor of ribonucleotide reductase has not been identified with certainty. Although NADPH does not react with the purified enzyme, it seems to be the ultimate electron donor for the reaction *in vivo*. Two independent systems which transfer hydrogen from NADPH to the B1 subunit of ribonucleotide reductase have been identified in *E. coli.*

The glutaredoxin system consists of glutathione, glutathione reductase, and glutaredoxin (88). In this system (Fig. 4.4a) the flavoprotein glutathione reductase catalyzes the hydrogen transfer from NADPH to glutathione. Subsequently, reduced glutathione reduces the disulfide in the active site of ribonucleotide reductase (Fig. 4.2) in a reaction catalyzed by the heat stable protein glutaredoxin (MW 12 000) (89, 90). Mutants deficient in glutaredoxin have not been found, but a mutant lacking glutathione synthetase has been isolated (69). It contains low levels of glutathione and has reduced specific

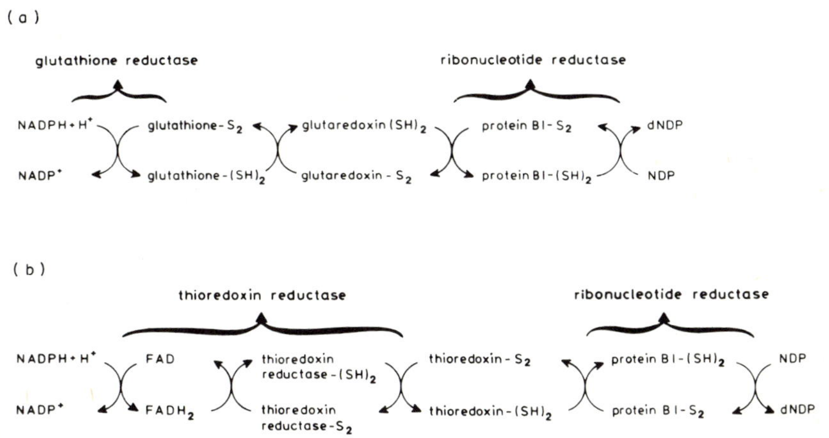

FIG. 4.4. Electron transport from NADPH to ribonucleoside diphosphates mediated by the glutaredoxin system (a), and the thioredoxin system (b) of *E. coli.* NDP and dNDP are ribo- and deoxyribonucleoside diphosphates, respectively.

activity of the B1 subunit of ribonucleotide reductase, suggesting that the glutaredoxin system is involved in ribonucleotide reduction in *E. coli.*

The thioredoxin system consists of thioredoxin reductase and thioredoxin. Thioredoxin is a heat stable protein with known amino acid sequence (MW 11 600). It contains only two cysteines, which are separated by two amino acids and located on a protrusion of the molecule (93). In the presence of ribonucleotide reductase two hydrogen atoms are transferred from the two cysteines to a disulfide in the B1 subunit (184). However, also other protein disulfides are reduced by thioredoxin (91).

Regeneration of the reduced form of thioredoxin is catalyzed by the flavoenzyme *thioredoxin reductase* (EC 1.6.4.5) (MW 66 000) with NADPH as hydrogen donor. The equilibrium of this reaction is far in the direction of thioredoxin reduction. Thioredoxin reductase consists of two identical or very similar subunits. The reaction is probably a ping pong mechanism, where NADPH donates reducing equivalents to the enzyme bound flavine adenine dinucleotide, which in turn reduces a cystine in the thioredoxin reductase molecule (186). Subsequently the two cysteins reduce the oxidized thioredoxin. The electron transport from NADPH via thioredoxin to ribonucleotides is shown in Fig. 4.4b. Alternatively, thioredoxin can be reduced nonenzymatically by low molecular weight compounds such as dihydrolipoamide (91).

It has been observed that in nearly all thioredoxin molecules *in vivo* were phosphorylated in one of the two cysteines (153). This would make the participation of this fraction of thioredoxin in oxidation-reduction reactions unlikely. However, in a more recent study (91a), no phosphorylation of thioredoxin was detected.

An *E. coli* mutant defective in thioredoxin reductase has been isolated. The mutant has no detectable thioredoxin reductase activity *in vitro*, but shows no defect in ribonucleotide reduction *in vivo*. Thus thioredoxin reductase seems not to be essential for ribonucleotide reduction (63).

Mutants of *E. coli* lacking thioredoxin (*trxA*, previously designated *tsnC*) have been isolated. The *trxA* gene is located around 84 minutes on the *E. coli* chromosome (*see* Appendix; 28, 123, 124). Mutants completely devoid of the thioredoxin protein show a "leaky" cysteine requirement. Thus, while thioredoxin may participate in cysteine biosynthesis, it is not essential for ribonucleotide reduction (92, 188). Accordingly thymine starvation which causes derepression of ribonucleotide reductase synthesis does not result in increased levels of thioredoxin or thioredoxin reductase (64).

These findings suggest that the glutaredoxin system is the principal physiological donor of reducing equivalents in ribonucleotide reduction. In support of this it has been shown that reduced glutaredoxin is almost 10-fold more active than reduced thioredoxin as reductant of ribonucleotide reductase (90).

3. Conversion of dCDP to dUMP (the dCTP pathway)

In enteric bacteria the majority of the cellular dUMP required for thymidine nucleotide synthesis is derived from a cytidine nucleotide via reduction of CDP. The dCDP thus formed is converted to dUMP through a series of reactions involving phosphorylation of dCDP, deamination of dCTP, and hydrolysis of dUTP to dUMP (Fig. 4.1). The enzymes catalyzing these reactions are described below.

Nucleoside diphosphokinase (EC 2.7.4.6). The phosphorylation of dCDP to dCTP is presumably catalyzed by nucleoside diphosphokinase. The *S. typhimurium* enzyme (MW 85 000) can use all naturally occurring ribo- and deoxyribonucleoside diphosphates as phosphate acceptors and all common nucleoside triphosphates as donors (73). The true substrates for the enzyme are the nucleotide-Mg^{2+} complexes, which show mutual competitive inhibition. The reaction proceeds by a ping-pong mechanism, in which the triphosphate donates its terminal phosphate to the enzyme as the first step. Subsequently the enzyme-bound phosphate is transferred to the recipient diphosphate.

A *S. typhimurium* strain mutated in the structural gene for nucleoside diphosphokinase (*ndk*) has been isolated (74). Crude extracts of the mutant contain only two per cent of the enzyme activity found in the parent, when assayed with each of the naturally occurring nucleoside diphosphates as acceptors and ATP as donor. The *ndk* gene is located at 55 units on the *Salmonella* linkage map (*see* Appendix; J. L. Ingraham, personal communication).

A nucleoside diphosphokinase with properties similar to the *Salmonella* enzyme has been found in *E. coli* (162, 166). 75% of this activity is membrane bound, and most of it is released from intact cells by a cold hypoosmotic shock. Based on this it was concluded that the enzyme is loosely bound to the outer surface of the cytoplasmic membrane (162). If this interpretation is correct, then the function of the enzyme in the synthesis of nucleic acid precursors is questionable. However, the observation that inhibition of the enzyme *in vivo* by desdanine results in cessation of growth, suggests an essential physiological function of nucleoside diphosphokinase (165, 166).

Two other enzymes displaying nucleoside diphosphate kinase activity have been found in *E. coli*, but the specific activity of both is low compared to the nucleoside diphosphokinase. The first activity is associated with DNA polymerase I (128), from which it cannot be separated. The second activity is associated with pyruvate kinase, whose concentration in the cell under aerobic conditions is low. Under anaerobic conditions, however, *E. coli* contains an increased level of pyruvate kinase concomitantly with a decreased level of nucleoside diphosphokinase. Thus, under these conditions pyruvate kinase may play a role in the synthesis of nucleoside triphosphates (167, 168).

Deoxycytidine triphosphate deaminase (EC 3.5.4.13). dCTP is further metabolized by two different routes. It serves as a substrate for DNA polymerase, and it is deaminated irreversibly to dUTP by dCTP deaminase. Purified dCTP deaminase from *S. typhimurium* (MW 82 000) is highly specific for Mg^{2+}-dCTP as substrate. It is half saturated at a dCTP concentration of 300 μM. In accordance with its function at a metabolic branch point the enzyme activity is regulated. It shows positive cooperativity towards dCTP and is feedback inhibited by dTTP, the ultimate product of the pathway. As seen in Fig. 4.5, dTTP affects the reaction by increasing the sigmoidicity of the dCTP saturation curve without affecting the maximal velocity. The only other nucleotide found to inhibit the enzyme is dUTP (12). Mutants deleted for the structural gene encoding dCTP deaminase (*dcd*) have been isolated in both *S. typhimurium* and *E. coli*. In both species the *dcd* gene is located close to the *his* cluster around 45 min on the chromosomes (*see* Appendix; 13, 142). Introduction of a *dcd* mutation results in sensitivity to 5-bromodeoxyuridine

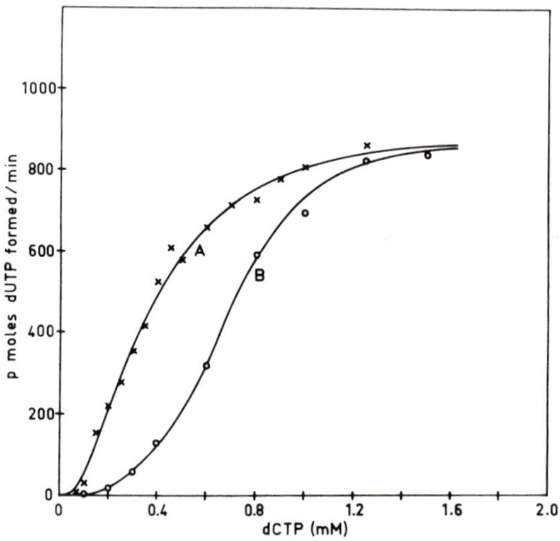

FIG. 4.5. Kinetics of dCTP saturation of dCTP deaminase from *S. typhimurium* in the absence (A) and in the presence (B) of 0.4 mM dTTP (12).

(66), but it does not cause any growth requirement, indicating that the dUDP-pathway for dUMP synthesis is capable of taking over the entire synthesis of dUMP (*see* below).

Deoxyuridine triphosphatase (EC 3.6.1.23). dUTP is broken down to dUMP and pyrophosphate in a reaction catalyzed by dUTPase (MW 64 500). The enzyme from *E. coli* appears to contain zinc ions and consists of four subunits of equal size (178). It is highly specific for dUTP and shows Michaelis–Menten kinetics with a K_m of 12 μM.

Mutants of *E. coli* with reduced levels of dUTPase have been isolated (86, 108). The structural gene for the enzyme (*dut*, synonymous with *dnaS*) is located close to the *pyrE* gene on the *E. coli* map around 80 min (Appendix). The most deficient *dut* mutant contains less than 2% of the parental dUTPase activity. This mutant shows a somewhat prolonged generation time, an increased sensitivity to 5-fluorodeoxyuridine, and an increased rate of recombination (Hyper-Rec) and spontaneous mutation. A *dut*-mutation will cause an increase in the dUTP pool and presumably a decrease in the dUMP pool. The low dUMP pool may explain the increased sensitivity to fluorodeoxyuridine due to less efficient competition with 5-fluorodUMP. The increased dUTP pool leads to an increased frequency of misincorporation of uracil for thymine in DNA. Excision-repair of these uracil moieties from the DNA leads to temporary breaks and gaps, explaining the Hyper-Rec phenotype of the mutants (*see also* p. 168).

A defective λ-phage containing the *dut* and *pyrE* genes has been isolated, and from that a detailed restriction map of this region of the *E. coli* chromosome has been obtained (2, 118).

4. Conversion of dUDP to dUMP (the dUDP Pathway)

About 25% of the cellular dUMP in enteric bacteria is derived via reduction of UDP to dUDP. The relative small contribution from this pathway to dUMP synthesis may be a reflexion of the fact that the K_m of ribonucleotide reductase for UDP is 10-fold higher than for CDP. Cells totally blocked in the main pathway for dUMP synthesis, i.e. *dcd* mutants, show a normal growth pattern. This shows that the capacity of the dUDP-pathway is sufficient to provide the cells with all the required dUMP.

Very little is at present known about the pathway by which dUDP is converted to dUMP. One likely possibility is that the conversion involves the sequential action of nucleoside diphosphokinase and dUTPase, where the last step is shared with the dCDP-pathway.

5. Conversion of dUMP to dTTP

Thymidylate synthase (EC 2.1.1.45). The initial step in the conversion of dUMP to dTTP is catalyzed by thymidylate synthase. In this reaction a methylene group and two reducing equivalents are transferred from 5,10-methylene tetrahydrofolate to dUMP. The products are dTMP and dihydrofolate (Fig. 4.6). Regeneration of tetrahydrofolate occurs in a reaction catalyzed by dihydrofolate reductase with NADPH as the hydrogen donor (Fig. 4.6).

Thymidylate synthase from *E. coli* K12 is a dimer (MW $2 \times 32\,000$) (80), which is stabilized and activated by thiol groups. The half saturation constants for dUMP and 5,10-methylene tetrahydrofolate are 10 μM and 14 μM, respectively. 5-fluorodUMP is a potent inhibitor of thymidylate synthase. In the presence of tetrahydrofolate, 5-fluoro dUMP is covalently and, for all practical purposes, irreversibly bound to the enzyme. Besides a rather weak product inhibition by dTMP no regulation of the enzymic activity or its synthesis has been observed.

Strains carrying mutations in the structural gene for thymidylate synthase, *thyA*, have been isolated. The *thyA* gene is located around 60 min on the *E. coli* and *S. typhimurium* chromosomes (Appendix). In addition to being thymine requiring, *thyA* mutants are resistant to the folate analogs aminopterin and trimethoprim, both of which inhibit dihydrofolate reductase competitively.

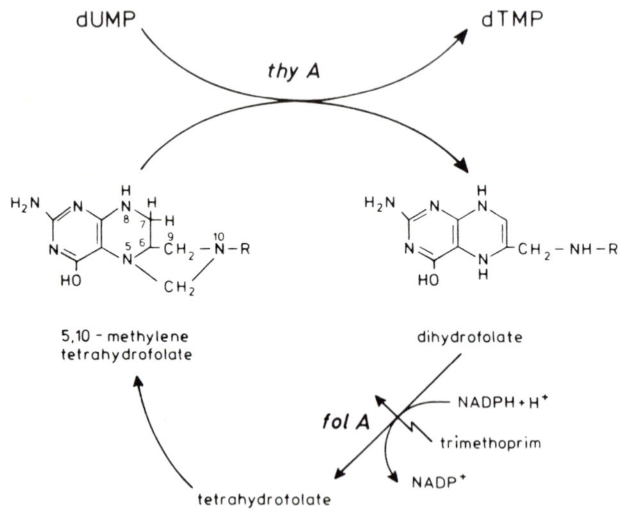

FIG. 4.6. dTMP synthesis and its relation to folic acid metabolism. *thyA*: thymidylate synthase. *folA*: dihydrofolate reductase.

The folate analog-resistant phenotype of *thyA* mutants is due to the fact that the thymidylate synthase reaction is the only known tetrahydrofolate-dependent reaction, where tetrahydrofolate is stoichiometrically oxidized to dihydrofolate. Thus, in a *thyA*$^+$ strain inhibition of dihydrofolate reductase will lead to depletion of tetrahydrofolate due to continued synthesis of dTMP by thymidylate synthase, even in the presence of exogenous thymine or thymidine. This, in turn, will result in cessation of growth, because tetrahydrofolate derivatives are required in a variety of biosynthetic pathways. Cells, in which thymidylate synthase is not functioning, do not oxidize tetrahydrofolate and are therefore resistant to trimethoprim (13).

Thymidylate kinase (EC 2.7.4.9). dTMP kinase catalyzes the phosphorylation of dTMP to dTDP with ATP as phosphate donor.

The *E. coli* enzyme (135) has an absolute requirement for divalent metal ions. Apparently the actual phosphate donor is a nucleotide-Mg^{2+} complex. The most effective donors are ATP and dATP with K_m-values of 1.2 mM, but other nucleoside triphosphates may serve as donors as well. The enzyme can use dTMP and dUMP as phosphate acceptors with K_m values of 240 μM and 10 μM, respectively. dUMP inhibits the phosphorylation of dTMP competitively ($K_i = 3$ mM). Mutants partially defective in the enzyme have been isolated in *E. coli* (Fuchs, personal communication).

The last step in the synthesis of dTTP is presumably catalyzed by nucleoside diphosphokinase (*see* p. 157).

6. Physiological studies on dTTP synthesis

Having summarized our present knowledge of the enzymology and the genetics of dTTP biosynthesis, we will discuss the function, the quantitative importance, and the regulation of dTTP biosynthesis *in vivo*. The discussion will largely be based on studies of the deoxyribonucleotide pools in *E. coli* and *S. typhimurium* mutants defective in one or more of the enzymes involved in dTTP synthesis.

*Mutants lacking thymidylate synthase (*thy *mutants)*. Mutants of *E. coli* and *S. typhimurium* lacking thymidylate synthase show an absolute growth requirement for thymine. The concentration of exogenous thymine required to satisfy the growth requirement of *thy* mutants is very high, i.e. 0.15 mM, as compared to the amount actually required for incorporation into DNA. Secondary mutations are easily obtained, which permit *thy* mutants to grow on 15 μM thymine. Such *low rhymine-requiring* mutants contain increased internal pools of deoxyribose 1-phosphate due to mutational blocks in deoxyribose 1-phosphate catabolism (*see* p. 182).

The internal dUMP concentration in *thy*$^{+}$ strains is too low to be measured accurately, but it is less than 10 μM. Thus, dUMP is probably present in concentrations close to the K_m-value of thymidylate synthase for dUMP, i.e. 10 μM. In mutants lacking thymidylate synthase the steady state concentration of dUMP is around 500 μM (Table 4.2) (132, 141). When present in such high concentration dUMP is continuously catabolized via deoxyuridine to uracil and deoxyribose 1-phosphate, thereby supplying the deoxyribose 1-phosphate required for the utilization of exogenous thymine (cf. Fig. 4.10 and Section III.A.). The uracil produced is reutilized by the cells for UMP synthesis (141).

The internal concentrations of dTTP, dCTP, and dGTP in exponentially growing *thy* mutants are dependent on the concentration of thymine supplied in the medium. In the presence of the minimal concentration of thymine compatible with balanced growth the steady state levels of dTTP and dGTP are only 20–30% of the levels found in the parental strain, whereas the dCTP level is increased 3–6 times. By increasing the exogenous concentration of thymine the steady state concentration of dTTP and dGTP rise, while that of dCTP falls progressively (Fig. 4.7). The dCTP and dGTP pools, but not the dTTP pool, may be normalized completely with high external concentrations of thymine. In low thymine-requiring strains the dTTP may be normalized by

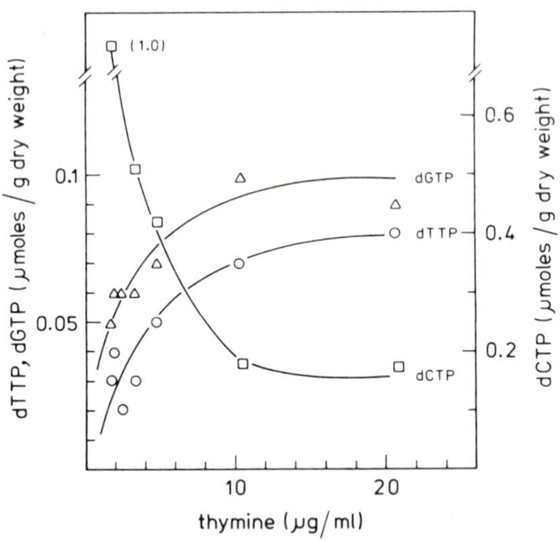

FIG. 4.7. The intracellular pools of dTTP (○——○), dGTP (△——△), and dCTP (□——□) in an *E. coli thyA,deoB* mutant (low thymine requirement), when grown exponentially in different concentrations of exogenous thymine. Note the difference in ordinates (K. F. Jensen, personal communication).

TABLE 4.2

Effect of thymine concentration on pyrimidine deoxynucleotide metabolism in *E. coli* mutants.

Exp.	Genotype	Additions to medium	Intracellular conc. (μM)[a] dTTP	dCTP	dUMP	Relative amount[b] of ribonucleoside diphosphate reductase	Relative activity[c] of dCTP deaminase
1	wild type	none	100	100	<10	1	1
2	*thyA,deoB*	thymine (16 μM)	30	400	500	5	10
3	*thyA,deoB*	thymine (160 μM)	50	100	400	1.3	1
4	*thyA*	thymine (160 μM)	20	600	700	nd[e]	14
5	*thyA,deoB*	none[d] (75′)	0	1000	800	6	17
6	*Δdcd*	none	30	2000	<10	nd	0

[a] The pool data shown are average values compiled from a large number of experiments with both *E. coli* 15 *thy* and *E. coli* K12, *thy* strains. The absolute values may vary considerably between strains. However, the relative values are quite constant and very similar to values obtained with *S. typhimurium* LT2 *thy* strains. Calculations of the intracellular concentrations are based on the assumption that the internal volume of a cell grown in minimal medium is 1.5 μm^3.

[b] From Filpula and Fuchs (56, 57), using *E. coli* K12 strains.

[c] Calculated from *in vitro* data obtained with purified dCTP deaminase of *S. typhimurium* (Fig. 4.5; ref. 12). The values given represent change in dCTP deaminase activity solely due to changes in the concentration of the substrate dCTP. The effect of changes in the negative effector dTTP has not been considered. However, they would only increase the differences shown.

[d] Thymine starvation for 75 min.

[e] nd = not determined.

the simultaneous addition of thymine and a deoxyribonucleoside (9, 94, 99, 132).

As discussed previously (p. 157), dUMP is synthesized via two different routes, one of which involves deamination of dCTP. It is therefore not surprising that the contribution from this pathway to dUMP synthesis varies with variations in the internal dCTP concentration. When a *thy* mutant of *S. typhimurium* is grown with 0.8 mM thymine in the medium, it contains a normal dCTP pool. Under these conditions 70% of its dUMP is synthesized via dCTP deamination and 30% via UDP reduction. However, when the same strain is grown in the presence of 0.16 mM thymine, where the dCTP pool is 5–10 fold above normal, 97% of the dUMP is derived from dCTP (141).

Variations in the concentration of exogenously supplied thymine also affect DNA replication in thymine-requiring strains. As shown by Pritchard and Zaritsky, the chromosomal replication time of *E. coli thy* mutants increases progressively as the thymine concentration in the medium is decreased. At thymine concentrations above the minimal concentration compatible with balanced growth these changes are not associated with significant changes in growth rate. In order to compensate for the increase in replication time the cells increase the frequency by which new replication forks are initiated (159, 210). Although it seems likely that the lowered dTTP pool is rate limiting for DNA replication fork movement during growth in low levels of thymine, the possibility remains that limitation for dGTP may also be involved (Fig. 4.7; 151).

Complete starvation of a thymine auxotroph for thymine results in cessation of DNA synthesis due to rapid depletion of the internal dTTP pool, the half-life of which is 1–2 min (138). This causes unbalanced growth, which eventually leads to loss of viability, "thymine-less death". Depletion of the dTTP pool is accompanied by significant changes in the deoxyribonucleotide metabolism of the cells (Table 4.2) as revealed by: (a) intracellular accumulation of large amounts of dCTP following a lag of about 15 min (140), (b) a 6-fold increase in the rate of synthesis of ribonucleotide reductase (56), and (c) excretion of large quantities of deoxyribose which can be detected in the medium provided the strain is blocked in deoxyribose 1-phosphate catabolism, i.e. is low thymine-requiring (20).

More detailed studies of the kinetics of dCTP accumulation during thymine starvation of an *E. coli thy* mutant showed that the expansion of the dCTP pool is accompanied by an increase in its *turnover*. After 75 min of thymine starvation, the dCTP pool has increased 10–20 fold. At this time the rate of synthesis and the rate of utilization of dCTP have increased 6.4 and 5.9 fold, respectively (140). The increased rate of synthesis may be explained by the observed increase in the ribonucleotide reductase level. Since no DNA synthesis occurs under these conditions, the high rate of utilization of dCTP must indicate that thymine starvation induces dCTP catabolism. In fact the rate by which deoxyribose is excreted during thymine starvation of low thymine-requiring *E. coli* strains may account for the total turnover of dCTP (140).

In an attempt to explain the apparent connection between dTTP, dCTP, and dUMP metabolism in thymine-requiring strains under steady state growth conditions (cf. Table 4.2), the following model is proposed. When the internal dTTP concentration falls below a certain level, which is probably not much below the level found in thy^+ strains, it becomes rate limiting for DNA replication thereby inducing new replication forks on the chromosome. According to Filpula and Fuchs any condition, which favours induction of new replication forks, promotes an increase in the rate of ribonucleotide reductase synthesis (57). Thus, as a result of the low dTTP pool the rate of CDP reduction will increase and dCTP will accumulate. Since the dCTP concentration in thy^+ strains (50–100 μM) is considerably below the half-saturating concentration of dCTP deaminase (Fig. 4.5), any increase in the dCTP pool will result in a nearly proportional increase in the rate by which it is deaminated. The dUTP produced is rapidly hydrolyzed to dUMP by dUTPase, and the dUMP is subsequently catabolized to uracil and deoxyribose 1-phosphate.

The decrease in the steady state levels of dGTP observed during growth in the presence of suboptimal levels of thymine (Fig. 4.7) may be a consequence of the decreased dTTP pools, since GDP reduction catalyzed by ribonucleotide reductase is specifically stimulated by dTTP (Table 4.1).

Mutants lacking dCTP deaminase (dcd). Enteric bacteria have the potential for synthesizing dUMP and thus thymidine nucleotides either from dCTP via deamination or directly from UDP by reduction. By differential labeling of the cytidine and uridine nucleotide pools of cytidine deaminase negative strains grown in the presence of both cytidine and uracil, it was shown that 75% of the cellular dTTP is derived from cytidine and 25% from uracil (141). Since these results were obtained with *cdd* mutants growing in the presence of cytidine, it was desirable to obtain independent evidence for the physiological importance of dCTP deamination in dUMP synthesis. Such evidence came from studies of *E. coli* and *S. typhimurium* mutants defective in dCTP deaminase activity (*dcd*). The following phenotypic traits are associated with the *dcd* mutation (13, 142, 144):

(1) Mutants deleted for the *dcd* gene are thymine prototrophic, showing that UDP-reduction can take over the entire production of dUMP for dTTP synthesis. This was confirmed by differential labeling of the cytidine and uridine nucleotide pools in a double mutant lacking both dCTP deaminase and cytidine deaminase.

(2) The intracellular dTTP concentration in mutants growing exponen-

tially in minimal medium is only 25–50% of that found in a *dcd*$^+$ strain. At the same time the dCTP pool is greatly enlarged, being 10–30 fold higher than in the parental strain (Table 4.2). Thus the introduction of a *dcd* mutation does interfere with dTTP synthesis.

(3) Exogenous thymidine or deoxyuridine, but not thymine, normalizes both the dTTP and the dCTP pools. The fact that deoxyuridine promotes normal pools indicates that the low dTTP pool in the mutant is due to shortage in the supply of dUMP.

(4) The mutants show an increased sensitivity towards 5-bromodeoxyuridine (66). This is probably due to less efficient competition by the low amounts of dTTP with 5-bromodUTP for incorporation into DNA.

These observations agree with the model proposed to explain the regulatory function of dTTP in dUMP synthesis in thymine requiring strains (p. 165). According to the model a mutant, which is entirely dependent on UDP-reduction for dUMP and thereby dTTP synthesis, should contain derepressed levels of ribonucleotide reductase. Under steady state growth conditions this may be accomplished by keeping the dTTP pool so low that the chromosomal replication time would be affected. As a result of increased ribonucleotide reductase activity the rate of dCTP synthesis will also be elevated. Thus the steady state level of dCTP in a mutant lacking dCTP deaminase would be expected to be considerably higher than the dCTP level in a thymine-requiring mutant grown under conditions where the dTTP pool is depressed to the same extent as in the *dcd* mutant. A comparison of exp. 2 and 6 of Table 4.2 shows that this is indeed observed.

Mutants defective in dUTPase (dut). With the discovery of dCTP deaminase and its physiological function it became apparent that dUTP is an intermediate in dTTP synthesis from cytidine nucleotides. Whether dUTP is also an intermediate in the UDP reduction pathway, i.e. in the conversion of dUDP to dUMP, is at present not known.

The conversion of dUTP to dUMP is catalyzed by dUTPase. Since DNA polymerase III of *E. coli* will use dUTP and dTTP as substrates with about equal efficiency (178), it is essential for the cell to keep the dUTP pool low relative to the dTTP pool to minimize uracil incorporation into DNA. At the same time the dUTP pool must be high enough to ensure a sufficient rate of dUMP formation via dUTPase.

Mutants of *E. coli* defective in dUTPase activity (*dut* mutants) have been isolated, but they are all leaky, i.e. *in vitro* they have from a few to fifty per cent residual dUTPase activity (86, 108, 191). After brief (^{3}H)-thymine pulses *dut* mutants accumulate *s*hort *O*kazaki *f*ragments (Sof phenotype), which are rapidly incorporated into high molecular weight DNA (190, 191). In addition *dut* mutants show an increased rate of recombination (Hyper–Rec) (191), and an increased rate of mutation (62).

Of particular interest for the present discussion is the fact that the *dut* mutants so far isolated are phenotypical Thy^+ and show no sign of being stressed in their ability to synthesize dTTP endogenously (Table 4.3). This may be explained by postulating that dUTPase is not obligatory for the conversion of dUTP to dUMP. Alternatively, the residual dUTPase activity of the mutants may suffice to satisfy the dUMP requirement of the cells. Measurements of the dUTP pools of different strains of *E. coli* seem to support this latter explanation. As shown in Table 4.3 the internal concentration of dUTP in a dut^+ strain is less than 0.3 μM, and thus significantly below the K_m-value of the *E. coli* dUTPase for dUTP (12 μM) (178). In a *dut* mutant (*dut-1*), which contains only a few percent residual dUTPase activity, the intracellular concentration of dUTP is increased to 9 μM (Table 4.3). Provided the K_m-value of the mutant dUTPase is not significantly different from that of the wild type enzyme, such a 30 fold increase in dUTP may ensure a normal flow of dUTP to dUMP catalyzed by the mutant enzyme.

E. coli contains a uracil-DNA N-glycosylase, which removes uracil moieties from DNA (112). The apyrimidinic sites created by this reaction are subsequently cleaved by an endonuclease (exonuclease III), and the resulting gaps repaired by a process which involves DNA polymerase I and DNA ligase (190, 202). Accordingly, mutants defective in uracil-DNA N-glycosylase, i.e. *ung* mutants, will retain any uracil incorporated into their DNA (47). By using *E. coli ung* mutants it has been shown that the DNA of an *ung,dut*$^+$ strain contains 1 uracil per 250–500 thymine residues, whereas the DNA of an *ung,dut* double mutant contains 1 uracil per 6–25 thymine residues (189, 200). Together with the pool measurements (Table 4.3) these results support the notion that the ratio between the dUTP and the dTTP concentrations in the cells determines the frequency of uracil incorporation into DNA. The

TABLE 4.3

Pool sizes of dUTP and dTTP in different mutants of *E. coli*.

	Pool sizes (n moles/g dry weight)		Intracellular conc.[a] (μM)		
Genotype	dUTP	dTTP	dUTP	dTTP	dUTP:dTTP
dut$^+$,*thy*$^+$	1[b]	121	0.3[b]	40	1:133[b]
dut-1,*thy*$^+$	26	274	8.7	91	1: 10
dut-1,*thy*$^-$	24	93	8.0	31	1: 4

[a] Calculated by assuming that 1 g bacterial dry weight corresponds to an intracellular volume of 3 ml.
[b] This is a maximal value due to the very low number of counts in the spot, i.e. two-fold above background.

transient appearance of short Okazaki fragments in a *ung*$^+$,*dut* strain (the Sof phenotype) is thus the result of excision of a large number of uracil moieties by uracil-DNA N-glycosylase. The increased number of breaks in the DNA may also explain the Hyper–Rec phenotype of the *dut* mutants as well as their increased mutation frequencies. Accordingly, *ung* mutations have been shown to suppress both the Sof as well as the Hyper–Rec phenotype of the *dut* mutants (190, 199, 200).

Introduction of the *dut-1* mutation into a thymine-requiring strain has no effect on the growth rate of the strain as measured turbidometrically, even though the dUTP:dTTP pool-ratio is 1:4 (Table 4.3, line 3) (Neuhard, unpublished exp.). This ratio implies that the *thy,dut-1* mutant should incorporate one uracil for every four thymine into DNA. Thus, the uracil excision-repair system is able to cope with at least this level of uracil misincorporation with no apparent effect on growth. However, it may be anticipated that such a strain will show a more pronounced Sof and Hyper–Rec phenotype, and will mutate with a higher frequency than the corresponding *thy*$^+$ strain.

Mutants lacking ribonucleoside diphosphate reductase (nrd). Addition of hydroxyurea to growing cultures of *E. coli* results in a rapid depletion of the intracellular pools of the four deoxyribonucleoside triphosphates (136). This agrees with the observation that purified ribonucleoside diphosphate reductase is inhibited by hydroxyurea (p. 152). Furthermore it suggests that only one enzyme, i.e. ribonucleoside diphosphate reductase, is responsible for the reduction of all four ribonucleotides to the corresponding 2′-deoxyribonucleotides *in vivo*. The definitive proof for this came with the isolation of *E. coli* mutants defective in either the B1 or the B2 subunit of ribonucleoside diphosphate reductase (*nrdA* and *nrdB*, respectively).

Enteric bacteria are devoid of deoxycytidine, deoxyadenosine, and deoxyguanosine kinase activity* implying that a tight *nrd* mutation is lethal for the cell. Thus, one approach was to look for *nrd* mutants among conditional *dna* mutants. This led to the finding that the *dnaF* gene is allelic with the *nrdA* gene which codes for the B1 subunit (67). A different approach was used by Fuchs and Neuhard (68), who based their selection on the observations (a) that the synthesis of dTTP in a *dcd* mutant is entirely dependent on UDP reduction (*see* p. 165), and (b) that purified ribonucleoside diphosphate reductase has a 10-fold lower K_m for UDP than for the other three nucleoside diphosphate substrates. Thus, in *dcd* strain a missense mutation in *nrd* which decreases the K_m-values of the enzyme towards its substrates would be expected primarily to interfere with UDP reduction. In a *dcd* strain such a

* Evidence for the existence of deoxycytidine kinase activity in certain strains of *E. coli* C has been presented (146).

mutation might create a growth requirement for deoxyuridine or thymidine. By using this approach, a mutant affecting the B2 protein of ribonucleoside diphosphate reductase (*nrdB*) was obtained. The *nrdB* mutant showed an increased sensitivity towards hydroxyurea (65).

Studies of a temperature sensitive *nrdA* mutant have shown that the deoxyribonucleoside triphosphate pools do *not* decrease significantly following a shift to the nonpermissive temperature (122). Thus the rapid cessation of DNA synthesis observed in this mutant following a shift in temperature to 42° is not caused by lack of precursors. In view of the growing body of evidence for the existence of DNA replication complexes involving deoxyribonucleotide synthesizing enzymes (40, 42, 119, 177, 193), the pool data may be interpreted in favour of the existence in *E. coli* of such a complex containing ribonucleoside diphosphate reductase as a component. The structural alterations occurring in the B1 protein of the temperature sensitive *nrdA* mutant at the nonpermissive temperature might inactivate the replication complex without abolishing completely the catalytic capacity of ribonucleoside diphosphate reductase.

B. *Bacillus*

The main features of endogenous dTTP synthesis in *B. subtilis* are depicted in Fig. 4.8, which also indicates some of the existing uncertainties in our present knowledge of the pathways involved. It should be noted that in contrast to *E. coli* and *S. typhimurium*, *Bacillus* has the capacity to phosphorylate deoxycytidine, deoxyadenosine, and deoxyguanosine to the corresponding 5′-monophosphates (195, 196).

1. Ribonucleotide reductase

The enzyme responsible for the reduction of ribonucleotides to deoxyribonucleotides in *B. subtilis* has not yet been characterized. Some evidence has been obtained indicating that uridine and cytidine nucleotide reduction occurs at the triphosphate level in a reaction which is insensitive to hydroxyurea and *un*affected by the addition of cobalamin (161). Since these results were obtained with crude extracts of *B. subtilis*, further characterization of this activity is required before any conclusion concerning the nature of the reaction can be drawn. Yau and Wachsman (208) have reported the presence of a hydroxyurea insensitive, but coenzyme B_{12} stimulated ribonucleotide reductase in crude extracts of *B. megaterium*.

Addition of nalidixic acid in concentrations which inhibit DNA synthesis, causes derepression of ribonucleotide reductase synthesis in *B. subtilis*, in a manner very similar to that found in *E. coli* (58).

The presence of deoxyribonucleoside kinases in *B. subtilis* makes it possible

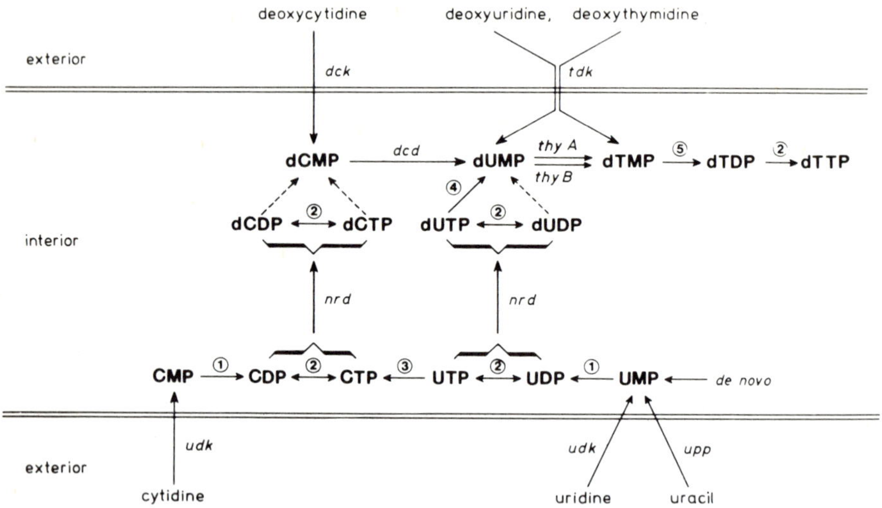

FIG. 4.8. Pyrimidine nucleotide interconversion in *B. subtilis*. The enzymes involved are identified by numbers or by their gene symbols, as follows: 1, UMP(CMP) kinase (197); 2, nucleoside diphosphokinase (173); 3, CTP synthetase; 4, dUTPase; 5, dTMP kinase; *dcd*, dCMP deaminase; *dck*, deoxycytidine kinase; *nrd*, ribonucleotide reductase (it is not known whether ribonucleoside di- or triphosphates are substrates); *tdk*, thymidine kinase; *thyA* and *B*, thymidylate synthase A and B; *udk*, uridine(cytidine)kinase; *upp*, uracil phosphoribosyltransferase. Broken arrows indicate hypothetical reactions.

to select mutants defective in endogenous synthesis of deoxyribonucleotides. *B. subtilis* mutants requiring deoxyribonucleosides for growth have been isolated by several groups (8, 161), but the mutants have not been characterized enzymatically or genetically. Thus, it is not known whether the mutants isolated by the different groups are allelic. Neuhard and Stauning (unpublished results) isolated deoxyribonucleoside-requiring mutants in a strain lacking deoxycytidine/cytidine deaminase and dCMP deaminase. The following phenotypic traits of these mutants indicate that they are deficient in ribonucleotide reduction: (a) They have an absolute growth-requirement for each of the four deoxyribonucleosides: deoxyadenosine, deoxyguanosine, deoxycytidine, *and* thymidine. The latter may be replaced by deoxyuridine. (b) Removal of any of the four deoxyribonucleosides from growing cultures of the mutants results in immediate cessation of DNA synthesis, while cell-mass continues to increase for several hours. After a lag of about 1 hour, the viability of the culture starts to decrease exponentially in a manner very similar to what is observed during "thymine—less death". (c) Starvation for any of the deoxyribonucleosides results in the specific depletion of the corresponding deoxyribonucleoside triphosphate from the acid-soluble pools of the mutants.

2. *dUMP synthesis*

Differential labeling of the cytosine and uracil nucleotide pools have shown that dUMP is synthesized by two different routes also in *B. subtilis*: one is via reduction of a uridine nucleotide, and the second via deamination of a deoxycytidine nucleotide. Under the experimental conditions employed, each of the pathways contributes about 50% of the total flow (131). In contrast to enteric bacteria, the latter pathway involves deamination of dCMP. A dCMP deaminase from *B. subtilis* has been highly purified (MW 170 000), and it was found to be an allosteric enzyme, which requires dCTP and Zn^{2+} for activity. In the presence of saturating concentrations of dCTP and Zn^{2+} the dCMP saturation curve is hyperbolic. The dCTP saturation curve is sigmoidal, the sigmoidicity being dependent on the Zn^{2+} and dCMP concentrations. In the presence of saturating concentrations of dCMP and Zn^{2+} the half saturation constant for dCTP is 1.4 μM (131). dTTP is a negative effector for the enzyme, the degree of inhibition observed being dependent on the concentration of dCTP.

The sedimentation constant $S^{\circ}_{20,w}$ of the enzyme is 6.5 S whether determined in the absence of effectors or in the presence of either dCTP, dTTP, or Zn^{2+} added separately. However, in the presence of both dCTP (or dTTP) *and* Zn^{2+}, the sedimentation constant is 6.9 S. This modest increase suggests that Zn^{2+} is required for the binding of the effectors to dCMP deaminase, and that this binding promotes a conformational change in the enzyme (Møllgaard, unpublished results).

B. subtilis mutants lacking dCMP deaminase (*dcd*) have been obtained. They have the following phenotypes (131): (a) In minimal medium the growth rates of the mutants are identical to that of the parent. (b) They contain 2–3 fold increased dCTP and dCMP pools, and significantly lowered dUMP, dTMP, and dTTP pools. (c) Differential labeling of the pyrimidine nucleotide pools reveals that dTTP is synthesized exclusively via uridine nucleotide reduction in *dcd* mutants. This indicates that dCMP deaminase is the only deoxycytidine nucleotide deaminase in *B. subtilis*, and that it is involved in dUMP synthesis *in vivo*. In addition these results show that uridine nucleotide reduction alone is capable of supplying sufficient dUMP for normal dTTP synthesis.

The enzyme responsible for dCMP synthesis from the primary product of cytidine nucleotide reduction is at present unknown, as is the final step in dUMP synthesis via uridine nucleotide reduction. The latter conversion may, however, involve the action of dUTPase, an enzyme which has been identified in *B. subtilis* (158).

3. *Thymidylate synthase*

B. subtilis appears to be unique in that it contains two thymidylate synthases, TSase A and TSase B. The enzymes have been separated by DEAE-cellulose chromatography, and both were shown to catalyze the reductive transfer of the methylene group of 5,10-methylenetetrahydrofolate to dUMP. As judged from *in vitro* experiments 90–95% of the total thymidylate synthase activity of wild-type *B. subtilis* is accounted for by TSase A. TSase B is more heat labile *in vitro* and has a 5-fold higher K_m for dUMP (20 μM) than TSase A (4 μM). Both enzymes are inhibited by the product dTMP and by 5-fluoro-dUMP. However, TSase A is about 10-fold more sensitive towards these inhibitors than TSase B (139). The molecular weight of both enzymes is about 60 000, a value close to what has been reported for thymidylate synthases from other organisms. TSase A consists of two subunits of similar size (Neuhard, unpublished experiments).

Two mutations in the unlinked genes *thyA* and *thyB* are required to create a thymine-requirement at 37° in *B. subtilis* (204, 209). At 46°, however, a mutation in *thyA* is sufficient to make the cells thymine-auxotrophic (Table 4.4), indicating that the *thyB* gene-product is either not formed or is inactivated at the high temperature (139). By subjecting extracts of wild type cells grown at 46°, as well as extracts of the two single mutants *thyA*,*thyB*$^+$ and *thyA*$^+$,*thyB* to DEAE cellulose chromatography, the *thyA* and *thyB* gene-products were identified as TSase A and TSase B, respectively.

The phenotypes of strains carrying the *thyA* and *thyB* mutations in different combinations are summarized in Table 4.4. As shown, a *thyA*$^+$,*thyB* mutant is phenotypically identical to the wild-type. The only means of identifying the *thyB*-genotype in a *thyA*$^+$ background is by chromatography of crude extracts and screening for the presence of two thymidylate synthases. In contrast, a *thyA*,*thyB*$^+$ mutant has a distinct phenotype: it shows a temperature-sensitive thymine-requirement and is partially resistant to trimethoprim in the presence of thymine at the permissive temperature (Table 4.4). The *thyA*,*thyB*$^+$ strains are further characterized by their high internal concentrations of dUMP and slightly decreased dTTP pools, when they are grown at 37° (Table 4.4). The high dUMP pool, corresponding to an internal concentration of about 30 μM as compared to less than 3 μM in the wild-type, may explain why a *thyA*,*thyB*$^+$ strain grows normally without added thymine at 37°C: the dUMP concentrations in a *thyA*$^+$ strain is so low that TSase B is probably not functioning, and TSase A is functioning below its maximal velocity. In the *thyA*,*thyB*$^+$ mutant, however, the dUMP concentration is well above the K_m value for TSase B and may therefore ensure an adequate supply of dTMP in such mutants.

It remains at present unexplained why TSase B, which seems to have no major function in dTMP synthesis in wild-type *B. subtilis* and which is

TABLE 4.4

Phenotypes of *B. subtilis thy* mutants[a].

Genotype	Thymine phenotype		Resistance to trimethoprim at 37°C[b]			Thymidylate synthase (nmol/min/mg)	Pools at 37°C[c] (nmol/g dry weight)	
	37°C	46°C	0.5 μg/ml	3 μg/ml	10 μg/ml		dUMP	dTTP
thyA⁺,*thyB*⁺	Thy⁺	Thy⁺	–	–	–	0.43	>10	220
thyA⁺,*thyB*	Thy⁺	Thy⁺	–	–	–	0.51	nd[d]	180
thyA,*thyB*⁺	Thy⁺	Thy⁻	++	+	–	0.02	90	150
thyA,*thyB*	Thy⁻	Thy⁻	++	++	++	0.005	460	40

[a] Taken in part from (139).
[b] – no growth, + weak growth, and ++ full growth after 24 hrs on solid medium containing 50 μg thymine per ml.
[c] Determined in cultures growing exponentially in a low-phosphate medium containing ^{32}P-orthophosphate. With the *thyA*,*thyB* mutant, the medium was supplied with 50 μg thymine per ml.
[d] Not determined.

temperature-sensitive *in vivo*, has been retained by the organism. It should be emphasized, however, that all the *in vitro* informations on the two thymidylate synthases have been obtained with 5,10-methylenetetrahydrofolate as substrate, although the physiological substrate probably is a derivative of one of the naturally occurring pteroylpolyglutamates (85).

4. Spores

Sporulation in *Bacillus* is initiated when vegetative cells cease exponential growth. During the first two hours of sporulation there is a postexponential doubling of cellular DNA content (111). Simultaneously the intracellular concentrations of all four deoxyribonucleoside triphosphates decrease, reaching 10–20% of their initial values two hours after initiation of sporulation (Neuhard, unpublished results). No further DNA synthesis is observed after this time. In the dormant spore the low levels of the deoxyribonucleoside triphosphates are maintained in accordance with the observation that ribonucleotide reductase activity is low in spores compared to vegetative cells. During spore germination DNA synthesis is correlated with the total deoxyribonucleoside triphosphate pool, which in turn seems to be dependent on the activity of ribonucleotide reductase (175).

C. *Lactobacillus*

Most *Lactobacillus* species have a growth requirement for a deoxyribonucleoside. If the medium is supplied with purines and pyrimidines, only one (anyone) deoxyribonucleoside is required. In *L. leichmannii* and *L. lactis* the deoxyribonucleoside requirement may be replaced by vitamin B_{12}. These observations indicate that *L. leichmannii* and *L. lactis* contain a coenzyme B_{12} dependent pathway for deoxyribonucleotide synthesis. Furthermore they imply that most *Lactobacillus* species contain pathways for scavenging deoxyribonucleosides. Accordingly it has been found, that lactobacilli contain deoxyribonucleoside kinase activities for all four deoxyribonucleosides (49, 157), as well as *trans*-N-deoxyribosylases catalyzing the transfer of deoxyribosyl groups between a variety of purine and pyrimidine bases (15) (cf. Chapter 3, p. 109).

1. Ribonucleotide reduction

The reduction of ribonucleotides in *L. leichmannii* is catalyzed by one single protein, which reduces all four naturally occurring ribonucleoside *tri*phosphates to the corresponding 2′-deoxyribonucleoside triphosphates with retention of the configuration at position 2′ (review: 186). The enzyme has no requirement for divalent metal ions, but has an absolute requirement for 5′-deoxy-5′-adenosylcobalamin (coenzyme B_{12}). Deoxyribonucleoside triphosphates are positive effectors, regulating both the specificity and the catalytical activity of the enzyme. *In vitro* the reducting equivalents can be donated by dithiols such as lipoic acid, dithiothreitol, and *E. coli* thioredoxin. The physiological donor is at present unknown, but a thioredoxin and a thioredoxin reductase have been isolated from *L. leichmannii.* They may constitute the hydrogen carriers between NADPH and the ribonucleoside triphosphate reductase *in vivo*.

The ribonucleoside triphosphate reductase from *L. leichmannii* is a monomer with a molecular weight of 76 000. In contrast to the *E. coli* ribonucleoside diphosphate reductase, it does not aggregate in the presence of magnesium ions or the effectors (29, 150). Binding studies have shown two nucleotide binding sites: the catalytic site, which is highly specific for the β-D-ribofuranose 5′-triphosphate moiety of the substrates, and an effector site, which binds both ribo- and deoxyribonucleoside triphosphates. The latter site is involved in the regulation of both substrate specificity and catalytic activity (29, 59, 117). *In vitro*, maximal stimulation of the reduction of GTP, CTP, ATP, and UTP is obtained by the addition of dTTP, dATP, dGTP, and dCTP, respectively. Indirect evidence indicates a second effector site specific for nucleoside triphosphates, which promotes coenzyme binding (59, 117). The binding of a nucleotide to one of the sites affects the affinities of the other sites giving a very complex regulatory pattern, which in addition depends on the presence or absence of magnesium ions. In contrast to what was found for the *E. coli* ribonucleotide reductase, no negative effector has been found for the *L. leichmannii* enzyme, and it is not inhibited by hydroxyurea.

The requirement for 5′-deoxy-5′-adenosylcobalamin is absolute, and the binding of the coenzyme to the enzyme requires the presence of both a nucleoside triphosphate and a hydrogen donor. The coenzyme is converted to a cobalamin- and a deoxyadenosyl-radical, and both participate actively in the catalytic process together with mercapto-groups on the enzyme (18, 106, 169, 182, 207).

L. leichmannii is not able to synthesize thymine nucleotides unless folate or thymine is added to the growth medium. If thymine is removed from cells growing without folate, the specific activity of ribonucleotide reductase increases ten-fold (14). A similar increase in specific activity is found in *L. casei* following inhibition of DNA synthesis by mitomycin C (58). Both of these observations indicate that ribonucleotide reductase synthesis in *Lactobacillus* is regulated in a manner similar to what has been found in *E. coli* (p. 154).

2. *dUMP synthesis*

L. acidophilus contains a dCMP deaminase which is activated by dCTP and feedback inhibited by dTTP (174). This suggests that dUMP is partially derived from a deoxycytidine nucleotide also in *Lactobacillus*. In addition dUMP may be synthesized directly from a uridine nucleotide by reduction of UTP to dUTP catalyzed by ribonucleoside triphosphate reductase. This latter pathway requires the participation of a dUTPase, which has not yet been identified in *Lactobacillus*.

3. Thymidylate synthase

Mutants of *L. casei* have been isolated, which contain 100-fold elevated levels of thymidylate synthase (48). From such mutants large quantities of pure thymidylate synthase are easily obtained, facilitating studies of its structure and reaction mechanism (Review: 60, 156). The potential use of specific inhibitors of thymidylate synthase in cancer therapy has further stimulated research on this enzyme (33).

Thymidylate synthase from *L. casei* has a molecular weight of 70 000 and consists of two identical subunits with known primary structure (120). The enzyme contains two binding sites for dUMP, one with high and one with low affinity (11, 32, 115). The low affinity site is probably not accessible initially, but becomes available for dUMP binding when the high affinity site is filled with dUMP and cofactor (32).

Binding experiments as well as kinetic studies suggest that the reaction is ordered and sequential (34, 37, 70, 72). The reaction is initiated by the binding of dUMP to the enzyme. Arginyl groups on the enzyme seem to participate in the binding through a strong electrostatic interaction with the phosphate moiety of the nucleotide (30). After binding of 5,10-methylentetrahydrofolate catalysis occurs, and finally dihydrofolate and dTMP are released in this order. Normal Michaelis–Menten kinetics are observed for both substrates, and the K_m-values for dUMP and 5,10-methylenetetrahydrofolate are 0,7 μM and 14 μM, respectively.

Nonenzymatic model studies as well as direct studies of the enzymatic reaction have led to a model for the catalytic mechanism of thymidylate synthase, which is shown in Fig. 4.9 (35, 155, 156). The reaction is initiated by a nucleophilic attack on the 6-position of the pyrimidine ring by the mercapto group of a cysteinyl residue in the catalytic site of the enzyme (16, 35, 36, 134, 201). Subsequently the methylene group of 5,10-methylenetetrahydrofolate binds to the 5-position of the pyrimidine ring giving a ternary complex, where the enzyme and methylenetetrahydrofolate are bound *trans* to each other across the 5,6-bond of the pyrimidine (98). After splitting of the covalent bond between tetrahydrofolate and the nucleotide, the reducing equivalents are transferred from the tetrahydrofolate to the pyrimidine base giving dihydrofolate and dTMP as products.

5-Fluoro-2′-deoxyuridylate (FdUMP) is an extremely potent inhibitor of thymidylate synthase. In the presence of 5,10-methylenetetrahydrofolate, thymidylate synthase binds two equivalents of FdUMP reversibly, forming a very stable, covalent, ternary complex with an association constant of 10^{13}M (16, 34, 36, 45, 70, 72, 109). Due to the fluoride atom in the 5-position, the catalytic sequence shown in Fig. 4.9 cannot continue beyond the ternary complex.

In most investigations on thymidylate synthase, 5,10-methylene-tetrahydrofolate is used as a substrate. However, a number of studies indicate that the physiological substrate is a polyglutamate derivative, probably the tetraglutamate, which is the most abundant folate compound in *L. casei* (7, 70, 71, 107).

D. General discussion

The studies on endogenous dTTP synthesis discussed above for the gram-negative bacteria. *E. coli* and *S. typhimurium*, and the gram-positive bacteria, *B. subtilis* and *Lactobacillus*, as well as less detailed studies on other microorganisms, allow some generalizations to be made concerning endogenous dTTP synthesis in microorganisms.

Deoxyribonucleotides are synthesized exclusively by reduction of the corresponding ribonucleotides. The reduction is catalyzed by a single enzyme, which is unspecific towards the base moiety, but highly specific towards the number of phosphates on the ribonucleotide substrates. The enzymes are allosterically regulated by nucleoside triphosphates, which regulate not only the rate of reduction, but also the specificity of the enzyme. In addition, the synthesis of the enzymes appears to be coupled to DNA-replication, since

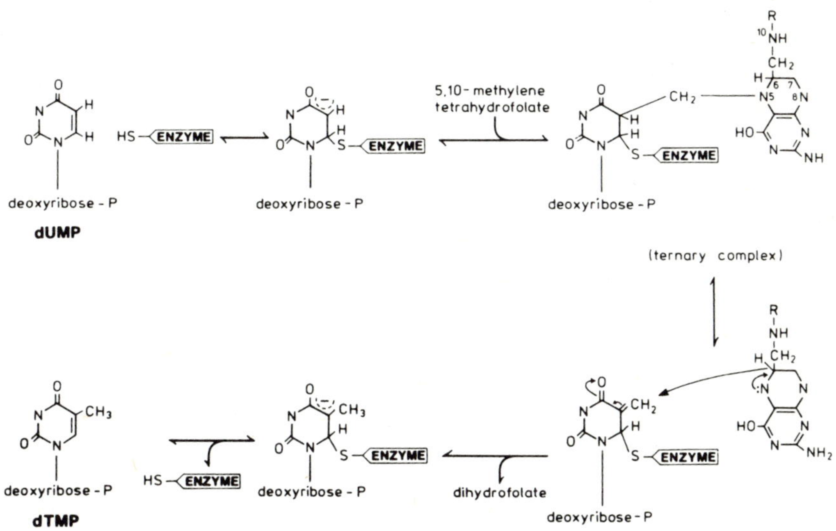

FIG. 4.9. Catalytic sequence for the thymidylate synthase reaction.

inhibition of DNA synthesis by different means results in increased ribonucleotide reductase levels (55, 58).

Two types of ribonucleotide reductases have been found. One, represented by the *E. coli* enzyme, uses ribonucleoside diphosphates as substrates, contains iron, and is inhibited by hydroxurea. The second type, represented by the enzyme from *L. leichmannii*, uses ribonucleoside triphosphates as substrates and requires coenzyme B_{12} for activity. The latter type does not contain iron and is not inhibited by hydroxyurea. A large number of microorganisms have been screened for the two types of ribonucleotide reductases, and no general pattern for their distribution has been found. Thus, among the green algae and the fungi, as well as among different procaryotic groups, both types of ribonucleotide reductases are found (31, 54, 55, 76, 77, 81, 116, 181). However, also other types of ribonucleotide reductases exist in microorganism as shown by the recent characterization of a deoxyadenosylcobalamin-dependent ribonucleoside *di*phosphate reductase from *Corynebacterium nephridii* (187).

Deoxyuridylate is synthesized from both deoxycytidine and deoxyuridine nucleotides. The pathway by which dUDP or dUTP is converted to dUMP, has not been established with certainty, but most likely a dUTPase participates. Although the presence of dUTPase has been established only in few microorganisms, it is probably ubiquitously distributed, since it is required for the exclusion of dUTP from the DNA precursor pool (p. 000). The second route for dUMP synthesis involves deoxycytidine nucleotide deamination catalyzed either by a dCTP deaminase or a dCMP deaminase. Both types of deaminases are allosterically regulated in accordance with their physiological function in thymidine nucleotide synthesis, i.e. they are regulated positively and negatively by dCTP and dTTP, respectively. So far dCTP deaminases have only been found in the gram-negative bacteria, *E. coli* and *S. typhimurium*, whereas dCMP deaminases are found in gram-positive bacteria such as *B. subtilis*, *L. acidophilus*, and *Staphylococcus aureus* (46) in the fungus *Neurospora crassa* (163), and in the protozoan parasites of the genus *Plasmodium* (176).

Thymidylate synthase catalyzes the conversion of dUMP to dTMP. This is the only known reaction, in which a methyl group is introduced into the 5-position of pyrimidine deoxyribonucleotides. Thymidylate synthase has been identified in all microorganisms examined (17, 83, 103, 121, 126, 137, 163) with the exception of certain members of the *Mycoplasma* group (130). *B. subtilis* is unique in that it contains two thymidylate synthases (p. 172). It is at present not known, if other *Bacillus* species are similar in that respect. All known thymidylate synthases are strongly inhibited by FdUMP, suggesting similar reaction mechanisms for all of them. Except for a weak product inhibition by dTMP, it appears that neither the activity nor the synthesis of the enzyme is regulated in procaryotes. In contrast, it has been reported that the synthesis of thymidylate synthase in *Tetrahymena pyriformis* is induced under certain growth conditions, and that this induction is mediated through cyclic GMP (41). Whether this will turn out to be of more general occurrence among eucaryotic microorganisms remains to be shown.

III. SYNTHESIS OF THYMIDINE NUCLEOTIDES FROM EXOGENOUS THYMINE AND THYMIDINE

Radioactive labeling procedures with either thymine or thymidine as the labeled precursors are crucial to the study of DNA synthesis and its regulation. It is essential to realize, however, that many factors, genetic as well as physiological, influence both the rate and the extent of thymine and thymidine utilization in microorganisms. In the following we will summarize our present knowledge of the mechanisms that control the conversion of exogenous thymine and thymidine to dTMP. Finally we will draw some conclusions concerning the application of radioactive thymine and thymidine in studies of DNA synthesis.

A. *Escherichia coli* and *Salmonella typhimurium*

1. Conversion of thymine to dTMP

The mechanism by which exogenous thymine is transported into the cytoplasma is at present unknown (cf. Chapter 6). Once inside the cell, thymine is converted to dTMP in two steps. First thymine reacts with deoxyribose 1-phosphate to form thymidine and inorganic phosphate, and subsequently thymidine is phosphorylated to dTMP (Fig. 4.10). No other pathways exist, since mutants blocked in either one of these steps are unable to convert thymine to dTMP. Although the properties of the enzymes and their genetic regulation are discussed in detail elsewhere in this book (Chapters 3 and 5), a few characteristics relevant to the present discussion will be presented below.

Thymidine phosphorylase. The enzyme is specific for deoxyribose 1-phosphate and will use uracil or thymine as substrates. It is inhibited by a number of ribo- and deoxyribonucleosides and by ribose-1-phosphate (19, 172, 205).

The expression of the thymidine phosphorylase gene *deoA* is controlled by at least two regulatory genes: the *cytR* and *deoR* genes. Deoxyribose 5-phosphate induces the synthesis of the enzyme by binding to the *deoR* gene product, whereas cytidine, and in the case of *S. typhimurium* also uridine, induce by binding to the *cytR* gene product. The latter induction is also dependent on cAMP and cAMP-binding protein (CRP).

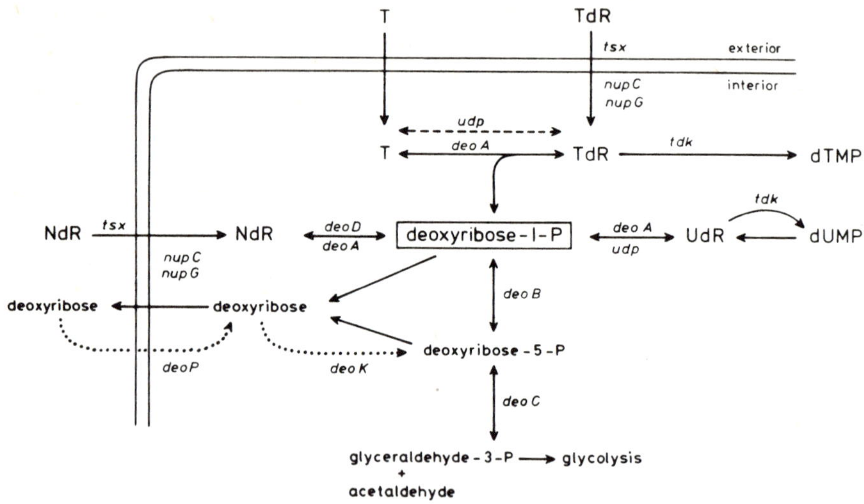

FIG. 4.10. Metabolism of thymine, thymidine and deoxyribose 1-phosphate in *E. coli* and *S. typhimurium*. N may be uracil, adenine, hypoxanthine or guanine. ----, route of minor importance; ·····, route exclusive for *S. typhimurium*. Enzymes or proteins are identified by their gene-symbols as follows: *deoA*, thymidine phosphorylase; *deoB*, phosphodeoxyribomutase; *deoC*, deoxyriboaldolase; *deoD*, purine nucleoside phosphorylase; *deoK*, deoxyribokinase; *deoP*, deoxyribose permease; *nupC*, nucleoside uptake; *nupG*, nucleoside uptake; *tdk*, thymidine kinase; *tsx*, phage T6 receptor and nucleoside uptake; *udp*, uridine phosphorylase.

Uridine phosphorylase. Purified uridine phosphorylase from *E. coli* shows weak activity towards deoxyuridine (6%) and thymidine (2%) (110). The synthesis of the enzyme is negatively controlled by the *cytR* repressor and positively by the cAMP–CRP system. Thus, like the enzymes encoded by the *deo*-operon, uridine phosphorylase is induced by cytidine, and in *S. typhimurium* also by uridine. When present at high levels uridine phosphorylase may contribute significantly to the thymidine phosphorylase activity of the cells (p. 182).

Thymidine kinase. Thymidine kinase catalyzes the phosphorylation of thymidine and deoxyuridine to the corresponding monophosphates. Also deoxyuridine derivatives halogenated on the 5-carbon of the pyrimidine ring may act as acceptors. The reaction is dependent on divalent metal ions. The activity of the enzyme is allosterically regulated (96, 97, 147, 148) i.e. it is inhibited by dTTP, and activated by dATP and dCTP as well as by the corresponding diphosphates.

Deoxyribose 1-phosphate catabolism. Deoxyribose 1-phosphate, the second substrate of the thymidine phosphorylase reaction, is rapidly catabolized by the sequential action of phosphodeoxyribomutase and deoxyriboaldolase (Fig. 4.10). The structural genes for the two enzymes, *deoB* and *deoC* together with the *deoA* and *deoD* genes encoding thymidine phosphorylase and purine nucleoside phosphorylase, constitute the *deo*-operon (Chapter 5). The expression of these four genes is controlled by the *deoR* and *cytR* coded repressor proteins.

All commonly used *E. coli* strains are devoid of deoxyribokinase activity, and hence cannot use deoxyribose as the sole carbon and energy source. In contrast, *S. typhimurium* contains a deoxyribokinase as well as a deoxyribose transport system, both of which are induced by deoxyribose (87).

2. *Utilization of thymine for DNA synthesis*

Thymine prototrophs. Thymine prototrophic strains of *E. coli* and *S. typhimurium* do not incorporate exogenous thymine into DNA, even though they contain the necessary enzymes. However, simultaneous addition of a deoxyribonucleoside results in extensive incorporation, provided the cells can degrade the nucleosides phosphorolytically. Incorporation continues as long as there is deoxyribonucleoside left in the medium. This suggests that the endogenous pool of deoxyribose 1-phosphate is the limiting compound in thymine utilization; a suggestion which has been supported by direct measurements of the internal deoxyribose 1-phosphate pools under different conditions (99).

E. coli double mutants, which contain a mutation that prevents deoxyribose 1-phosphate catabolism (*deoB*) and a mutation which results in constitutive synthesis of thymidine phosphorylase (*deoR*), will incorporate some thymine into DNA, corresponding to 10% of the incorporation observed in a thymine auxotrophic strain. Mutants containing only one of these mutations do not show any incorporation (99).

Thymine auxotrophs. Mutants of *E. coli* and *S. typhimurium* defective in thymidylate synthase, have acquired the ability to utilize exogenous thymine, although not very effectively. The thymine concentration required to support normal growth of *thy* mutants is about 150 μM. From such *high thymine-requiring* strains secondary mutants capable of growing on 15 μM thymine, i.e. *low thymine-requirers*, are easily obtained. The mutations responsible for the low thymine requirement are in either of the genes *deoB* or *deoC*, both of which code for enzymes involved in deoxyribose 1-phosphate catabolism (Fig. 4.10). Therefore, these mutants will have a higher endogenous pool of deoxyribose 1-phosphate, which allows a more efficient utilization of thymine for thymidine synthesis (99). This implies that the thymine requirement of a *thy* strain should increase under conditions, where deoxyribose 1-phosphate catabolism is stimulated. In fact double mutants of the genotypes *thy*, *deoR* or *thy*, *cytR*, in which the deoxyribose 1-phosphate catabolizing enzymes are synthesized constitutively, have become *super-high thymine-requiring*, and will only grow when supplied with more than 0.3 mM thymine (1).

The size of the deoxyribose 1-phosphate pool is not the only factor that determines the efficiency of thymine utilization in *thy* strains. Also the level of thymidine phosphorylase is of importance. When thymine requiring strains are plated on media containing 5–10 μM thymine, *super-low thymine requirers* are selected with a high frequency. They turn out to be mutated in *deoB* and to contain an additional mutation in either the *deoR* or the *cytR* gene, resulting in high levels of thymidine phosphorylase. In fact, upon detailed analyses many strains, which are claimed to be low thymine-requirers, appear to be super-low thymine-requirers (1, 133).

A *thy* mutant which in addition is deficient in thymidine phosphorylase (*deoA*) requires *thymidine* for growth. However, very high concentrations of thymine (0.5 mM) will support the growth of such mutants due to the weak thymidine phosphorylase activity exhibited by uridine phosphorylase (p. 181). Accordingly, mutations in either the *cytR* gene or in the operator of the *udp* gene which result in high levels of uridine phosphorylase will enable a *thy*, *deoA* mutant to grow on 150 μM thymine (25, 129).

Finally it should be emphasized that ribonucleosides are potent inhibitors of thymine utilization due to their inhibitory effect on thymidine phosphorylase either directly or via the ribose 1-phosphate produced by their phosphorolytic cleavage.

In Table 4.5 we have listed some of the mutations known to affect the thymine-requirement of *thy* strains.

Deoxyribose 1-phosphate synthesis. The introduction of a *thy* mutation does not lead to changes in the levels of the deoxyribose 1-phosphate catabolizing enzymes or of thymidine phosphorylase. Thus, the increase in the deoxyribose 1-phosphate pool found in *thy* mutants is caused by an increase in the rate of synthesis of deoxyribose 1-phosphate.

Beacham and Pritchard used thymine-requiring mutants lacking phosphorylases for thymidine, purine nucleosides, and uridine in various combinations to show that deoxyribose 1-phosphate is derived from deoxyuridine, and that both thymidine phosphorylase and uridine phosphorylase may participate in the phosphorylytic cleavage of this compound (10). This suggests that deoxyribose 1-phosphate arises through the breakdown of dUMP. This is supported by the observation that the endogenous pool of dUMP in thymine requiring strains is 50–100 fold higher than in Thy$^+$ strains (Table 4.2). Therefore, the following sequences of events seem to take place

TABLE 4.5

Thymine-requirement of different *thyA* mutants of *E. coli* and *S. typhimurium*.

Genotype	Thymine phenotype	Thymine conc. required for growth
thyA	high requirer	150 μM
thyA,deoB or *deoC*	low requirer	15 μM
thyA,cytR or *deoR*	super high requirer	300 μM
thyA,deoB,cytR or *deoR*	super low requirer	2–5 μM
thyA,deoA	super high requirer	>500 μM
thyA,deoA,udp	thymidine requirer	—
thyA,deoA,cytR	high requirer	150 μM
thyA,dut	thymidine requirer	—
thyA,dcd	thymidine requirer	—
thyA,deoC,deoB,nrdA$_{ts}$[a]	"high requirer"	>16 μM[b]

[a] The *nrdA* mutation confers temperature sensitivity to the ribonucleoside diphosphate reductase B1 subunit. The growth test was performed at 37°C where the protein is partially inactivated.
[b] Growth was only tested on 16 μM thymine. At this concentration the *nrdA*$^+$ parent grows well, while the mutant does not grow. With 16 μM thymine + deoxyadenosine both strains grow well.

following the introduction of a block in thymidylate synthase: The dTTP pool decreases and dUMP accumulates. As discussed previously (p. 165) a drop in the dTTP pool causes an increase in the rate of dUMP formation via the CDP-reduction pathway, which promotes the accumulation of dUMP. The resulting high internal concentration of dUMP will stimulate its own hydrolysis to deoxyuridine, which in turn gives rise to the deoxyribose 1-phosphate required for thymine utilization. This implies that the utilization of thymine for dTTP synthesis will largely be dependent on the size of the dUMP pool and hence on the rate of dUMP synthesis. With a given concentration of exogenous thymine the system will reach a steady state, where the size of the dTTP pool is determined by the availability of deoxyribose 1-phosphate, through regulation of dUMP synthesis. An increase in the concentration of exogenous thymine will cause an immediate swelling of the dTTP pool, which in turn will decrease the rate of deoxyribose 1-phosphate formation. As a result a new steady state will be reached in which the dTTP pool is somewhat higher than before. This is in accordance with the finding that the dTTP pool of exponentially growing *thy* strains increases with increasing concentration of thymine in the medium (p. 162 and Table 4.2).

Studies on mutants defective in dUMP synthesis have confirmed that dUMP is an intermediate in the synthesis of deoxyribose 1-phosphate by thymine-requiring strains. Thus, thymine-requiring strains with additional mutations in one of the genes encoding either ribonucleotide reductase (*nrd*), dCTP deaminase (*dcd*), or dUTPase (*dut*), require much higher concentrations of thymine for growth, than does the parental *thy* strains; in fact *thy,dut*, and *thy,dcd* mutants do not grow properly unless they are supplied with thymidine in the growth medium (Table 4.5) (13, 67; Neuhard, J., unpublished observations).

dUMPase. The enzyme responsible for the conversion of dUMP to deoxyuridine has not yet been identified. However, a dUMPase has been found both in *E. coli* (192) and *S. typhimurium* (Uerkwitz, personal communication), and the enzymes have been separated from the known periplasmic phosphatases and from dUTPase. The only nucleotides that are cleaved by the enzymes are dUMP, dTMP, and UMP. Whether these enzymes are involved in deoxyribose 1-phosphate synthesis from dUMP cannot be decided until mutants defective in this activity have been isolated and characterized.

3. *Utilization of thymidine for DNA synthesis*

The conversion of extracellular thymidine to dTMP is depicted in Fig. 4.10. In *E. coli* at least two transport systems are known, which may actively transport thymidine across the cytoplasmic membrane. Mutants in each of these systems have been isolated and characterized (*nupC* and *nupG*). In addition an outer membrane protein coded for by the *tsx* gene facilitates the passage of a variety of nucleosides including thymidine through the outer membrane (82). Thus, *tsx* mutants exhibit a low rate of thymidine incorporation at low thymidine concentrations (less than 3 μM) (127). The expression of the transport systems as well as the *tsx* gene are controlled by the same regulatory systems which control the nucleoside catabolizing enzymes (cf. Chapter 6).

Once thymidine is inside the cytoplasmic membrane, it may be phosphorylated to dTMP by thymidine kinase or it may be degraded to thymine and deoxyribose 1-phosphate by thymidine phosphorylase and, to a lesser degree, by uridine phosphorylase.

Thymidine is readily utilized for DNA synthesis by both Thy^+ and Thy^- strains. In Thy^+ strains, however, the incorporation ceases after a short time due to the rapid breakdown of thymidine to thymine. Since thymidine induces thymidine phosphorylase, no significant increase in the amount of thymidine incorporated into DNA is observed when the thymidine concentration in the medium is increased. Any condition, which favors the induction of thymidine phosphorylase or uridine phosphorylase, will decrease the amount of thymidine incorporated. In the extreme case, i.e. in *deoR* mutants which contain 40–50 fold increased levels of thymidine phosphorolytic activity, exogenous thymidine is not incorporated into DNA. Addition of uridine, a potent inhibitor of thymidine phosphorylase, promotes thymidine incorporation (205). However, as soon as the uridine is broken down by the cells, the incorporation ceases. Alternative ways of getting efficient thymidine incorporation over extended periods of time include: (a) to employ mutants defective in thymidine phosphorylase (52), and (b) to use dTMP as exogenous precursor (171). Utilization of dTMP requires prior dephosphorylation, which seems to be the rate-limiting step. Thus dTMP addition causes a slow feeding of thymidine which seems to favor phosphorylation over degradation. With labeled dTMP as precursor a Thy^+ strain derives 50% of the thymine residues in the DNA from exogenous dTMP, the rest being derived from endogenous synthesis. A similar contribution from exogenous thymidine is observed in mutants lacking thymidine phosphorylase. In both cases the decrease in endogenous thymidine nucleotide synthesis appears to be caused by a swelling of the dTTP pool, which in turn decreases the availability of dUMP for thymidylate synthase (26, 27).

4. Practical applications

At this stage it is appropriate to discuss briefly the implication of the observations presented in the previous sections for the use of radioactive thymine or thymidine as precursors for DNA-labeling. In this context it is important to realize that most thymine-requiring mutants, unless they have been kept on high concentrations of thymine throughout, have acquired secondary mutations in the *deoB* or *deoC* genes and possibly also in the *deoR* or *cytR* genes. Thus they have become either low or super-low thymine-requiring. When such mutants are made Thy^+ and used for further strain constructions, these additional mutations persist, but are in most cases unrecognized. A large number of "well established" strains, which at one time were thymine-requirers, contain mutations affecting the utilization of thymine or thymidine.

Long-term labeling of DNA. For the general labeling of DNA, the most reproducible results are obtained by employing thymine-requiring strains and labeled thymine as precursor. In that way the specific activity of the newly synthesized DNA is controlled by the specific activity of the precursor. By using mutants with a super-low thymine requirement (p. 182) one can economize with the labeled precursor. If a Thy^+ strain, for some reason, has to be used, the most extensive and reproducible labeling is obtained with dTMP as precursor or, provided the cells lack thymidine phosphorylase, with thymidine. Labeling with thymine in combination with a deoxyribonucleoside or with thymidine in combination with uridine may also be used, however, the amount of label going into DNA will be highly dependent on the nucleoside degrading capacity of the strains, and thereby on the presence of mutations, which affects this capacity, i.e. mutations in the *cytR*,*deoR*,*deoB*,*deoC*,*upd*, and *deoD* genes.

Independent of the labeling method employed, the specific activity of the newly synthesized DNA of a Thy^+ strain cannot be deduced from the specific activity of the added precursor due to ongoing endogenous dTTP synthesis.

Pulse labeling of DNA. Studies on the mechanism of DNA replication, recombination, and DNA repair necessitate experiments which measure short time incorporation of labeled precursors of high specific activities into nascent DNA. In such experiments the labeled precursor must be rapidly and efficiently converted to the deoxyribonucleoside triphosphate even when used in low concentrations. Therefore, the precursor of choice for such experiments is thymidine, which is actively concentrated by the nucleoside transport systems, and which is efficiently converted to dTTP. In certain strains it may cause swelling of the dTTP pool, thereby ensuring a more rapid incorporation of the label into DNA. Thymine does not offer these advantages, unless the strain to be labeled is a super-low thymine-requiring mutant (p. 182).

In the interpretation of thymidine pulse-labeling experiments it is of utmost importance to realize that artefacts may arise due to known or unknown genetic differences between bacterial strains. As pointed out by McKeown and coworkers (127) many strains used in studies of DNA metabolism carry mutations in the *tsx* gene. Since the *tsx* gene product is involved in the transport of thymidine through the outer membrane (82), such mutants show a severely reduced capacity for thymidine uptake at low concentrations.

Another very important factor affecting the incorporation of thymidine into DNA, is the intracellular level of thymidine phosphorylase. The presence of mutations which influence this level may give the impression of altered levels of DNA synthesis. This is particularly obvious in strains harboring mutations in the *deoR* or *cytR* genes.

Mutations in *deoC* gene, coding for deoxyriboaldolase, may also influence the level of thymidine phosphorylase, since such mutants are unable to degrade deoxyribose 5-phosphate, the endogenous inducer of thymidine phosphorylase. In a wild-type genetic background a *deoC* mutation does not cause increased thymidine phosphorylase levels. However, the presence of a *deoC* mutation in strains, which have an increased turnover of deoxyribonucleotides, causes accumulation of deoxyribose 5-phosphate and induction of thymidine phosphorylase (132). Many mutants used in studies of the mechanism of DNA replication contain mutations, which interfere with the velocity of DNA replication. Since such mutants may contain elevated levels of ribonucleotide reductase (p. 154), they may be expected to have a high turnover of deoxyribonucleotides. If the same strains contain a *deoC* mutation, this will result in endogenous induction of thymidine phosphorylase. In thymidine pulse-labeling experiments this may appear as a lowered level of DNA synthesis compared to a non mutant strain.

An additional problem in pulse labeling experiments is that the actual amount of newly synthesized DNA cannot be computed from the specific activity of the precursor used, not even in a *thy* mutant, due to the dilution of the label by the preexisting pool of unlabeled thymidine nucleotides. A more detailed discussion of this problem, and how it may be solved, has been presented elsewhere (3, 143).

B. Other microorganisms

As emphasized in the previous section the capacity of a given organism to utilize exogenously supplied thymine or thymidine for DNA synthesis is dependent on several factors. In the case of thymine, the cell must possess thymidine phosphorylase, thymidine kinase, and a source of deoxyribose 1-phosphate, whereas only thymidine kinase is required for thymidine utilization.

The successful isolation of thymine-requiring mutants of a variety of bacteria such as *Enterobacter aerogenes* (83), *Proteus mirabilis* (149), *Micrococcus radiodurans* (113), *Serratia marcescens* (95), *Bacillus megaterium* (194), *Bacillus thuringiensis* (78), and *Bacillus subtilis* (139) shows that these organisms can utilize exogenous thymine. Since these mutants showed a high thymine requirement, which by secondary mutations could be transformed into a low thymine requirement, they appear to utilize external thymine by a mechanism very similar to that found in *E. coli* and *S. typhimurium*. This has been confirmed for *B. subtilis* and *B. thuringiensis*, where the low thymine requirement is caused by mutations in either the deoxyriboaldolase gene or the phosphodeoxyribomutase gene (78, 164). Thymine-requiring mutants of *Streptococcus faecalis* seem to differ from this general pattern in that they appear to be low thymine-requiring even when isolated on high concentrations of thymine (4). This may indicate that this organism lacks a deoxyribose 1-phosphate catabolizing system.

The procedure generally employed for isolating mutants defective in thymidylate synthase is to select for resistance to inhibitors of dihydrofolate reductase, i.e. aminopterin or trimethoprim, in the presence of thymine or thymidine. The selection is based on the fact that thymidylate synthase is the only C_1-transfer reaction in which tetrahydrofolate is simultaneously oxidized to dihydrofolate. Therefore, dihydrofolate reductase is required for the regeneration of tetrahydrofolate. In the presence of aminopterin or trimethoprim the continued action of thymidylate synthase will deplete the intracellular tetrahydrofolate pool, thereby inhibiting the biosynthesis of serine, methionine, dTMP, and purine nucleotides. In addition, the lack of tetrahydrofolate prevents the synthesis of N-formyl-methionyl-$tRNA_f^{Met}$, which in many procaryotes is required for the initiation of protein synthesis. *B. subtilis*, *S. faecalis*, and probably certain other gram-positive bacteria can initiate protein synthesis with non-formylated methionyl-$tRNA_f^{Met}$ under conditions, where tetrahydrofolate is not available (5, 38, 39, 170). These bacteria will therefore grow in the presence of folate analogs, provided serine, methionine, thymine, and a purine base are added to the medium. Thus in order to select for thymine auxotrophs of these organisms in the presence of trimethoprim and thymine it is necessary to use a medium, which is devoid of either purines, serine, or methionine.

Certain microorganisms are unable to convert thymine to dTMP due to lack of one of the enzymes required, i.e. thymidine phosphorylase or thymidine kinase:

(1) *Thymidine phosphorylase deficient organisms. Pseudomonas acidovorans* is remarkably poor in its capacity to metabolize nucleosides (105). It contains thymidine kinase, but is deficient in thymidine phosphorylase activity. In keeping with this, mutants lacking thymidylate synthase are obligate thymidine-requiring, i.e. thymine cannot support growth of these mutants (104). Very high concentrations of thymidine, i.e. 1 mM, are required to satisfy the growth requirement, suggesting that *Ps. acidovorans* is devoid of an efficient thymidine transport system. Obligate thymidine auxotrophs have also been isolated in *Diplococcus pneumoniae* (61) and *Rhizobium lupini* (125). Since these mutants require less than 0.1 mM thymidine for growth, they may contain a more efficient thymidine transport system than *Ps. acidovorans*. Even though wild type strains of *Ps. acidovorans* and *R. lupini* contain thymidine kinase, they do not incorporate exogenous thymidine into DNA. This indicates that dTTP is synthesized more efficiently from dUMP via thymidylate synthase than from exogenous thymidine.

(2) *Thymidine kinase deficient organisms.* Several fungi, including *Saccharomyces cerevisiae*, *Neurospora crassa*, and *Aspergillus nidulans* are devoid of thymidine kinase (79), which precludes any specific labeling of the DNA of these eucaryotic microorganisms with either thymine or thymidine. In addition it makes the selection of thymidylate synthase deficient mutants of these organisms quite complicated.

In *S. cerevisiae* these problems were largely overcome, when selection procedures were devised that yield mutants permeable to dTMP (*tup*-mutants) from which dTMP requiring mutants were subsequently obtained (21, 22). The selection of *tup* mutants is accomplished by employing conditions where endogenous dTMP synthesis is prevented by inhibiting dihydrofolate reductase and thereby creating a tetrahydrofolate deficiency. Only mutants, which are permeable to dTMP, are able to grow on media containing aminopterin, sulfanilamide, amino acids, adenine, and dTMP. By varying the concentration of dTMP in the selections, mutants with different dTMP-requirements are obtained (53). The *tup* mutations so far studied define at least four different genetic loci (21, 203). Mutants with a specific dTMP requirement for growth may be selected from the permeable strains on the same medium as used for the *tup* selections but lacking adenine. The rationale for this selection is that only mutants blocked in thymidylate synthase will be able to keep a sufficient pool of tetrahydrofolate for purine biosynthesis under conditions, where tetrahydrofolate synthesis is inhibited by aminopterin and sulfanilamide. The mutants thus obtained all map at the *tmp1* locus, which defines the structural gene for thymidylate synthase (17, 114). Thus by employing mutants of the genotype *tmp1,tup* it has become possible to use the same DNA labeling techniques for *S. cerevisiae* as for *E. coli*, except that the precursor must be dTMP instead of thymine or thymidine.

In *N. crassa*, *As. nidulans*, and the yeast *Rhodotorula glutinis* [$2^{14}C$]-labeled thymidine and thymine are incorporated into both RNA and DNA. This observation led to the discovery of an oxidative pathway by which thymidine and thymine are converted into uracil (84, 206; Shaffer *et al.* (1979) Abstr. Int. Congress of Biochem. Toronto). The conversion involves the oxidation of thymidine in the 2′-position to thymine ribonucleoside, followed by hydrolytic cleavage of the glycosyl bond. Thymine is subsequently oxidized via 5-hydroxymethyluracil, and 5-formyluracil to uracil-5-carboxylic acid, which is decarboxylated to yield uracil (p. 115).

The cyanobacterum *Anacystis nidulans* as well as several *Neisseria*, *Moraxella*, and *Acinetobacter* species are deficient in both thymidine phosphorylase and thymidine kinase (100, 101, 102, 211). Specific DNA-labeling may, however, be obtained in some of these organisms by employing low concentrations of deoxyadenosine (160). This indicates that these organisms contain deoxyadenosine kinase activity and are devoid of adenosine deaminase and adenine nucleoside phosphorylase activities. Some evidence has been obtained, indicating that *An. nidulans* has the capacity to demethylate thymine and thymidine (75, 160).

ACKNOWLEDGEMENTS

Research from our laboratory has been supported by grants from the Danish Natural Science Research Council, the Carlsberg Foundation, and NATO.

REFERENCES

1. Ahmad, S. I., and Pritchard, R. H. (1971). A regulatory mutant affecting the synthesis of enzymes involved in the catabolism of nucleosides in *Escherichia coli. Molec. Gen. Genet.* **111**, 77–83.
2. An, G., Justesen, J., Watson, R. J., and Friesen, J. D. (1979). Cloning the *spoT* gene of *Escherichia coli*: Identification of the *spoT* gene product. *J. Bacteriol.* **137**, 1100–1110.
3. Anderson, M. L. M. (1978). Size distribution of short chain DNA in two strains of *Escherichia coli. In* "DNA synthesis. Present and future" (I. Molineux and M. Kohiyama, *Eds.*), pp. 309–319. Plenum Press, New York.
4. Andrew, M. H. E. (1973). Use of trimethoprim to obtain thymine-requiring mutants of *Streptococcus faecalis. J. Gen. Microbiol.* **74**, 195–199.
5. Arnold, H-H. (1977). Initiation of protein synthesis in *Bacillus subtilis* in the presence of trimethoprim or aminopterin. *Biochim. Biophys. Acta* **476**, 76–87.
6. Atkin, C. L., Thelander, L., Reichard, P., and Lang, G. (1973). Iron and free radical in ribonucleotide reductase. Exchange of iron and Mössbauer spectroscopy of the protein B2 sub-unit of the *E. coli* enzyme. *J. Biol. Chem.* **248**, 7464–7472.
7. Baugh, C. M., Braverman, E., and Nair, M. G. (1974). The identification of poly-γ-glutamyl chain lengths in bacterial folates. *Biochemistry* **13**, 4952–4957.
8. Bazill, G. W., and Karamata, D. (1972). Temperature sensitive mutants of *Bacillus subtilis* defective in deoxyribonucleotide synthesis. *Molec. Gen. Genet.* **117**, 19–29.
9. Beacham, I. R., Beacham, K., Zaritsky, A., and Pritchard, R. H. (1971). Intracellular thymidine triphosphate concentrations in wild type and in thymine requiring mutants of *Escherichia coli* 15 and K12. *J. Mol. Biol.* **60**, 75–86.
10. Beacham, I. R., and Pritchard, R. H. (1971). The role of nucleoside phosphorylase in the degradation of deoxyribonucleosides by thymine-requiring mutants of *E. coli. Molec. Gen. Genet.* **110**, 289–298.
11. Beaudette, N. V., Langerman, N., Kisliuk, R. L., and Gaumont, Y. (1977). A calorimetric study of the binding of 2′-deoxyuridine-5′-phosphate and 5-fluoro-2′-deoxyuridine-5′-phosphate to thymidylate synthetase. *Arch. Biochem. Biophys.* **179**, 272–278.
12. Beck, C. F., Eisenhardt, A. R., and Neuhard, J. (1975). Deoxycytidine triphosphate deaminase of *Salmonella typhimurium*. Purification and characterization. *J. Biol. Chem.* **250**, 609–616.
13. Beck, C. F., Neuhard, J., and Thomassen, E. (1977). Thymidine-requiring mutants of *Salmonella typhimurium* that are defective in deoxyuridine 5′-phosphate synthesis. *J. Bacteriol.* **129**, 305–316.
14. Beck, W. S., and Hardy, J. (1965). Requirement of ribonucleotide reductase for cobamide coenzyme, a product of ribosomal activity. *Proc. Natl. Acad. Sci. U.S.A.* **54**, 286–293.
15. Beck, W. S., and Levin, M. (1963). Purification, kinetics and repression control of bacterial *trans*-N-deoxyribosylase. *J. Biol. Chem.* **238**, 702–709.
16. Bellisario, R. L., Maley, G. F., Galivan, J. H., and Maley, F. (1976). Amino acid sequence at the FdUMP binding site of thymidylate synthetase. *Proc. Natl. Acad. Sci. U.S.A.* **73**, 1848–1852.
17. Bisson, L., and Thorner, J. (1977). Thymidine 5′-monophosphate-requiring mutants of *Saccharomyces cerevisiae* are deficient in thymidylate synthase. *J. Bacteriol.* **132**, 44–50.

18. Blakley, R. L., Orme-Johnson, W. H., and Bozdeck, J. M. (1979). Mechanism of *Lactobacillus leichmannii* ribonucleotide reductase studied with Coα-[α-(Aden-9-yl)]-Coβ-adenosylcobamide (Pseudocoenzyme B_{12}) as coenzyme. *Biochemistry* **18**, 2335–2339.
19. Blank, J. G., and Hoffee, P. A. (1975). Purification and properties of thymidine phosphorylase from *Salmonella typhimurium Arch. Biochem. Biophys.* **168**, 259–265.
20. Breitman, T. R., and Bradford, R. M. (1964). The induction of thymidine phosphorylase and excretion of deoxyribose during thymine starvation. *Biochem. Biophys. Res. Commun.* **17**, 786–791.
21. Brendel, M. (1976). A simple method for the isolation and characterization of thymidylate uptaking mutants in *Saccharomyces cerevisiae. Molec. Gen. Genet.* **147**, 209–215.
22. Brendel, M., Fäth, W. W., and Laskowski, W. (1975). Isolation and characterization of mutants of *Saccharomyces cerevisiae* able to grow after inhibition of dTMP synthesis. *In* "Methods in cell biology" (D. M. Prescott, *Ed.*), Vol. 11, pp. 287–294. Academic Press, New York.
23. Brown, N. C., and Reichard, P. (1969). Ribonucleoside diphosphate reductase. Formation of active and inactive complexes of proteins B1 and B2. *J. Mol. Biol.* **46**, 25–38.
24. Brown, N. C., and Reichard, P. (1969). Role of effector binding in allosteric control of ribonucleoside diphosphate reductase. *J. Mol. Biol.* **46**, 39–55.
25. Buxton, R. S. (1975). Genetic analysis of thymidine-resistant and low-thymine-requiring mutants of *Escherichia coli* K-12 induced by bacteriophage Mu-1. *J. Bacteriol.* **121**, 475–484.
26. Cannon, W. D., and Breitman, T. R. (1967). Control of deoxynucleotide biosynthesis in *Escherichia coli.* I. Decrease of pyrimidine deoxynucleotide biosynthesis *in vivo* in the presence of deoxythymidylate. *Biochemistry* **6**, 810–816.
27. Cannon, W. D., and Breitman, T. R. (1968). Control of deoxynucleotide biosynthesis in *Escherichia coli.* II. Effect of deoxythymidylate on the biosynthesis of both deoxynucleotides and ribonucleotide reductase. *Arch. Biochem. Biophys.* **127**, 534–542.
28. Chamberlin, M. (1974). Isolation and characterization of prototrophic mutants of *Escherichia coli* unable to support the intracellular growth of T7. *J. Virol.* **14**, 509–516.
29. Chen, A. K., Bhan, A., Hopper, S., Abrams, R., and Franzen, J. S. (1974). Substrate and effector binding to ribonucleoside triphosphate reductase of *Lactobacillus leichmannii. Biochemistry* **13**, 654–661.
30. Cipollo, K. L., and Dunlap, R. B. (1978). Essential arginyl residues in thymidylate synthetase. *Biochem. Biophys. Res. Commun.* **81**, 1139–1144.
31. Cowles, J. R., Evans, H. J., and Russell, S. A. (1969). B_{12} coenzyme-dependent ribonucleotide reductase in *Rhizobium* species and the effects of cobalt deficiency on the activity of the enzyme. *J. Bacteriol.* **97**, 1460–1465.
32. Danenberg, K. D., and Danenberg, P. V. (1979). Evidence for a sequential interaction of the subunits of thymidylate synthetase. *J. Biol. Chem.* **254**, 4345–4348.
33. Danenberg, P. V. (1977). Thymidylate synthetase—a target enzyme in cancer chemo-therapy. *Biochim. Biophys. Acta* **473**, 73–92.
34. Danenberg, P. V., and Danenberg, K. D. (1978). Effect of 5,10-methylenetetrahydrofolate on the dissociation of 5-fluoro-2′-deoxyuridylate from thymidylate synthetase: Evidence for an ordered mechanism. *Biochemistry* **19**, 4018–4024.

35. Danenberg, P. V., and Heidelberger, C. (1976). The effect of Raney nickel on the covalent thymidylate synthetase-5-fluoro-2′-deoxyuridylate-5,10-methylene-tetrahydrofolate complex. *Biochemistry* **15**, 1331–1337.
36. Danenberg, P. V., Langenbach, R. J., and Heidelberger, C. (1974). Structures of reversible and irreversible complexes of thymidylate synthetase and fluorinated pyrimidine nucleotides. *Biochemistry* **13**, 926–929.
37. Daron, H. H., and Aull, J. L. (1978). A kinetic study of thymidylate synthase from *Lactobacillus casei*. *J. Biol. Chem.* **253**, 940–945.
38. Delk, A. S., and Rabinowitz, J. C. (1974). Partial nucleotide sequence of a prokaryotic initiator tRNA that functions in its non-formylated form. *Nature (London)* **252**, 106–109.
39. Delk, A. S., Romeo, J. M., Nagle, D. P., and Rabinowitz, J. C. (1976). Biosynthesis of ribothymidine in the transfer RNA of *Streptococcus faecalis* and *Bacillus subtilis*. A methylation of RNA involving 5,10-methylenetetrahydro-folate. *J. Biol. Chem.* **251**, 7649–7656.
40. Dermody, J. J., Bourguignon, G. J., Fogelson, P. D., and Sternglanz, R. (1974). Nalidixic acid-sensitive and resistant modes of DNA replication in *Escherichia coli*. *Biochem. Biophys. Res. Commun.* **61**, 1340–1347.
41. Dickens, M. S., Lucas-Lenard, J., and Roth, J. S. (1975). Induction of thymidylate synthetase activity in *Tetrahymena* by cyclic guanosine monophosphate. *Biochem. Biophys. Res. Commun.* **67**, 1319–1325.
42. Döbeln, U. von (1976). Nucleoside diphosphate kinase activity associated with ribonucleotide reductase. *Biochem. Biophys. Res. Commun.* **72**, 1160–1168.
43. Döbeln, U. von (1977). Ribonucleotide reductase from *Escherichia coli*. Identification of allosteric effector sites by chromatography on immobilized effectors. *Biochemistry* **16**, 4368–4371.
44. Döbeln, U. von, and Reichard, P. (1976). Binding of substrates to *Escherichia coli* ribonucleotide reductase. *J. Biol. Chem.* **251**, 3616–3622.
45. Donato, H., Aull, J. L., Lyon, J. A., Reinsch, J. W., and Dunlap, R. B. (1976). Formation of ternary complexes of thymidylate synthetase as followed by absorbance, fluorescence, and circular dichroic spectra and gel electrophoresis. *J. Biol. Chem.* **251**, 1303–1310.
46. Duncan, B. K., Diamond, G. R., and Besman, M. J. (1972). Regulation of enzymatic activity through subunit interaction. A possible example. *J. Biol. Chem.* **247**, 8136–8138.
47. Duncan, B. K., Rochstroh, P. A., and Warner, H. R. (1978). *Escherichia coli* K-12 mutants deficient in uracil-DNA glycosylase. *J. Bacteriol.* **134**, 1039–1045.
48. Dunlap, R. B., Harding, N. G. L., and Huennekens, F. M. (1971). Thymidylate synthetase from amethopterin-resistant *Lactobacillus casei*. *Biochemistry* **10**, 88–97.
49. Durham, J. P., and Ives, D. H. (1971). The metabolism of deoxyribonucleosides in *Lactobacillus acidophilus*: regulation of deoxyadenosine, deoxycytidine, deoxyguanosine and deoxythymidine kinase activities by nucleotides. *Biochim. Biophys. Acta* **228**, 9–25.
50. Ehrenberg, A., and Reichard, P. (1972). Electron spin resonance of the iron-containing protein B2 from ribonucleotide reductase. *J. Biol. Chem.* **247**, 3485–3488.
51. Eriksson, S., Sjöberg, B.-M., Hahne, S., and Karlström, O. (1977). Ribonucleoside diphosphate reductase from *Escherichia coli*. An immunological assay and a novel purification from an overproducing strain lysogenic for phage *λdnrd*. *J. Biol. Chem.* **252**, 6132–6138.

52. Fangman, W. L. (1969). Specificity and efficiency of thymidine incorporation in *Escherichia coli* lacking thymidine phosphorylase. *J. Bacteriol.* **99**, 681–687.
53. Fäth, W. W., and Brendel, M. (1976). Isolation and properties of yeast mutants with highly efficient thymidylate utilization. *Z. Naturforsch.* **31c**, 467–478.
54. Feller, W., and Follmann, H. (1976). Ribonucleotide reductase activity in green algae. *Biochem. Biophys. Res. Commun.* **70**, 752–758.
55. Feller, W., Schimpff-Weiland, G., and Follmann, H. (1980). Deoxyribonucleotide biosynthesis in synchronous algae cells. *Eur. J. Biochem.* **110**, 85–92.
56. Filpula, D., and Fuchs, J. A. (1977). Regulation of ribonucleoside diphosphate reductase synthesis in *E. coli.* Increased enzyme synthesis as a result of inhibition of deoxynucleic acid synthesis. *J. Bacteriol.* **130**, 107–113.
57. Filpula, D., and Fuchs, J. A. (1978). Regulation of the synthesis of ribonucleoside diphosphate reductase in *Escherichia coli*: Specific activity of the enzyme in relationship to perturbations of DNA replication. *J. Bacteriol.* **135**, 429–435.
58. Filpula, D., and Fuchs, J. A. (1979). Increased synthesis of ribonucleotide reductase after deoxyribonucleic acid inhibition in various species of bacteria. *J. Bacteriol.* **139**, 694–696.
59. Follmann, H., and Hogenkamp, H. P. C. (1971). Interaction of ribonucleotide reductase with ribonucleotide analogs. *Biochemistry* **10**, 186–192.
60. Friedkin, M. (1973). Thymidylate synthetase. *Adv. Enzymol.* **38**, 235–292.
61. Friedman, L. R., and Ravin, A. W. (1972). Genetic and biochemical properties of thymidine-dependent mutants of Pneumococcus. *J. Bacteriol.* **109**, 459–461.
62. Frisch, S. M., Couch, J. L., and Glaser, D. A. (1978). Mutator activity of a short Okazaki fragment mutant of *Escherichia coli. J. Bacteriol.* **134**, 1192–1194.
63. Fuchs, J. A. (1977). Isolation of an *Escherichia coli* mutant deficient in thioredoxin reductase. *J. Bacteriol.* **129**, 967–972.
64. Fuchs, J. A. (1977). Coordinate control of the synthesis of ribonucleoside diphosphate reductase components in *Escherichia coli. J. Bacteriol.* **130**, 957–959.
65. Fuchs, J. A., and Karlström, H. O. (1973). A mutant of *Escherichia coli* defective in ribonucleosidediphosphate reductase. 2. Characterization of the enzymatic defect. *Eur. J. Biochem.* **32**, 457–462.
66. Fuchs, J. A., and Karlström, O. (1976). Mapping of *nrdA* and *nrdB* in *Escherichia coli* K-12. *J. Bacteriol.* **128**, 810–814.
67. Fuchs, J. A., Karlström, H. O., Warner, H. R., and Reichard, P. (1972). Identification of the gene product defective in a *dnaF* mutant of *E. coli. Nature. New Biol.* **238**, 69–71.
68. Fuchs, J. A., and Neuhard, J. (1973). A mutant of *Escherichia coli* defective in ribonucleosidediphosphate reductase. 1. Isolation of the mutant as a deoxyuridine auxotroph. *Eur. J. Biochem.* **32**, 451–456.
69. Fuchs, J. A., and Warner, H. R. (1975). Isolation of an *E. coli* mutant deficient in glutathione synthesis. *J. Bacteriol.* **124**, 140–148.
70. Galivan, J. H., Maley, F., and Baugh, C. M. (1976). Demonstration of separate binding sites for the folate coenzymes and deoxynucleotides with inactivated *Lactobacillus casei* thymidylate synthetase. *Biochem. Biophys. Res. Commun.* **71**, 527–534.
71. Galivan, J., Maley, F., and Baugh, C. M. (1977). Protective effect of the pteroylpolyglutamates and phosphate on the proteolytic inactivation of thymidylate synthetase. *Arch. Biochem. Biophys.* **184**, 346–354.
72. Galivan, J. H., Maley, G. F., and Maley, F. (1976). Factors affecting substrate binding in *Lactobacillus casei* thymidylate synthetase as studied by equilibrium dialysis. *Biochemistry* **15**, 356–362.

73. Ginther, C. L., and Ingraham, J. L. (1974). Nucleoside diphosphokinase of *Salmonella typhimurium*. *J. Biol. Chem.* **249**, 3406–3411.
74. Ginther, C. L., and Ingraham, J. L. (1974). Cold-sensitive mutant of *Salmonella typhimurium* defective in nucleoside diphosphokinase. *J. Bacteriol.* **118**, 1020–1026.
75. Glaser, V. M., Al-Nuri, M. A., Groshev, V. V., and Shestakov, S. V. (1973). The labelling of nucleic acids by radioactive precursors in the blue-green algae *Anacystis nidulans* and *Synechocystis aquatilis Sanv. Arch. Mikrobiol.* **92**, 217–226.
76. Gleason, F. K. (1979). Control of ribonucleotide reductase in *Synechococcus Sp.*, a unicellular cyanobacterium. *Arch. Microbiol.* **123**, 15–21.
77. Gleason, F. K., and Wood, J. M. (1976). Ribonucleotide reductase in blue-green algae: Dependence on adenosylcobalamin. *Science* **192**, 1343–1344.
78. Grigorieva, T. M., Smirnov, V. V., and Sukhodolets, V. V. (1978). Thymine auxotrophs and secondary mutations decreasing the growth requirement for thymine in *Bacillus thuringiensis*. *Genetika* **14**, 1310–1318.
79. Grivell, A. R., and Jackson, J. F. (1968). Thymidine kinase: Evidence for its absence from *Neurospora crassa* and some other micro-organisms, and the relevence of this to the specific labelling of deoxyribonucleic acid. *J. Gen. Microbiol.* **54**, 307–317.
80. Haertle, T., Wohlrab, F., and Guschlbauer, W. (1979). Thymidylate synthetase from *Escherichia coli* K 12. Purification and dependence of kinetic properties on sugar conformation and size of the 2′ substituent. *Eur. J. Biochem.* **102**, 223–230.
81. Hamilton, F. D. (1974). Ribonucleotide reductase from *Euglena gracilis*. A 5′-deoxyadenosylcobalamin-dependent enzyme. *J. Biol. Chem.* **249**, 4428–4434.
82. Hantke, K. (1976). Phage T6-colicin K receptor and nucleoside transport in *Escherichia coli*. *FEBS Letts.* **70**, 109–112.
83. Harrison, A. P. (1965). Thymine incorporation and metabolism by various classes of thymine-less bacteria. *J. Gen. Microbiol.* **41**, 321–333.
84. Hayaishi, O., Nozaki, M., and Abbott, M. T. (1975). Oxygenases: Dioxygenases. *In* "The Enzymes" (P. D. Boyer, *Ed.*), Vol. XII B, pp. 119–189. Academic Press, New York.
85. Hintze, D. N., and Farmer, J. L. (1975). Identification of poly-γ-glutamyl chain lengths in folates of *Bacillus subtilis*. *J. Bacteriol.* **124**, 1236–1239.
86. Hochhauser, S. J., and Weiss, B. (1978). *E. coli* mutants deficient in deoxyuridine triphosphatase. *J. Bacteriol.* **134**, 157–166.
87. Hoffee, P. (1968). The 2-deoxyribose gene-enzyme complex in *Salmonella typhimurium*. I. Isolation and enzymatic characterization of 2-deoxyribose negative mutants. *J. Bacteriol.* **95**, 449–457.
88. Holmgren, A. (1976). Hydrogen donor system for *Escherichia coli* ribonucleoside-diphosphate reductase dependent upon glutathione. *Proc. Natl. Acad. Sci. U.S.A.* **73**, 2275–2279.
89. Holmgren, A. (1979). Glutathione-dependent synthesis of deoxyribonucleotides. Purification and characterization of glutaredoxin from *Escherichia coli*. *J. Biol. Chem.* **254**, 3664–3671.
90. Holmgren, A. (1979). Glutathione-dependent synthesis of deoxyribonucleotides. Characterization of the enzymatic mechanism of *Escherichia coli* glutaredoxin. *J. Biol. Chem.* **254**, 3672–3678.
91. Holmgren, A. (1979). Thioredoxin catalyses the reduction of insulin disulfides by dithiothreitol and dihydrolipoamide. *J. Biol. Chem.* **254**, 9627–9632.
91a. Holmgren, A. and Fagerstedt, M. (1982). The *in vivo* distribution of oxidised and

reduced thioredoxin in *Escherichia coli*. *J. Biol. Chem.* **257**, 6926–6930.

92. Holmgren, A., Ohlson, I., and Grankvist, M.-L. (1978). Thioredoxin of *E. coli*. Radioimmunological and enzymatic determinations in wild type cells and mutants defective in phage T7 DNA replication. *J. Biol. Chem.* **253**, 430–436.
93. Holmgren, A., Söderberg, B.-O., Eklund, H., and Bränden, C.-I. (1975). Three-dimensional structure of *E. coli* thioredoxinS_2 to 2.8 Å resolution. *Proc. Natl. Acad. Sci. U.S.A.* **72**, 2305–2309.
94. Hosono, R., and Kuno, S. (1975). An inhibitory effect of thymidine on its own conversion to nucleotide in *Escherichia coli*. *Eur. J. Biochem.* **57**, 177–179.
95. Ishibashi, M., Sugino, Y., and Hirota, Y. (1964). Chromosomal location of thymine and arginine genes in *Escherichia coli* and an F′ incorporating them. *J. Bacteriol.* **87**, 554–561.
96. Iwatsuki, N., and Okazaki, R. (1967). Mechanism of regulation of deoxythymidine kinase of *Escherichia coli*. I. Effect of regulatory deoxynucleotides on the state of aggregation of the enzyme. *J. Mol. Biol.* **29**, 139–154.
97. Iwatzuki, N., and Okazaki, R. (1967). Mechanism of regulation of deoxythymidine kinase of *Escherichia coli*. II. Effect of temperature on the enzyme activity and kinetics. *J. Mol. Biol.* **29**, 155–165.
98. James, T. L., Pogolotti, A. L., Ivanetich, K. M., Wataya, Y., Lam, S. S. M., and Santi, D. V. (1976). Thymidylate synthetase: Fluorine-19 nmr characterization of the active site peptide covalently bound to 5-fluoro-2′-deoxyuridylate and 5,10-methylenetetrahydrofolate. *Biochem. Biophys. Res. Commun.* **72**, 404–410.
99. Jensen, K. F., Leer, J. C., and Nygaard, P. (1973). Thymine utilization in *Escherichia coli* K12 on the role of deoxyribose 1-phosphate and thymidine phosphorylase. *Eur. J. Biochem.* **40**, 345–354.
100. Jyssum, S. (1971). Utilization of thymine, thymidine, and TMP by *Neisseria meningitidis*. 2. Lack of enzymes for specific incorporation of exogenous thymine, thymidine and TMP into DNA. *Acta Path. Microbiol. Scand. Section B* **79**, 778–788.
101. Jyssum, S. (1974). Search for thymidine phosphorylase, nucleoside deoxyribosyltransferase, and thymidine kinase in genus *Neisseria*. *Acta Path. Microbiol. Scand. Section B* **82**, 53–56.
102. Jyssum, S., and Bovre, K. (1974). Search for thymidine phosphorylase, nucleoside deoxyribosyltransferase, and thymidine kinase in *Moraxella*, *Acinetobacter*, and allied bacteria. *Acta Path. Microbiol. Scand. Section B* **82**, 57–66.
103. Kelln, R. A., and Warren, R. A. J. (1973). Studies on the biosynthesis of α-putrescinylthymine in bacteriophage Ø W-14-infected *Pseudomonas acidovorans*. *J. Virol.* **12**, 1427–1433.
104. Kelln, R. A., and Warren, R. A. J. (1973). Obligate thymidine auxotrophs of *Pseudomonas acidovorans*. *J. Bacteriol.* **113**, 510–511.
105. Kelln, R. A., and Warren, R. A. J. (1974). Pyrimidine metabolism in *Pseudomonas acidovorans*. *Can. J. Microbiol.* **20**, 427–433.
106. Kim, J. J., Abrams, R., and Franzen, J. S. (1977). Influence of effectors on the rate of reaction of reduced ribonucleoside triphosphate reductase with N-ethylmaleimide. *Arch. Biochem. Biophys.* **182**, 674–682.
107. Kisliuk, R. L., Gaumont, Y., and Baugh, C. M. (1974). Polyglutamyl derivatives of folate as substrates and inhibitors of thymidylate synthetase. *J. Biol. Chem.* **249**, 4100–4103.
108. Konrad, E. B., and Lehman, I. R. (1975). Novel mutants of *Escherichia coli* that accumulate very small DNA replicative intermediates. *Proc. Natl. Acad. Sci. U.S.A.* **72**, 2150–2154.

109. Leary, R. P., Beaudette, N., and Kisliuk, R. L. (1975). Interaction of deoxyuridylate with thymidylate synthetase. *J. Biol. Chem.* **250**, 4864–4868.
110. Leer, J. C., Hammer-Jespersen, K., and Schwartz, M. (1977). Uridine phosphorylase from *Escherichia coli.* Physical and chemical characterization. *Eur. J. Biochem.* **75**, 217–224.
111. Leighton, T., Khachatourians, G., and Brown, N. (1975). The role of semiconservative DNA replication in bacterial cell development. *In* "DNA Synthesis and Its Regulation" (M. Goulian., P. Hanawalt and C. F. Fox, *Eds.*), pp. 677–687. W. A. Benjamin, Inc. Menlo Park, California, U.S.A.
112. Lindahl, T. (1979). DNA glycosylases, endonucleases for apurinic/apyrimidinic sites, and base excision-repair. *In* "Progress in Nucleic Acid Research and Molecular Biology" (W. E. Cohn, *Ed.*), Vol. 22, pp. 135–192. Academic Press, New York.
113. Little, I. G., and Hanawalt, P. C. (1973). Thymineless death and ultraviolet sensitivity in *Micrococcus radiodurans. J. Bacteriol.* **113**, 233–240.
114. Little, I. G., and Haynes, R. H. (1979). Isolation and characterization of yeast mutants auxotrophic for 2′-deoxythymidine 5′-monophosphate. *Molec. Gen. Genet.* **168**, 141–151.
115. Lockshin, A., and Danenberg, P. V. (1979). Thymidylate synthetase and 2′-deoxyuridylate form a tight complex in the presence of pteroyltriglutamate. *J. Biol. Chem.* **254**, 12285–12288.
116. Lowdon, M., and Vitols, E. (1973). Ribonucleotide reductase activity during the cell cycle of *Saccharomyces cerevisiae. Arch. Biochem. Biophys.* **158**, 177–184.
117. Ludwig, W., and Follmann, H. (1978). The specificity of ribonucleoside triphosphate reductase. Multiple induced activity changes and implications for deoxyribonucleotide formation. *Eur. J. Biochem.* **82**, 393–403.
118. Lundberg, L., Nyman, P. O., and Karlström, O. (1979). Isolation and Characterization of plasmids carrying the *dut* gene of *Escherichia coli. Acta Chem. Scand. B* **33**, 597–598.
119. Lunn, C. A., and Pigiet, V. (1979). Characterization of a high activity form of ribonucleoside diphosphate reductase from *Escherichia coli. J. Biol. Chem.* **254**, 5008–5014.
120. Maley, G. F., Bellisario, R. L., Guarino, D. U., and Maley, F. (1979). The primary structure of *Lactobacillus casei* thymidylate synthetase. III. The use of 2-(2-nitro phenylsulfenyl)-3-methyl-3-bromoindolanine and limited tryptic peptides to establish the complete amino acid sequence of the enzyme. *J. Biol. Chem.* **254**, 1301–1304.
121. Mandelbaum-Shavit, F. (1976). Resistance of *Pedicoccus cerevisiae* to amethopterin as a consequence of changes in enzymatic activity and cell permeability. I. Dihydrofolate reductase, thymidylate synthetase and formyltetrahydrofolate synthetase in amethopterin-resistant and -sensitive strains of *Pedicoccus cerevisiae. Biochim. Biophys. Acta* **428**, 664–673.
122. Manwaring, J. D., and Fuchs, J. A. (1979). Relationship between deoxyribonucleoside triphosphate pools and deoxyribonucleic acid synthesis in an *nrdA* mutant of *Escherichia coli. J. Bacteriol.* **138**, 245–248.
123. Mark, D. F., Chase, J. W., and Richardson, C. C. (1977). Genetic mapping of *trxA*, a gene affecting thioredoxin in *Escherichia coli* K12. *Molec. Gen. Genet.* **155**, 145–152.
124. Mark, D. F., and Richardson, C. C. (1976). *Escherichia coli* thioredoxin: A subunit of bacteriophage T7 DNA polymerase. *Proc. Natl. Acad. Sci. U.S.A.* **73**, 780–784.

125. Mattes, R. (1974). Investigation of the nucleoside metabolism in *Rhizobium lupini*: Obligatory thymidine phosphorylase deficiency. *Hoppe-Seylers. Z. Physiol. Chem.* **355**, 1229.
126. McCuen, R. W., and Sirotnak, F. M. (1975). Thymidylate synthetase from *Diplococcus pneumoniae*. Properties and inhibition by folate analogues. *Biochim. Biophys. Acta* **384**, 369–380.
127. McKeown, M., Kahn, M., and Hanawalt, P. (1976). Thymidine uptake and utilization in *E. coli*: A new gene controlling nucleoside transport. *J. Bacteriol.* **126**, 814–822.
128. Miller, L. K., and Wells, R. D. (1971). Nucleoside diphosphokinase activity associated with DNA polymerases. *Proc. Natl. Acad. Sci. U.S.A.* **68**, 2298–2302.
129. Mironov, A. S., and Sukhodolets, V. V. (1979). Promoter-like mutants with increased expression of the *Escherichia coli* uridine phosphorylase structural gene. *J. Bacteriol.* **137**, 802–810.
130. Mitchell, A., and Finch, L. R. (1977). Pathways of nucleotide biosynthesis in *Mycoplasma mycoides* subsp. *mycoides*. *J. Bacteriol.* **130**, 1047–1054.
131. Møllgaard, H., and Neuhard, J. (1978). Deoxycytidylate deaminase from *Bacillus subtilis*. Purification, characterization, and physiological function. *J. Biol. Chem.* **253**, 3536–3542.
132. Munch-Petersen, A. (1970). Deoxyribonucleoside catabolism and thymine incorporation in mutants of *Escherichia coli* lacking deoxyriboaldolase. *Eur. J. Biochem.* **15**, 191–202.
133. Munch-Petersen, A., Nygaard, P., Hammer-Jespersen, K., and Fiil, N. (1972). Mutants constitutive for nucleoside-catabolizing enzymes in *Escherichia coli* K12. *Eur. J. Biochem.* **27**, 208–215.
134. Munroe, W. A., Lewis, C. A., and Dunlap, R. B. (1978). Kinetics of sulfhydryl group modification of thymidylate synthetase: A proposal for activation of catalytical cysteinyl residues. *Biochem. Biophys. Res. Commun.* **80**, 355–360.
135. Nelson, D. J., and Carter, C. E. (1969). Purification and characterization of thymidine 5′-monophosphate kinase from *Escherichia coli B. J. Biol. Chem.* **244**, 5254–5260.
136. Neuhard, J. (1967). Studies on the acid-soluble nucleotide pool in *Escherichia coli*. IV. Effect of hydroxyurea. *Biochim. Biophys. Acta* **145**, 1–6.
137. Neuhard, J., Maltman, K. L., and Warren, R. A. J. (1980). Bacteriophage ØW-14 infected *Pseudomonas acidovorans* synthesizes hydroxymethyldeoxyuridine triphosphate. *J. Virol.* **34**, 347–353.
138. Neuhard, J., and Munch-Petersen, A. (1966). Studies on the acid soluble nucleotide pool in thymine-requiring mutants of *Escherichia coli* during thymine starvation. II. Changes in the amounts of deoxycytidine triphosphate and deoxyadenosine triphosphate in *Escherichia coli* 15 $T^-A^-U^-$. *Biochim. Biophys. Acta* **114**, 61–71.
139. Neuhard, J., Price, A. R., Schack, L., and Thomassen, E. (1978). Two thymidylate synthetases in *Bacillus subtilis*. *Proc. Natl. Acad. Sci. U.S.A.* **75**, 1194–1198.
140. Neuhard, J., and Thomassen, E. (1971). Turnover of the deoxyribonucleoside triphosphates in *Escherichia coli* 15T during thymine starvation. *Eur. J. Biochem.* **20**, 36–43.
141. Neuhard, J., and Thomassen, E. (1971). Deoxycytidine triphosphate deaminase: Identification and function in *Salmonella typhimurium*. *J. Bacteriol.* **105**, 657–665.
142. Neuhard, J., and Thomassen, E. (1976). Altered deoxyribonucleotide pools in P2

eductants of *Escherichia coli* K-12 due to deletion of the *dcd* gene. *J. Bacteriol.* **126**, 999–1001.

143. O'Donovan, G. A. (1978). Thymidine metabolism in bacteria (and "how, or how not, to label DNA"). *In* "DNA synthesis. Present and future" (I. Molineux and M. Kohiyama, *Eds.*), pp. 219–253. Plenum Press, New York.
144. O'Donovan, G. A., Edlin, G., Fuchs, J. A., Neuhard, J., and Thomassen, E. (1971). Deoxycytidine triphosphate deaminase: Characterization of an *Escherichia coli* mutant deficient in the enzyme. *J. Bacteriol.* **105**, 666–672.
145. O'Donovan, G. A., and Neuhard, J. (1970). Pyrimidine metabolism in microorganisms. *Bacteriol. Rev.* **34**, 278–343.
146. Ohkawa, M. (1979). Abnormal metabolism of thymidine nucleotides and phosphorylation of deoxycytidine in *Escherichia coli* C *thy*$^-$,*ura*$^-$ mutant. *Eur. J. Biochem.* **100**, 165–173.
147. Okazaki, R., and Kornberg, A. (1964). Deoxythymidine kinase of *Escherichia coli.* I. Purification and some properties of the enzyme. *J. Biol. Chem.* **239**, 269–274.
148. Okazaki, R., and Kornberg, A. (1964). Deoxythymidine kinase of *Escherichia coli.* II. Kinetics and feedback control. *J. Biol. Chem.* **239**, 275–284.
149. Okubadejo, O. A., and Maskell, R. M. (1973). Thymine-requiring mutants of *Proteus mirabilis* selected by co-trimoxazole in vivo. *J. Gen. Microbiol.* **77**, 533–535.
150. Panagon, D., Orr, M. D., Dunstone, J. R., and Blakley, R. L. (1972). A monomeric, allosteric enzyme with a single polypeptide chain. Ribonucleotide reductase of *Lactobacillus leichmannii. Biochemistry* **11**, 2378–2388.
151. Pato, M. (1979). Alterations of deoxyribonucleoside triphosphate pools in *Escherichia coli*: Effects on deoxyribonucleic acid replication and evidence for compartementation. *J. Bacteriol.* **140**, 518–524.
152. Petersson, L., Gräslund, A., Ehrenberg, A., Sjöberg, B-M., and Reichard, P. (1980). The iron center in ribonucleotide reductase from *Escherichia coli. J. Biol. Chem.* **255**, 6706–6712.
153. Pigiet, V., and Conley, R. R. (1978). Isolation and characterization of phosphothioredoxin from *Escherichia coli. J. Biol. Chem.* **253**, 1910–1920.
154. Platz, A., and Sjöberg, B-M. (1980). Construction and characterization of hybrid plasmids containing the *Escherichia coli nrd* region. *J. Bacteriol.* **143**, 561–568.
155. Plese, P. C., and Dunlap, R. B. (1977). Sulfhydryl group modification of thymidylate synthetase and its effect on activity and ternary complex formation. *J. Biol. Chem.* **252**, 6139–6144.
156. Pogolotti, A. L., and Santi, D. V. (1977). The catalytic mechanism of thymidylate synthetase. *In* "Bioorganic Chemistry" (E. E. Van Tamelen, *Ed.*), Vol. 1, pp. 277–311. Academic Press, New York.
157. Powell, J. W., and Wachsman, J. T. (1973). Evidence for four deoxynucleoside kinase activities in extracts of *Lactobacillus leichmannii. Applied Microbiol.* **25**, 869–872.
158. Price, A. R., and Frato, J. (1975). *Bacillus subtilis* deoxyuridinetriphosphatase and its bacteriophage PBS2-induced inhibitor. *J. Biol. Chem.* **250**, 8804–8811.
159. Pritchard, R. H., and Zaritsky, A. (1970). Effect of thymine concentration on the replication velocity of DNA in a thymineless mutant of *Escherichia coli. Nature (London)* **226**, 126–131.

160. Restaino, L., and Frampton, E. W. (1975). Labelling the deoxyribonucleic acid of *Anacystis nidulans*. *J. Bacteriol.* **124**, 155–160.
161. Rima, B. K., and Takahashi, I. (1978). Deoxyribonucleoside-requiring mutants of *Bacillus subtilis*. *J. Gen. Microbiol.* **107**, 139–145.
162. Roisin, M. P., and Kepes, A. (1978). Nucleoside diphosphate kinase of *Escherichia coli*, a periplasmic enzyme. *Biochim. Biophys. Acta* **526**, 418–428.
163. Rossi, M., and Woodward, D. O. (1975). Enzymes of deoxythymidine triphosphate biosynthesis in *Neurospora crassa* mitochondria. *J. Bacteriol.* **121**, 640–647.
164. Rumyantseva, E. V., Sukhodolets, V. V., and Smirnov, Y. V. (1979). Isolation and characterization of mutants for genes of nucleoside catabolism in *Bacillus subtilis*. *Genetika* **25**, 594–604.
165. Saeki, T., Hori, M., and Umezawa, H. (1972). Cyclamidomycin (Desdanine), an inhibitor of nucleoside diphosphokinase of *Escherichia coli*. *J. Antibiotics* **25**, 343–349.
166. Saeki, T., Hori, M., and Umezawa, H. (1974). Kinetic studies on the inhibition of nucleoside diphosphate kinase by desdanine. *J. Biochem.* **76**, 623–629.
167. Saeki, T., Hori, M., and Umezawa, H. (1974). Pyruvate kinase of *E. coli*. Its role in supplying nucleoside triphosphates in cells under anaerobic conditions. *J. Biochem.* **76**, 631–637.
168. Saeki, T., Hori, M., and Umezawa, H. (1975). Effect of desdanine on nucleoside diphosphate kinase and pyruvate kinase of *Escherichia coli*. *J. Antibiotics* **28**, 974–981.
169. Sando, G. N., Blakley, R. L., Hogenkamp, H. P. C., and Hoffmann, P. J. (1975). Studies on the mechanism of adenosylcobalamin-dependent ribonucleotide reduction by the use of analogs of the coenzyme. *J. Biol. Chem.* **250**, 8774–8779.
170. Schmidt, W., Arnold, H-H., and Kersten, H. (1977). Tetrahydrofolate-dependent biosynthesis of ribothymidine in transfer ribonucleic acids of gram-positive bacteria. *J. Bacteriol.* **129**, 15–21.
171. Schwan, H., and Holldorf, A. W. (1975). Effective utilization of exogenous deoxythymidine 5′-monophosphate for DNA synthesis in enterobacteria. *FEBS Letts.* **57**, 179–182.
172. Schwartz, M. (1971). Thymidine phosphorylase from *Escherichia coli*. Properties and kinetics. *Eur. J. Biochem.* **21**, 191–198.
173. Sedmark, J., and Ramaley, R. (1971). Purification and properties of *Bacillus subtilis* nucleoside diphosphokinase. *J. Biol. Chem.* **246**, 5365–5372.
174. Sergott, R. C., Debeer, L. J., and Bessman, M. J. (1971). On the regulation of a bacterial deoxycytidylate deaminase. *J. Biol. Chem.* **246**, 7755–7758.
175. Setlow, P. (1973). Deoxyribonucleic acid synthesis and deoxynucleotide metabolism during bacterial spore germination. *J. Bacteriol.* **144**, 1099–1107.
176. Sherman, I. W. (1979). Biochemistry of Plasmodium (Malarial parasites). *Microbiol. Rev.* **43**, 453–495.
177. Shivakumar, A. G., and Dubnau, D. (1978). Plasmid replication in *dna* ts mutants of *Bacillus subtilis*. *Plasmid* **1**, 405–416.
178. Shlomai, J., and Kornberg, A. (1978). Deoxyuridine triphosphatase of *Escherichia coli*. Purification, properties, and use as reagent to reduce uracil incorporation into DNA. *J. Biol. Chem.* **253**, 3305–3312.
179. Sjöberg, B-M., Reichard, P., Gräslund, A., and Ehrenberg, A. (1977). Nature of

the free radical in ribonucleotide reductase from *Escherichia coli*. *J. Biol. Chem.* **252**, 536–541.

180. Sjöberg, B-M., Reichard, P., Gräslund, A., and Ehrenberg, A. (1978). The tyrosine free radical in ribonucleotide reductase from *Escherichia coli*. *J. Biol. Chem.* **253**, 6863–6865.
181. Stutzenberger, F. (1974). Ribonucleotide reductase of *Pithomyces chartarum*: Requirement for B_{12} coenzyme. *J. Gen. Microbiol.* **81**, 501–503.
182. Tamao, Y., and Blakley, R. L. (1973). Direct spectrophotometric observation of an intermediate formed from deoxyadenosylcobalamin in ribonucleotide reduction. *Biochemistry* **12**, 24–34.
183. Thelander, L. (1973). Physicochemical characterization of ribonucleoside diphosphate reductase from *Escherichia coli*. *J. Biol. Chem.* **248**, 4591–4601.
184. Thelander, L. (1974). Reaction mechanism of ribonucleoside diphosphate reductase from *E. coli*. Oxidation-reduction-active disulfides in the B1 subunit. *J. Biol. Chem.* **249**, 4858–4862.
185. Thelander, L., Larsson, B., Hobbs, J., and Eckstein, F. (1976). Active site of ribonucleoside diphosphate reductase from *E. coli*: Inactivation of the enzyme by 2′-substituted ribonucleoside diphosphates. *J. Biol. Chem.* **251**, 1398–1405.
186. Thelander, L., and Reichard, P. (1979). Reduction of ribonucleotides. *Ann. Rev. Biochem.* **48**, 133–158.
187. Tsai, P. K., and Hogenkamp, H. P. C. (1980). The purification and characterization of an adenosylcobalamin-dependent ribonucleoside diphosphate reductase from *Corynebacterium nephridii*. *J. Biol. Chem.* **255**, 1273–1278.
188. Tsang, M.G-S., and Schiff, J. A. (1978). Assimilatory sulfate reduction in an *Escherichia coli* mutant lacking thioredoxin activity. *J. Bacteriol.* **134**, 131–138.
189. Tye, B-K., Chien, J., Lehman, I. R., Duncan, B. K., and Warner, H. R. (1978). Uracil incorporation: A source of pulse-labeled DNA fragments in the replication of the *Escherichia coli* chromosome. *Proc. Natl. Acad. Sci. U.S.A.* **75**, 233–237.
190. Tye, B-K., and Lehman, I. R. (1977). Excission-repair of uracil misincorporated in DNA as a result of a defect in dUTPase. *J. Mol. Biol.* **117**, 293–306.
191. Tye, B-K., Nyman, P-O., Lehman, I. R., Hochhauser, S., and Weiss, B. (1977). Transient accumulation of Okazaki fragments as a result of uracil incorporation into nascent DNA. *Proc. Natl. Acad. Sci. U.S.A.* **74**, 154–157.
192. Uerkvitz, W., Karlström, O., and Munch-Petersen, A. (1973). A deoxyuridine monophosphate phosphatase detected in mutants of *Escherichia coli* lacking alkaline phosphatase and 5′-nucleotidase. *Molec. Gen. Genet.* **121**, 337–346.
193. Viswanathan, G., and Noronha, J. M. (1979). Studies on a membrane-bound and solubilized ribonucleotide reductase preparation from *Escherichia coli* TAU^-. *Biochim. Biophys. Acta* **567**, 325–330.
194. Wachsman, J. T., Kemp, S., and Hogg, L. (1964). Thymineless death in *Bacillus megaterium*. *J. Bacteriol.* **87**, 1079–1086.
195. Wachsman, J. T., and Morgan, D. D. (1971). Deoxynucleoside kinases of *Bacillus megaterium* KM. *J. Bacteriol.* **105**, 787–792.
196. Wachsman, J. T., and Morgan, D. D. (1973). Demonstration of a deoxycytidine kinase activity in extracts of *Bacillus megaterium*. *Applied Microbiol.* **25**, 506–508.
197. Waleh, N. S., and Ingraham, J. L. (1976). Pyrimidine ribonucleoside monophosphokinase and the mode of RNA turnover in *Bacillus subtilis*. *Arch. Microbiol.* **110**, 49–54.
198. Warner, H. R. (1973). Properties of ribonucleoside diphosphate reductase in nucleotide permeable cells. *J. Bacteriol.* **115**, 18–22.

199. Warner, H. R., and Duncan, B. K. (1978). *In vivo* synthesis and properties of uracil-containing DNA. *In* "DNA synthesis. Present and future" (I. Molineux and M. Kohiyama, *Eds.*), pp. 267–279. Plenum Press, New York.
200. Warner, H. R., Duncan, B. K., Garrett, C., and Neuhard, J. (1981). Synthesis and metabolism of uracil-containing deoxyribonucleic acid in *Escherichia coli. J. Bacteriol.* **145**, 687–695.
201. Wataya, Y., Matsuda, A., and Santi, D. V. (1980). Interaction of thymidylate synthetase with 5-nitro-2′-deoxyuridylate. *J. Biol. Chem.* **255**, 5538–5544.
202. Weiss, B., Rogers, S. G., and Taylor, A. F. (1978). The endonuclease activity of exonuclease III and the repair of uracil-containing DNA in *Escherichia coli. In* "DNA Repair Mechanisms" (P. C. Hanawalt., E. C. Friedberg and C. F. Fox, *Eds.*), pp. 191–194. Academic Press, New York.
203. Wickner, R. B. (1975). Mutants of *Saccharomyces cerivisiae* that incorporate deoxythymidine 5′-monophosphate into DNA *in vivo. In* "Methods in Cell Biology (D. M. Prescott, *Ed.*), Vol. 11, pp. 295–302. Academic Press, New York.
204. Wilson, M. C., Farmer, J. L., and Rothman, F. (1966). Thymidylate synthesis and aminopterin resistance in *Bacillus subtilis. J. Bacteriol.* **92**, 186–196.
205. Womack, J. E. (1977). Inhibition of thymidine phosphorylase *in vivo* provides a rapid method for switching DNA labelling. *Molec. Gen Genet.* **158**, 11–15.
206. Wondrach, L. M., Hsu, C. A., and Abbott, M. T. (1978). Thymine 7-hydroxylase and pyrimidine deoxyribonucleoside 2′-hydroxylase activities in *Rhodotorula glutinis. J. Biol. Chem.* **253**, 6511–6515.
207. Yamada, R., Tamao, Y., and Blakley, R. L. (1971). Degradation of 5′-deoxyadenosylcobalamin by ribonucleoside triphosphate reductase and binding of degradation products to the active center. *Biochemistry* **10**, 3959–3968.
208. Yau, S., and Wachsman, J. T. (1973). The *Bacillus megaterium* ribonucleotide reductase: evidence for a B_{12} coenzyme requirement. *Mol. Cell. Biochem.* **1**, 101–105.
209. Zahler, S. A., Benjamin, L. G., Glatz, B. S., Winter, P. F., and Goldstein, B. J. (1976). Genetic mapping of *alsA,alsR,thyA,kauA*, and *citD* markers in *Bacillus subtilis. In* "Microbiology-1976" (D. Schlessinger, *Ed.*), pp. 35–43. Am. Soc. Microbiol., Washington, DC. U.S.A.
210. Zaritsky, A., and Pritchard, R. H. (1971). Replication time of the chromosome in thymineless mutants of *Escherichia coli. J. Mol. Biol.* **60**, 65–74.
211. Øvrebø, S., and Kleppe, K. (1973). Pyrimidine Metabolism in *Acinetobacter calcoaceticus. J. Bacteriol.* **116**, 331–336.

5

Nucleoside catabolism

KARIN HAMMER-JESPERSEN

I. Introduction 203
II. Nucleoside catabolism in *Escherichia coli* and *Salmonella typhimurium* 204
A. Pathways of nucleoside catabolism 204
B. Properties of nucleoside catabolizing enzymes 206
C. Isolation and characterization of mutants defective in nucleoside catabolism 212
D. Induction of nucleoside catabolizing enzymes 218
E. Regulatory units in nucleoside catabolism 220
F. Mechanism of the regulation of the deo-operon . . . 224
G. The physiological role of nucleoside catabolizing enzymes 235
III. Other microorganisms 238
A. Pathways of nucleoside catabolism 238
B. *Bacillus* 241
C. Lactic acid 244
D. *Saccharomyces cerevisiae* 245
E. *Neurospora crassa* 249

I. INTRODUCTION

The splitting of the glycosylic bond of a nucleoside is the central reaction in nucleoside catabolism. By this reaction the nucleoside structure, as it occurs in nucleotides and nucleic acids, is destroyed, thus liberating the purine or pyrimidine base and the corresponding pentose sugar. Two different reactions for nucleoside cleavage have been found in nature. They are shown in Fig. 5.1 using inosine as a substrate.

Reaction A is an irreversible hydrolytic cleavage of the glycosylic bond, and it is catalyzed by a nucleoside hydrolase. Reaction B is a reversible phosphorolytic cleavage of the glycosylic bond catalyzed by a nucleoside phosphorylase.

The pathways of nucleoside catabolism which are treated in this chapter

METABOLISM OF NUCLEOTIDES, NUCLEOSIDES AND NUCLEOBASES IN MICROORGANISMS
ISBN: 0-12-510580-0

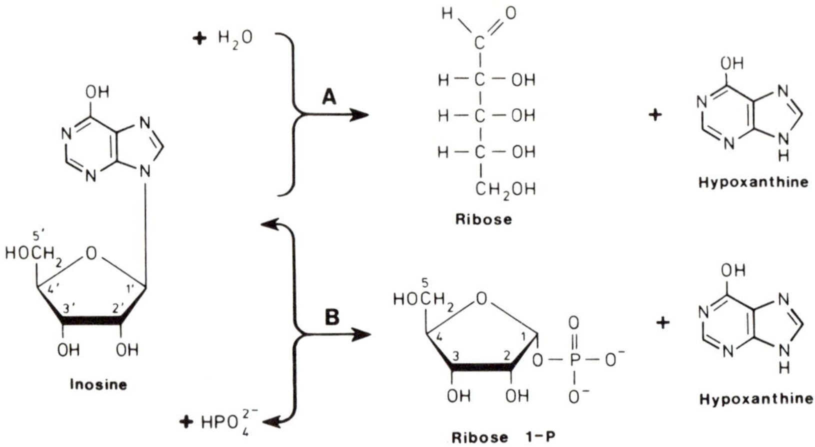

FIG. 5.1. Nucleoside degradation. Reaction A is catalyzed by a nucleoside hydrolase. Reaction B is catalyzed by a nucleoside phosphorylase.

cover: (a) the reactions which transform nucleosides into substrates for nucleoside phosphorylases or hydrolases, (b) the reactions for the cleavage of the glycosylic bond and (c) the reactions involved in the conversion of the pentose moiety into intermediates of glycolysis or the pentose phosphate shunt.

Most organisms are able to incorporate the free purine or pyrimidine bases into nucleic acids. These reactions are discussed in Chapters 2 and 3 in this book.

Some microorganisms—mainly fungi—are able to cleave the ring structure of the nucleobases. These pathways of purine and pyrimidine catabolism have been reviewed by Vogels and Van Der Drift (144), and they will not be discussed in the present chapter.

II. NUCLEOSIDE CATABOLISM IN *ESCHERICHIA COLI* AND *SALMONELLA TYPHIMURIUM*

A. Pathways of nucleoside catabolism

Four different types of enzymatic reactions are involved in the nucleoside catabolism: (i) deaminations catalyzed by adenosine or cytidine deaminase, (ii) phosphorolytic cleavage of the nucleosides catalyzed by a nucleoside phosphorylase, and (iii) conversion of pentose 1-phosphate into pentose 5-phosphate catalyzed by phosphopentomutase. Finally an aldolase reaction is required for cleavage of deoxyribose 5-phosphate. Fig. 5.2 shows the

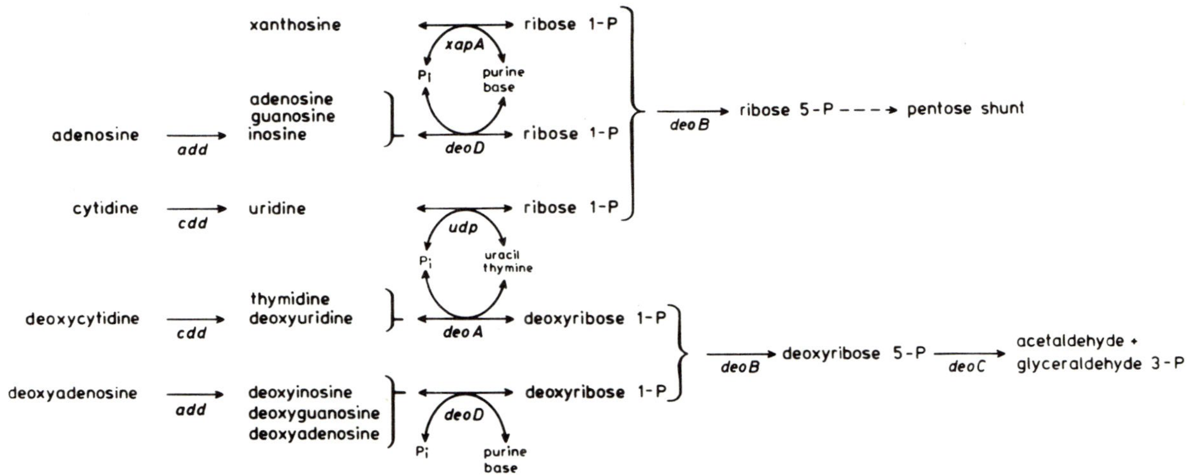

FIG. 5.2. Pathways of nucleoside catabolism in *E. coli*. The enzymes are shown by the 3-letter code used to designate their respective structural genes. The eight enzymes are: adenosine deaminase (*add*); cytidine deaminase (*cdd*); xanthosine phosphorylase (*xapA*); purine nucleoside phosphorylase (*deoD*); uridine phosphorylase (*udp*); thymidine phosphorylase (*deoA*); phosphopentomutase (*deoB*), and deoxyriboaldolase (*deoC*).

pathways of nucleoside catabolism in *Escherichia coli*. The deamination step is not needed for all the nucleosides, and the deoxyriboaldolase reaction is not needed for ribonucleoside catabolism. In contrast the nucleoside phosphorylase and the mutase reactions are necessary for the catabolism of all nucleosides.

Altogether eight different enzymes are involved in the pathways in Fig. 5.2. The presence of these enzymes ensures that ribonucleosides and deoxyribonucleosides may serve as sole carbon sources for *E. coli*. The same enzymes except xanthosine phosphorylase are present in *Salmonella typhimurium*.

B. Properties of nucleoside catabolizing enzymes

All the eight nucleoside catabolizing enzymes except xanthosine phosphorylase have been purified to near homogeneity. Some of the reported properties of these enzymes are given in Table 5.1. The homologous enzymes from *S. typhimurium* and *E. coli* appear to have very similar properties.

1. Nucleoside deaminases

Both ribonucleosides and deoxyribonucleosides are substrates for adenosine deaminase and cytidine deaminase, yielding free ammonia and the corresponding hypoxanthine and uracil nucleosides as reaction products (28, 107). Both reactions are irreversible, and very potent inhibitors of both enzymes have been found. Coformycin inhibits adenosine deaminase ($K_I = 3.0 \times 10^{-8}$ M) and tetrahydrouridine inhibits cytidine deaminase ($K_I = 2.4 \times 10^{-7}$ M). These inhibitors appear to act as transition state analogs and they have been used to inhibit the enzymes *in vivo* in *E. coli* (27, 44, 107, and P. Nygaard, personal communication).

2. Nucleoside phosphorylases

Four different nucleoside phosphorylases have been demonstrated in *E. coli*. Two are specific for pyrimidine nucleosides and two are specific for purine nucleosides. All four catalyze the phosphorolysis of the nucleoside substrates, the products being the free nucleobase and the corresponding pentose 1-phosphate (Fig. 5.1). The reactions are reversible and syntheses of the nucleosides are favoured at pH 7.4 with a K_{eq} of 54 for inosine and deoxyinosine (108) and a K_{eq} of 32 for uridine (18). Arsenate will substitute for phosphate in the reactions. The four enzymes are all composed of subunits (Table 5.1).

Purine nucleoside phosphorylase. The enzymes from *E. coli* and *S. typhimurium* are hexamers and have very similar properties (67, 118, Tables 5.1 and 5.2). Hypoxanthine, guanine, adenine and their respective nucleosides are substrates for the enzymes, whereas xanthine and xanthosine are not. Detailed substrate specificity and inhibitor studies have been performed with the *E. coli* enzyme. A few compounds have been found to be inhibitors but not substrates, namely 7-deazaadenosine and 7-deazainosine (29).

Steady state kinetic analysis suggests that purine nucleoside phosphorylase from both organisms works by a sequential mechanism, where nucleoside and phosphate bind randomly to the enzyme, while the products must dissociate from the enzyme in ordered sequence: first the purine base, then the pentose 1-phosphate. Normal Michaelis–Menten kinetics were observed for all substrates except for phosphate. The data indicate negative cooperativity in the binding of phosphate to the enzyme, and a Hill coefficient of 0.47 was calculated. The proposed reaction mechanism was confirmed by binding studies: nucleosides, phosphate and pentose 1-phosphate are able to bind to the free enzyme, while purine bases are not. The binding data also indicate one binding site per subunit for each ligand (65, 67).

Xanthosine phosphorylase. The enzyme has been found in *E. coli* but not in *S. typhimurium*. The substrate specificity is like that of purine nucleoside phosphorylase, with the difference that xanthosine phosphorylase uses xanthosine as substrate but not adenosine or deoxyadenosine (Table 5.2). The K_m for phosphate is high (8.4 mM), but low for pentose 1-phosphate (0.04 mM). The gel filtration behaviour of the enzyme is consistent with a molecular weight of 205 000 ($\pm 10\%$) and indicates that some dissociation to half size molecules may take place (47).

Uridine phosphorylase. The enzyme from *E. coli* is an octamer consisting of identical subunits. The preferred substrate is uridine but the enzyme is also active towards 5-substituted uridine derivatives. Uridine phosphorylase has some activity towards the corresponding pyrimidine deoxyribonucleosides. The enzyme shows no activity towards nucleosides containing cytosine or purines. The K_m values for phosphate and ribose 1-phosphate are high (Table 5.3 and 86). A sequential reaction mechanism with random addition of uridine and phosphate was found for the enzyme from *E. coli* B (81).

Thymidine phosphorylase. The enzymes from *E. coli* and *S. typhimurium* are both dimers consisting of identical subunits and they have similar properties (17, 53, 127, 128, Tables 5.1 and 5.3). The enzyme has specificity for deoxyuridine and several 5-substituted deoxyuridine derivatives, including thymidine, while deoxyribonucleosides containing purines or cytosine do not serve as substrates. The enzyme is specific for the 2′deoxyribose part of the nucleoside and has very low activity towards uridine (Table 5.3). Detailed substrate specificity and inhibitor studies have been performed with the

TABLE 5.1

Properties of nucleoside catabolizing enzymes.

		Molecular weight		Kinetic constants			
Enzyme and organism	Gene symbol	Holoenzyme	Subunit	K_m (mM)	Substrate	Spec. act.[a] μmol/(min × mg)	Ref.
Adenosine deaminase *E. coli*	*add*	29 000		0.075 0.040	adenosine deoxyadenosine	277 277	107
Cytidine deaminase *E. coli*	*cdd*	73 000 54 000		0.167 0.085	cytidine deoxycytidine	127 344	27, 59
Purine nucleoside phosphorylase *E. coli*	*deoD*	140 000	23 700	0.07 0.12 0.10	inosine phosphate deoxyribose 1-phosphate	195	65, 67
Purine nucleoside phosphorylase *S. typhimurium*	*deoD*	140 000	23 700	0.09 0.10 0.10	inosine phosphate deoxyribose 1-phosphate	301	65, 67, 118
Xanthosine phosphorylase *E. coli*	*xapA*	205 000		0.56 8.4 0.04	inosine phosphate deoxyribose 1-phosphate		47

Uridine phosphorylase *E. coli*	*udp*	176 000	22 000	0.15 7.3 1.4	uridine phosphate ribose 1-phosphate	133; 183[b]	86
Thymidine phosphorylase *E. coli*	*deoA*	94 000	47 000	0.89 0.35 0.3	thymidine phosphate deoxyribose 1-phosphate	122	127, 128
Thymidine phosphorylase *S. typhimurium*	*deoA*	94 000	47 000	2.1 2.3	thymidine phosphate	465	17, 53
Phosphopentomutase *E. coli*	*deoB*	45 000	45 000	0.014 0.043	deoxyribose 1-phosphate ribose 1-phosphate	310; 370[b]	43, 85
Deoxyriboaldolase *E. coli*	*deoC*	32 000[c] 50 000[c]	27 737		deoxyribose 5-phosphate	72; 245[b]	141
Deoxyriboaldolase *S. typhimurium*	*deoC*	57 000	28 000	0.10 3.5 0.6	deoxyribose 5-phosphate acetaldehyde glyceraldehyde 3-phosphate	58.3	51, 55

[a] Specific activity of the purified protein is given in units per mg protein (determined by the Lowry method).
[b] Specific activity is given as units per mg protein (determined by weight measurements).
[c] Gelfiltration in phosphate buffer gives MW 50 000 while gelfiltration in Tris-HCl-EDTA gives MW 32 000.

TABLE 5.2
Substrate specificity of purine nucleoside phosphorylases.

Substrate	Relative activity (%) Purine nucleoside[a] phosphorylase from *E. coli* K12	*S. typhimurium*	Xanthosine[b] phosphorylase *E. coli* K12
Deoxyinosine	100	88	100
Deoxyguanosine	74	100	88
Inosine	46	82	67
Guanosine	48	81	47
Xanthosine	⩽1	⩽1	24
Adenosine	61	78	⩽1
Deoxyadenosine	61	78	⩽1
Deoxyuridine	<1	<1	<0.1

[a] Data taken from reference 67.
[b] Data taken from reference 47.

TABLE 5.3
Substrate specificity of pyrimidine nucleoside phosphorylases.

Substrate	Relative activity (%) Uridine[a] phosphorylase *E. coli* K12	Thymidine phosphorylase *E. coli* K12[b]	*S. typhimurium*[c]
Thymidine	2	100	100
Deoxyuridine	6	175	100
Thymidine ribonucleoside	19	0.5	
Uridine	100	1.2	0
Deoxycytidine	<0.1	<0.01	0
Cytidine	<0.1		
5-Fluoro-2′-deoxyuridine	14	118	
5-Bromo-2′-deoxyuridine	27	53	40
5-Bromo-uridine	69	3.5	
Deoxyadenosine	<1	<0.1	0
Inosine	<1		

[a] Data taken from reference 86.
[b] K. Hammer-Jespersen, unpublished; determined at pH 7.1 in the presence of 5 mM nucleoside and 10 mM phosphate using a thymidine phosphorylase with a specific activity of 313 units/E_{280}.
[c] Data taken from references 17 and 53.

enzyme from *E. coli* B (9, 10). Almost no changes are allowed in the deoxyribose part of the nucleoside, only the 5′hydroxyl group is not necessary for activity. Thymine derivatives with certain hydrofobic side chains in the 5 or the 6 position are very potent inhibitors (10). Uridine and adenosine are inhibitors for the enzyme from *E. coli* K 12, competing with thymidine. K_i values of 1.1 mM and 3.0 mM were found for uridine and adenosine and $K_m = 0.9$ mM for thymidine (21, 98).

When whole cells were used 10–20 fold lower K_i and K_m values were determined (98). These apparently lower K_m and K_i values found *in vivo* are probably due to the presence of active nucleoside transport systems in the *E. coli* cell (*see* Chapter 6). The inhibition of thymidine phosphorylase by uridine *in vivo* has been utilized in a method for DNA labelling (147, and Chapter 4).

Kinetic analysis of the enzyme from *E. coli* K 12 indicates that the reaction mechanism is a sequential type with phosphate as the first substrate to bind to the enzyme and deoxyribose 1-phosphate as the last product to leave the enzyme (127).

3. Phosphopentomutase

This enzyme catalyzes the conversion of both ribose 1-phosphate and deoxyribose 1-phosphate to the corresponding pentose 5-phosphates. Phosphopentomutase is therefore needed for catabolism of all nucleosides. The enzyme has a very low K_m for both substrates. It is the only nucleoside catabolizing enzyme which is dependent upon cofactors for activity. Ribose 1.5 diphosphate is required for phosphopentomutase activity in a similar way as glucose 1.6 diphosphate for phosphoglucomutase activity. Furthermore a metal ion, either Mn^{2+}, Co^{2+} or Ni^{2+}, is an absolute requirement for activity. Addition of EDTA to the metal activated enzyme does not inhibit the phosphopentomutase activity instantaneously, indicating that a metallo-enzyme is formed and not a metallo-substrate complex (43, 85). Titration of the pure enzyme with $CoCl_2$ showed that addition of two Co^{2+} ions per molecule of phosphopentomutase is necessary for full activity (Karin Hammer-Jespersen, unpublished).

The enzyme is inhibited by phosphate, sulfate and several metal ions. Also nucleoside triphosphates are inhibitors (43). Inhibition of phosphopentomutase *in vivo* has been observed in cells infected with T4 phages (101).

Four different forms of the enzyme have been separated according to charge. They have identical molecular weights, amino acid composition, and N-terminal amino acid, indicating that they are products of the same gene. One of the two major forms was formed from the other by deamidation *in vitro*, but it is not known whether a similar conversion occurs *in vivo*. The two minor forms are believed also to constitute enzyme forms with modified peptide chains (85).

The enzyme from *S. typhimurium* has similar cofactor requirements and inhibition pattern as the *E. coli* enzyme (54).

4. Deoxyriboaldolase

The enzyme from *S. typhimurium* has been purified to homogeneity. It is a dimer with identical subunits (2 × 28 500) (55). The enzyme belongs to the class I aldolases: during the reaction a lysine residue from the enzyme forms a Schiffs base intermediate with the substrate acetaldehyde, and the enzymatic reaction does not require metal ions (51). The C-terminal tyrosine residue is essential for the catalytic activity but not for substrate binding (55). Genetic studies showed that missense mutations, which result in an inactive protein, were clustered in the proximal end of the structural gene, *deoC*. This indicates that the N-terminal part of the protein is directly involved in the catalytic function (63, 64).

Deoxyriboaldolase is specific for deoxyribose 5-phosphate, no activity is seen with ribose 5-phosphate or deoxyribose 1-phosphate. Deoxyribose, however, is a poor substrate (0.5% of the rate seen with deoxyribose 5-phosphate) (51).

The enzyme has also been purified to homogeneity from *E. coli*. The size of the peptide chain was found to be the same as in *S. typhimurium*, but gel filtration indicated that the enzyme could exist both as a monomer and as a dimer in *E. coli* (141). The amino acid sequence of the N-terminal end of the protein has been determined. This information together with the DNA sequence of the *deoC*-containing part of the *deo* operon has made it possible to determine the entire amino acid sequence of the *deoC* gene product (141).

C. Isolation and characterization of mutants defective in nucleoside catabolism

Mutants in the structural genes encoding the nucleoside catabolizing enzymes have been isolated and the genes mapped in *E. coli* (*see* Appendix). Six of the eight genes have similar locations on the genetic map of *S. typhimurium*. Xanthosine phosphorylase, however, is not found in this organism and the *add* gene has not been located.

None of the nucleoside catabolizing enzymes seem to be essential for the cell. Mutants deleted in the *deo*-genes, the *add* gene and the *cdd* gene have been isolated (25, 68, 151 and J. Josephsen, personal communication). Furthermore phage Mu and/or λ has been inserted into the *deo*-genes, the *udp*, *cdd* and *xap* genes (23, 24, 25, 26, 69, and B. Mygind, personal communication). Positive selection procedures have been designed for mutants in each of the eight genes. They are summarized in Table 5.4 and described below.

TABLE 3.4

Selection of mutants defective in genes coding for nucleoside catabolizing enzymes.

Gene	Gene product	Required genetic background for selection	Selection procedure Growth in presence of:	Other mutants obtained by the selection	Reference
add	adenosine deaminase	*purA* or *purB*	deoxyadenosine (16 μM) as source of adenine		56, 68
cdd	cytidine deaminase		fluorodeoxycytidine (20 μM) + uridine (80 μM) + uracil (180 μM)	*tdk*	69, 76, 105
deoA	thymidine phosphorylase	*deoC*	thymidine (4 mM)	*deoB*	1, 25, 119
		upp	fluorouracil (19 μM) + deoxyadenosine (2 mM)	*deoD,tdk*	1, 24
deoB	phosphopentomutase	*deoC*	thymidine (4 mM)	*deoA*	1, 25
		thyA	low thymine (16 μM)	*deoC*	20, 97
deoC	deoxyriboaldolase	*thyA*	low thymine (16 μM)	*deoB*	19, 54, 97
deoD	purine nucleoside phosphorylase	*deoC*	deoxyadenosine (4 mM)	*deoB*	102, 119
		upp	fluorouracil (19 μM) + deoxyadenosine (2 mM)	*deoA,tdk*	1, 25
		upp	fluorouracil (19 μM) + adenosine (2 mM)	*(udp),udk*	2
udp	uridine phosphorylase	*upp,deoA*	fluorouracil (19 μM) + adenosine (2 mM)	*deoD,udk*	111
xapA	xanthosine phosphorylase	*upp,deoD,xapR*	fluorouracil (19 μM) + deoxyguanosine (2 mM)	*xapR,deoA,tdk,nupG*	26

1. Mutations in the add *gene*

Deoxyadenosine in low concentrations cannot serve as source of adenine in *E. coli* and *S. typhimurium* wild type cells due to rapid deamination by adenosine deaminase. *Add* mutants have therefore been selected by plating an adenine requiring strain (*purA* or *purB*) with 16 μM deoxyadenosine as the only adenine source (56, 68). Another phenotype of the *add* mutants is the growth inhibition by high (4 mM) concentrations of adenosine and deoxyadenosine (48). This inhibition also explains why the *add* mutants cannot grow on these nucleosides as sole carbon source (Table 5.5), in spite of the presence of alternative pathways for their degradation (through purine nucleoside phosphorylase, Fig. 5.2).

2. Mutations in the cdd *gene*

Selection of mutants lacking cytidine deaminase is based upon the conversion of the non toxic nucleoside analog 5-fluorodeoxycytidine to the toxic nucleotide 5-fluorodeoxyuridinemonophosphate through the following reactions:

$$\text{5-fluorodeoxycytidine} \xrightarrow{1} \text{5-fluorodeoxyuridine} \xrightarrow{2} \text{5-fluorodeoxyuridinemonophosphate}$$

Reaction 1 and 2 are catalyzed by cytidine deaminase and thymidine kinase respectively. Mutants resistant to 5-fluorodeoxycytidine are therefore mutated in either *cdd*, *tdk* or nucleoside transport genes. Uridine and uracil are included in the selection medium to minimize toxic effects of 5-fluorodeoxyuridine and 5-fluorouracil produced by the surrounding wild type cells. *cdd* mutants are thus characterized by their resistance to 5-fluorodeoxycytidine and their sensitivity to 5-fluorodeoxyuridine (15, 44, 69, 76). Furthermore they are unable to grow on cytidine or deoxycytidine as sole carbon sources (Table 5.5).

3. Mutations in the deo *genes*

DeoA, *deoB*, *deoC*, and *deoD* mutants may be isolated in several different ways. The three main selection principles are described below. Mutations in the four *deo* genes occur with the same frequencies i.e. 10^{-6}, and genetic evidence so far indicates that only one gene exists for each of the four enzymes. The order of the four *deo* genes has been determined by transductions and found to be *deoCABD* (1, 62, 151).

Selection of mutants. (1) Selection for low thymine requirement. Mutants lacking thymidylate synthetase (*thyA*-mutants) require high concentrations

TABLE 5.5

Growth of *E. coli* and *S. typhimurium* mutants on nucleosides as sole carbon source.

Genotype	Growth											
	adenosine	inosine	guanosine	xanthosine[a]	cytidine	uridine	deoxycytidine	deoxyuridine	thymidine	deoxyadenosine	deoxyinosine	deoxyguanosine
add	−	+	+	+	+	+	+	+	+	−	+	+
cdd	+	+	+	+	−	+	−	+	+	+	+	+
deoA	+	+	+	+	+	+	−	−	−	+	+	+
deoB	−	−	−	−	−	−	−	−	−	−	−	−
deoC	+	+	+	+	+	+	−	−	−	−	−	−
deoD	−	−	−	+	+	+	+	+	+	−	−	−
udp	+	+	+	+	−	−	+	+	+	+	+	+
xapA	+	+	+	−	+	+	+	+	+	+	+	+
wild type	+	+	+	+	+	+	+	+	+	+	+	+

[a] *S. typhimurium* does not grow on xanthosine as carbon source.

of thymine (160 μM) for growth. Strains with secondary mutations which permit growth at low concentrations of thymine (16 μM) can readily be isolated. They were originally designated *tlr* (thymine low requirement) by Alikhanian *et al.* (7). Later it was shown that the catabolism of deoxyribonucleosides is blocked in *tlr* mutants and that they are deficient in either phosphopentomutase (*deoB*) or deoxyriboaldolase (*deoC*) (19, 54, 97 and Chapter 4).

(2) Resistance to deoxyribonucleosides. The *deoC* mutants are sensitive to all deoxyribonucleosides except deoxycytidine due to accumulation of high internal concentrations of deoxyribose 5-phosphate (5–12 mM) (13, 88, 99). Therefore when *deoC* mutants are spread on glucose minimal media in the presence of high concentrations of deoxyribonucleoside, only mutants which cannot form the toxic deoxyribose 5-phosphate will grow. Thus in the presence of thymidine *deoA* and *deoB* mutants will be selected, while in the presence of deoxyadenosine *deoD* and *deoB* mutants will appear (1, 25, 102, 119) (Table 5.4).

(3) Resistance to fluorouracil in the presence of a nucleoside. Mutants which lack uracil phosphoribosyltransferase (*upp*) are resistant to fluorouracil, but if a nucleoside is added to the growth medium they are sensitized to fluorouracil. In the presence of deoxyadenosine the following reactions occur:

$$\text{deoxyadenosine} + P_i \xrightarrow[deoD]{} \text{deoxyribose 1-phosphate} + \text{adenine};$$

$$\text{fluorouracil} + \text{deoxyribose 1-phosphate} \xrightarrow[deoA]{} \text{fluorodeoxyuridine} \xrightarrow[tdk]{} \text{fluorodUMP}$$

The toxicity of fluorouracil under these conditions is due to the intracellular formation of fluorodUMP by the sequential action of purine nucleoside phosphorylase (*deoD*), thymidine phosphorylase (*deoA*) and thymidine kinase (*tdk*). *DeoA* and *deoD* mutants have been obtained in this way (1, 25, and Table 5.4).

The phenotypes of the mutants. As described above the *deo* mutants have different phenotypes with respect to thymine requirement in *thyA* mutants, inhibition by deoxyribonucleosides, and resistance to fluorouracil + nucleoside (*see also* Chapters 3 and 4). The four different types of *deo* mutants may also be distinguished by their growth pattern on nucleosides as sole carbon source. *E. coli* and *S. typhimurium* contain a ribokinase (4, 8). Thus the presence of nucleoside hydrolases would make it possible for the cells to utilize ribonucleosides as carbon sources without participation of phosphopentomutase. As shown in Table 5.5 *deoB* mutants cannot grow on nucleosides as carbon sources thereby excluding the existence of a ribonucleoside hydrolase in *E. coli* and *S. typhimurium*. *S. typhimurium* contains a deoxyribokinase (52). The phenotype of the *deoB* mutants therefore excludes the presence of a deoxyribonucleoside hydrolase in *S. typhimurium*.

DeoC mutants cannot grow on any deoxyribonucleoside, but they will grow on all ribonucleosides. *DeoD* mutants cannot grow on purine nucleosides, except on xanthosine. *DeoA* mutants cannot grow on any pyrimidine deoxyribonucleosides.

4. Mutations in the udp *gene*

The selection of uridine phosphorylase mutants is based on the same principle as the selection of *deoA* mutants. A *upp* parent strain is used and the bacteria are sensitized to fluorouracil by addition of adenosine. By the sequential action of purine nucleoside phosphorylase (*deoD*), uridine phosphorylase (*udp*) and uridine kinase (*udk*), fluoro-UMP is formed (cf. Chapter 4). *udp* and *deoD* mutants have been obtained in this way (111). It has been recommended to use a *deoA,upp* parent strain for the isolation of *udp* mutants (111). The *udp* mutants cannot utilize pyrimidine ribonucleosides as carbon source (Table 5.5).

5. Mutations in the xapA *gene*

Xanthosine phosphorylase is not present in measurable amounts in wild type *E. coli* cells (47), this is in contrast to the enzyme levels of the seven other nucleoside catabolizing enzymes. Furthermore the substrate specificity of xanthosine phosphorylase is overlapping that of purine nucleoside phosphorylase (47, and Table 5.2). A strain expressing xanthosine phosphorylase constitutively and lacking purine nucleoside phosphorylase (*xapR,deoD*) may therefore be used for selection of *xapA* mutants. An additional introduction of a *upp* mutation makes it possible to select *xapA* mutants as strains resistant to fluorouracil in the presence of deoxyguanosine (26). The selection principle is the same as described for the selection of *deoD* and *deoA* mutants. Mutants defective in xanthosine phosphorylase cannot grow on xanthosine as sole carbon source, although they grow normally on all the other nucleosides given in Table 5.5. Thus the substrate specificity of xanthosine phosphorylase cannot be deduced from the phenotype of the *xapA* mutants. The substrate specificity of the other seven enzymes may in general be deduced from the growth pattern of the corresponding mutants on nucleosides as carbon source.

D. Induction of nucleoside catabolizing enzymes

The synthesis of all eight nucleoside catabolizing enzymes is inducible. The induction pattern is complex and five different regulatory systems are involved (Table 5.6).

*1. The Deo-Enzymes**

Deoxyribose 5-phosphate induction. The synthesis of all four Deo-enzymes is induced by deoxyribonucleosides. The true inducing compound is deoxyribose 5-phosphate. The main evidence for this is the observation that a deoxyribonucleoside cannot induce the Deo-enzymes in a mutant blocked in the pathway leading to formation of deoxyribose 5-phosphate from the added deoxyribonucleoside (20, 48, 119). Addition of thymidine to a thymidine phosphorylase (*deoA*) mutant does not induce the three other Deo-enzymes, whereas deoxyadenosine will do so. The induction is noncoordinate (48, 119). As shown in Table 5.6 the fold of induction of deoxyriboaldolase and thymidine phosphorylase is higher than that of phosphopentomutase which in turn is higher than that of purine nucleoside phosphorylase. The induction is not subject to catabolite repression i.e. it is independent of the cAMP level in the cell (45, 46, 48).

Cytidine induction. The four Deo-enzymes are induced by cytidine and also in this case the induction is noncoordinate. However, the fully induced Deo-enzyme levels are lower in presence of cytidine than in presence of a deoxyribonucleoside (102, 106). In *S. typhimurium* both cytidine and uridine serve as inducers (106) whereas in *E. coli* only cytidine is inducer (44).

The cytidine induction is sensitive to glucose repression: the induced levels are much lower in glucose grown cells than in glycerol grown cells (45, 48, 106). Thus the synthesis of the Deo-enzymes during cytidine inducing conditions is dependent upon the presence of cAMP and cAMP receptor protein (CRP). This conclusion has been confirmed using an *in vitro* system for coupled transcription and translation of the *deo*-genes (143). Adenosine induces as cytidine but only in mutants partially blocked in adenosine catabolism i.e. *deoD* or *add* mutants (102, 106).

Inosine, guanosine induction. Inosine and guanosine induce two of the four Deo-enzymes, namely phosphopentomutase and purine nucleoside phosphorylase. The fold of induction is low but significant (3–5, *see* Table 5.6). The true inducing compounds seem to be nucleosides. Thus guanosine induces phosphopentomutase in a mutant blocked in guanosine metabolism (the

* The term Deo-enzymes is an abbreviation for the enzymes encoded by the four deo genes: *deoC*, *deoA*, *deoB*, and *deoD*.

TABLE 5.6

Induction of the nucleoside catabolizing enzymes.[a]

Inducer	Regulatory gene	Glucose repression	Fold of Induction: Thymidine phosphorylase and deoxyriboaldolase	Phosphopento-mutase	Purine nucleoside phosphorylase	Uridine phosphorylase	Cytidine deaminase	Adenosine deaminase	Xanthosine phosphorylase
Deoxyribose 5-phosphate	*deoR*	—	60	9	6	—	—	—	—
Cytidine Adenosine[b] Uridine[c]	*cytR*	+	10	3	2	20	80	—	—
Inosine Guanosine	nd	—	—	5	3	—	—	—	—
Adenine[d] Hypoxanthine[d]	nd	—	—	—	—	—	—	100 20	—
Xanthosine[d]	*xapR*	+	—	—	—	—	—	—	1000

[a] Numbers give fold of induction (maximal); + means glucose repression; — no effect; nd = not determined.
[b] Adenosine induces only in *add* or *deoD* mutants.
[c] Uridine is inducing in *S. typhimurium* not in *E. coli*.
[d] Only in *E. coli*.

strain is devoid of guanosine kinase and purine nucleoside phosphorylase, K. Hammer-Jespersen, unpublished). The induction is not sensitive to catabolite repression, indicating that the inosine, guanosine regulated synthesis of phosphopentomutase and purine nucleoside phosphorylase is independent of CRP and cAMP (48, 119).

2. Cytidine deaminase and uridine phosphorylase

These enzymes which are both involved in cytidine catabolism are inducible by cytidine (Table 5.6). The induction follows the same pattern as the cytidine induction of the Deo-enzymes i.e. cytidine is the true inducing compound but also adenosine may serve as inducer (in mutants blocked in adenosine catabolism) (44, 102, 106). In *S. typhimurium* also uridine acts as inducer (106). The induction is sensitive to catabolite repression implying that CRP and cAMP is necessary for the expression of uridine phosphorylase and cytidine deaminase. This has been confirmed by use of mutants defective in adenylate cyclase or CRP (46).

3. Adenosine deaminase

In wild type *E. coli* adenosine deaminase is induced by adenine and hypoxanthine (114). Also the corresponding nucleosides will cause induction, but only after cleavage to the free nucleobases, as shown by the failure of the nucleosides to induce in a *deoD* strain. The induction of adenosine deaminase is independent of the cAMP-CRP complex (P. Nygaard, personal communication).

In *S. typhimurium* the synthesis of adenosine deaminase appears not to be inducible (114 and P. Nygaard, personal communication).

4. Xanthosine phosphorylase

Xanthosine phosphorylase is normally not present in measurable amounts in *E. coli*. The synthesis of the enzyme is induced only by xanthosine, the fold of induction being at least 1000 (47). The failure of other purine nucleosides to induce xanthosine phosphorylase explains the lack of growth of a *deoD* mutant on these nucleosides as carbon source. In fact the enzyme was discovered in pseudorevertants of *deoD* mutants on inosine as carbon source. These mutants were shown to contain high levels of xanthosine phosphorylase (26).

E. Regulatory units in nucleoside catabolism

From the induction pattern in Table 5.6 it is seen that five regulatory systems (regulons) are involved in the expression of nucleoside catabolizing enzymes in *E. coli*. One would expect to find five regulatory proteins in *E. coli* responding to the effectors: deoxyribose 5-phosphate, cytidine, inosine/guanosine, adenine/hypoxanthine, and xanthosine. The genes coding for three of these regulatory proteins namely *deoR*,*cytR* and *xapR* have been identified and mapped (Appendix). The regulatory proteins belonging to the adenosine deaminase regulon and to the inosine/guanosine regulon have not yet been identified. An interesting feature is that some of the structural genes belong to several different regulons.

1. The DEO R regulon*

The *deoR* gene product regulates the expression of the four *deo*-genes as well as the expression of two nucleoside transport systems (the G-system and the *tsx* receptor protein, cf. Chapter 6). The regulatory gene is not located near any of the structural genes (*see* Appendix). The *deoR* gene encodes a repressor protein as indicated by the high levels of the Deo-enzymes in (i) mutants containing a temperature sensitive *deoR* allele (45); (ii) mutants with phage λ inserted into the *deoR* gene (K. Hammer-Jespersen, unpublished). Using merodiploids it has been shown that $deoR^+$ is dominant over *deoR* (K. Hammer-Jespersen, unpublished).

DEO R regulated promoters are not dependent upon the cAMP-CRP complex. In contrast it has been found that the synthesis of the Deo-enzymes is stimulated in cells lacking the cAMP-CRP complex (46).

Selection of deoR Mutants (Table 5.7)

(1) Thymidylate synthetase mutants (*thyA*) require 160 μM thymine for growth. When *thyA* mutants are plated on 2.4 μM thymine secondary mutants appear, some of which have simultaneously acquired mutations in the *deoB* and *deoR* genes (2, 102, and Chapter 4). The combination of the block in the catabolism of deoxyribose 1-phosphate (*deoB* mutation) and the high level of thymidine phosphorylase (*deoR* mutation) ensures that a *thyA*,*deoB*,*deoR* mutant can grow in the presence of only 2.4 μM thymine. By this selection also *cytR deoB* and $deoO^c$ *deoB* mutants have been isolated (102, 3). The selection may be modified in the following way: in the presence of uridine the thymine requirement of a *thyA deoB* strain is increased, due to inhibition of thymidine phosphorylase by uridine, and only mutants with raised levels of thymidine phosphorylase will survive on plates with 16 μM thymine and 2 mM uridine (16).

(2) When *E. coli* or *S. typhimurium* is plated on inosine as sole carbon source fast growing colonies arise spontaneously; these are either *deoR* or *cytR* mutants (102, 106). The faster growth of the *deoR* mutant on inosine may be explained by the fact that it contains higher levels of purine nucleoside phosphorylase and phosphopentomutase than does a wild type strain.

* The *deoR* gene product is designated DEO R.

TABLE 5.7
Selection of regulatory mutants in nucleoside catabolism.

Regulatory gene	Required genetic background	Selection principle	Other mutants obtained by the selection	Reference
deoR	*thyA*	very low thymine (2.4 μM)	*deoB,cytR* or *deoB,deoO*[c]	23, 16, 102
		fast growth on inosine	*cytR*	102, 106
cytR	*thyA*	very low thymine (2.4 μM)	*deoB,deoR* or *deoB,deoO*[c]	102
		fast growth on inosine	*deoR*	102, 106
		fast growth on uridine[a]		102
	thyA deoA	thymine (160 μM)	*udpO*[c]	22, 133
xapR[a]	*deoD*	growth on inosine		26

[a] Not in *S. typhimurium*.

Isolation of *cytR* mutants by this procedure is probably mainly due to a selection for an increased transport of inosine, since the levels of purine nucleoside phosphorylase and phosphopentomutase are only slightly increased in the *cytR* mutants.

2. *The CYT R* regulon*

CYT R regulates the synthesis of six nucleoside catabolizing enzymes (Table 5.6) and the expression of three nucleoside transport systems (G-system, C-system, and the *tsx* gene product, cf. Chapter 6). Furthermore the CYT R regulated promoters are dependent upon CRP and cAMP (45, 46, 82, 100, 106).

Cytidine, adenosine and in *S. typhimurium* also uridine serve as effectors for CYT R (44, 106). Various structural analogs of cytidine and uridine were tested as effectors for the CYT R protein in *S. typhimurium*. Analogs of the nucleosides with substitutions in 5 and 5′ position are effective as inducers, but substitutions in the 6, 2′, and 3′ positions are not allowed. 5′-O-methylcytidine, 5′-O-methyluridine, N^4-methyl- and N_3-methylcytidine are gratuitous inducers (80).

The *cytR* gene encodes a repressor protein as indicated by the high levels of the Deo-enzymes, cytidine deaminase and uridine phosphorylase in the following mutants: (a) strains containing amber mutations in the *cytR* gene (45); (b) mutants with phage λ inserted into the *cytR* gene (K. Hammer-Jespersen, unpublished). Using merodiploids it has been shown that *cytR*$^+$ is dominant over *cytR* (102).

Selection of cytR mutants. Several positive selections for *cytR* mutants have been designed which make use of the increased levels of thymidine phosphorylase, uridine phosphorylase, or nucleoside transport systems in the *cytR* mutants (Table 5.7).

(1) The two selections employed for isolation of *deoR* mutants may be applied to obtain *cytR* mutants also.

(2) A *deoA thyA* mutant cannot grow on thymine but requires thymidine for growth because of the lack of thymidine phosphorylase (*see* Chapter 4). Uridine phosphorylase, however, has a slight activity towards thymidine (Table 5.3). Thus, if the level of this enzyme is increased by a *cytR* mutation the resulting mutant (*cytR deoA thyA*) grows in the presence of 160 μM thymine (22, 133).

(3) When *E. coli* cells are plated on uridine as sole carbon source fast growing colonies arise spontaneously, these are *cytR* mutants (102). This selection is based on the fact that uridine does not induce the synthesis of uridine phosphorylase and phosphopentomutase, which are both required for

* The *cytR* gene product is designated CYT R.

growth under these conditions. The selection cannot be used in *S. typhimurium* where uridine does induce these enzymes.

3. The XAP R* regulon

XAP R regulates the synthesis of xanthosine phosphorylase in *E. coli*, but it does not affect the synthesis of any of the other enzymes involved in nucleoside catabolism (Table 5.6). Xanthosine is the only effector found for the $xapR^{+}$ gene product (47). XAP R seems to be needed for the expression of xanthosine phosphorylase i.e. XAP R is probably an activator protein. This conclusion is based on the low frequency with which mutants with high levels of xanthosine phosphorylase are isolated and on the phenotypes of these mutants (26). One class of *xapR* mutants were clearly shown to harbour XAP R protein with changed effector specificity: in these strains xanthosine phosphorylase was induced by inosine and deoxyinosine as well as by xanthosine. In merodiploid strains $xapR/xapR^{+}$ intermediate levels of xanthosine phosphorylase were found (26). Further studies are needed to decide whether XAP R has both repressor and activator functions like the *araC* regulatory protein.

xapR mutants are isolated as pseudorevertants of *deoD* mutants Table 5.7. On inosine as carbon source the selection principle is high levels of xanthosine phosphorylase in presence of inosine (for further explanation *see* D 4).

F. Mechanism of the regulation of the *deo*-operon

1. A model for the deo-operon regulation

The *deo* operon contains four structural genes (*deoC,A,B* and *D*). All four genes code for enzymes responsible for the catabolism of deoxyribonucleosides; in addition the two distal gene products participate in the catabolism of ribonucleosides (Fig. 5.2 and 5.3). Transcription of all four structural genes into tetracistronic mRNA molecules (mRNA-1a and mRNA-1b) is initiated at two adjacent promoter regions *deoP* and *cytP* located to the left of the *deoC* gene, the order being *deoP cytP deoC*. Initiations from *deoP* are inhibited by DEO R. The corresponding site on the DNA is called *deo* operator (*deoO*). Initiations from *cytP* are dependent upon the presence of CRP and cAMP, and are inhibited by CYT R, when this is bound to the *cyt* operator site (*CYTO*). The synthesis of the tetracistronic mRNA-1 is thus subject to a dual negative control, but also a positive control (CRP + cAMP) is involved. The two distal genes, *deoB* and *deoD* (encoding phosphopentomutase and purine nucleoside phosphorylase) are furthermore transcribed into a dicistronic mRNA molecule (mRNA-2). The synthesis of mRNA-2 is initiated from a promoter (P2) located just before the *deoB* gene. Inosine and guanosine induce the synthesis of mRNA-2 but the corresponding regulatory protein(s) is not identified (Fig.

* The *xapR* gene product is designated XAP R.

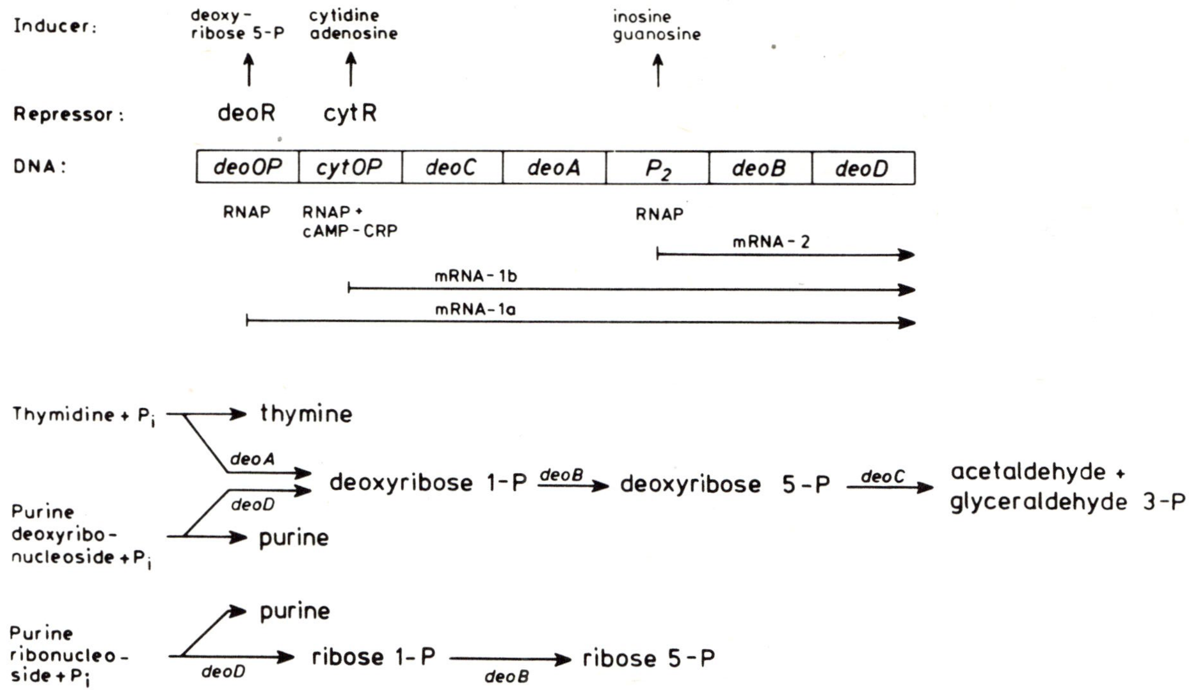

FIG. 5.3. Model for regulation of the *deo* operon. The Deo enzymes, deoxyriboaldolase, thymidine phosphorylase, phosphopentomutase and purine nucleoside phosphorylase, are coded for by *deoC*, *deoA*, *deoB*, and *deoD*, respectively, and located in a cluster counterclockwise to *thr*. One operator, *deoO*, binds the *deoR* repressor, the other one, *cytO*, binds the *cytR* repressor. For a proper binding of RNA polymerase to the *cyt* promoter (*cytP*) cAMP and CRP are needed, but initiation at the *deoP* is independent of the cAMP-CRP complex. Deoxyriboaldolase and thymidine phosphorylase are induced by deoxyribose-5-phosphate and cytidine or adenosine. Similarly, phosphodeoxyribomutase and purine nucleoside phosphorylase are induced by these components, but in addition inosine and guanosine give rise to an induction of these enzymes from a promoter, located at the left side of the *deoB* gene (P2). Cytidine and adenosine inactivate CYT R, whereas deoxyribose 5-phosphate inactivates DEO R. The *cytR* and *deoR* genes are not linked to the *deo* operon. RNAP, RNA polymerase.

5.3). The final levels of *deoB* and *deoD* products in the cell therefore reflect a combination of events taking place at three different promoter regions on the DNA: *deoP*, *cytP*, and *P2*. The levels of deoxyriboaldolase and thymidine phosphorylase (encoded by *deoC* and *deoA*) are regulated from two promoter sites: *deoP* and *cytP*.

2. In vivo *evidence for the model*

Regulatory mutants. The effects of *cytR* and *deoR* mutations on the Deo-enzyme levels are seen in Tables 5.6 and 5.8. The levels of deoxyriboaldolase and thymidine phosphorylase encoded by the *deoC* and *deoA* genes are increased 10 fold by a *cytR* mutation, 60 fold by a *deoR* mutation and about 120 fold when the cell is devoid of both CYT R and DEO R. The corresponding increase in the levels of phosphopentomutase and purine nucleoside phosphorylase is always smaller, and the regulation of the four Deo-enzymes is non-coordinate. The enzyme levels in the *deoR cytR* mutant are higher than the sum of the enzyme levels in the *deoR* mutant and the *cytR* mutant (Table 5.8), indicating a cooperative action of the DEO R and CYT R repressors. In the *cytR* mutant and in the *cytR deoR* mutant the Deo-enzymes are subject to catabolite repression, i.e. the Deo-enzyme levels are about four fold lower when the mutants have been grown on glucose as carbon source instead of glycerol. In contrast the Deo-enzyme levels in the wild type strains and the *deoR* mutants are almost the same whether the cells have grown on glucose or glycerol (45). These results indicate that the main part of the Deo-enzyme levels in the *cytR deoR* mutant originates from a promoter which needs CRP and cAMP i.e. the *cyt* promoter. DEO R should in accordance with this hypothesis be able to repress the expression from *cytP* (line 2 compared to line 3 and 4 in Table 5.8).

It must be emphasized that *cytP* and *deoP* are strong promoters. From Table 5.9 it is estimated that the Deo-enzymes in a *cytR deoR* mutant constitute about 7% of the total protein. Moreover the four Deo-enzymes seem to be synthesized in approximately equimolar amounts in the *deoR cytR* mutants.

The CYT R proteins of *E. coli* and *S. typhimurium* are probably closely related, since a merodiploid *cytR* mutant of *S. typhimurium* carrying the *E. coli cytR*$^+$ allele on an episome has repressed enzyme levels. The Deo-enzymes, cytidine deaminase and uridine phosphorylase are in this strain inducible by cytidine but not by uridine, proving that uridine is not effector for the *E. coli* CYT R (J. Neuhard, unpublished).

Units of transcription. The noncoordinate regulation of the Deo-enzymes mentioned above indicates that phosphopentomutase and purine nucleoside

TABLE 5.8

Enzyme levels in different mutants.

		Enzyme Activity (units/mg protein)[a]				
	Genotype	Deoxyriboaldolase (*deoC*)	Thymidine phosphorylase (*deoA*)	Phosphodeoxy-ribomutase (*deoB*)	Purine nucleoside phosphorylase (*deoD*)	Uridine phosphorylase (*udp*)
1	wild type	0.02	0.05	0.07	0.18	0.08
2	*cytR*	0.15	0.64	0.23	0.33	2.4
3	*deoR*	1.4	3.5	1.2	2.0	0.09
4	*cytR deoR*	2.3	9.4	2.1	3.6	2.5
5	*deoA*::(Mu)	0.02	0.001	0.05	0.13	0.06
6	*cytR deoR* *deoA*::(Mu)	2.0	0.020[b]	0.06	0.09	2.2

[a] The enzyme levels are measured after exponential growth with glycerol as carbon source. 1 unit = 1 μmol product formed per minute at 37°C.

[b] This level of activity is due to the high levels of uridine phosphorylase in the *cytR* mutant (compare line 5).

TABLE 5.9

Gene-protein relationship in the *deo*-operon of *E. coli* K12.

Gene symbol	Enzyme	Molecular weight	Specific activity (units/mg)	Base pairs per gene[a]	Enzyme in *deoR cytR* strain[b]	
					% of total protein	Numbers of subunits per cell
deoC	deoxyriboaldolase	1 × 27 737	245	777	1.00	41 200
deoA	thymidine phosphorylase	2 × 47 000	300	1281	3.12	79 600
deoB	phosphopentomutase	1 × 45 000	370	1206	0.57[c]	14 580[c]
deoD	purine nucleoside phosphorylase	6 × 23 700	195	663	1.80	91 100

[a] The sum of the four *deo*-genes, 3927 base pairs, corresponds to 2.6×10^6 daltons of DNA.

[b] These numbers are calculated from the enzyme levels shown in Table 5.8 line 4, and the specific activity of the pure proteins (this table).

[c] The apparent low amount of phosphopentomutase is probably an artefact due to inhibition of this enzyme activity in crude extracts.

phosphorylase belong to a transcriptional unit different from that of deoxyriboaldolase and thymidine phosphorylase. Furthermore the separate induction of phosphopentomutase and purine nucleoside phosphorylase by inosine and guanosine strongly implies that the *deoB*,*deoD* genes constitute a separate transcriptional unit (48, 119).

The direction and the length of the transcripts from the *deo*-operon was determined by use of polar mutants. Insertion of phage Mu into the *deoA* gene reduced the expression from the *deoB* and *deoD* genes drastically in a *deoR cytR* strain (Table 5.8 and 24). The same results were obtained by introduction of polar amber mutations into the *deoC* gene (6). These results indicate a clockwise transcription of the *deo*-operon into a tetracistronic mRNA-1. Also in other strains with increased expression of the *deo*-operon (i.e. *cytR* and *deoR* mutants) this polarity effect on the distal gene products was seen (6, 24). However, in a wild type background ($deoR^+$ $cytR^+$) the polar mutations in the *deoA* and the *deoC* genes show no significant polarity in the expression of the *deoB* and the *deoD* genes (Table 5.8 line 5 compared with line 1). The results with polar *deoC* and *deoB* mutations in a $cytR^+$ $deoR^+$ strain indicate the existence of two separate operons (*deoC deoA* and *deoB deoD*) both being transcribed clockwise (12, 22).

The combined results with polar mutations in the *deo*-operon may be explained by the model of overlapping transcription of the *deoB* and the *deoD* genes into a tetracistronic mRNA-1 and a dicistronic mRNA-2 species (Fig. 5.3), assuming that the main part of the *deoB* and *deoD* gene products in a $cytR^+$ $deoR^+$ strain originates from mRNA-2.

The internal promoter (P2) must be located after the polar mutations in the *deoA* gene. The levels of phosphopentomutase and purine nucleoside phosphorylase were the same in all the *deoA* : Mu mutants, whether the strains were $cytR^+$ or $cytR^-$, $deoR^+$ or $deoR^-$. Furthermore the synthesis of the enzymes was still inducible by inosine and guanosine. It was therefore concluded that the synthesis of mRNA-2 is not regulated by DEO R or by CYT R but by another factor which responds to inosine and guanosine (24).

Operator mutants. It is possible to isolate operator constitutive mutants ($deoO^c$) by the very low thymine selection (Table 5.7). So far one such mutant has been described (3). This operator mutation is a deletion covering part of the regulatory region and the *deoC* gene. In this mutant the expression of thymidine phosphorylase (*deoA*) is increased 20-fold compared to a wild type strain and the *deo*-operon expression is insensitive to glucose repression. The Deo-enzyme levels were increased two-fold by removal of DEO R, while removal of CYT R had no effect (5). From these results it is concluded that the deletion mutant has lost the *cyt* promoter and most of the *deoR* operator site on the DNA, suggesting the gene order *deoP,cytP,deoC* on the *E. coli* chromosome.

3. In vitro *evidence for the model*

The *in vitro* system developed by Zubay (152) has been used to investigate more specifically the molecular mechanisms which are involved in gene expression from the *deo*-operon. The Zubay system allows coupled transcription and translation of a specific added DNA template using a crude cell-free extract (S-30) as source of enzymes and ribosomes. The S-30 extracts must be prepared from strains devoid of any Deo-enzyme activities. As DNA templates λ phages or plasmids carrying the *deo* operon or parts of it have been used (74, 142).

By separating the transcription process from the translation process in the Zubay system it was demonstrated that both the DEO R and the CYT R regulation of the *deo* operon occurs at the transcriptional level (134).

DEO R regulation. Of a number of sugar phosphates tested only deoxyribose 5-phosphate was effector for DEO R *in vitro* (P. Valentin-Hansen, personal communication). The deoxyribose 5-phosphate induction is independent of addition of cAMP to the *in vitro* system, confirming that the cAMP–CRP complex is not necessary for transcription from *deoP*. Deoxyribose 5-phosphate cannot induce the Deo-enzymes when S-30 extracts prepared from *deoR* mutants are used. Addition of wild type extract to the S-30 extract from the *deoR* mutant represses the enzyme synthesis and it is now inducible by deoxyribose 5-phosphate. This confirms the repressor nature of the *deoR* gene product. Furthermore this kind of mixing experiments provide an assay for DEO R.

By using increasing concentrations of *deo*-DNA in the *in vitro* system the amount of DEO R in the S-30 extract is gradually titrated out by the added DNA (Fig. 5.4). These experiments are in accordance with *deoR* encoding a repressor protein. The repressor titrations and the S-30 mixing experiments allowed an estimate of the dissociation constant for the *deo* DNA-DEO R complex of 1.0×10^{-10} M. This is of the same order of magnitude as the K_D for the interaction of *lac*-repressor with *lac*-operator, when the high ionic strength in the Zubay system—corresponding to 0.34 M NaCl—is taken in consideration (116, 143). From the same experiments it was calculated that an *E. coli* cell contains at least 20 molecules of DEO R (143).

The degree of induction was measured at several different nonsaturating deoxyribose 5-phosphate concentrations. From these results a $K_d = 2 \times 10^{-6}$ M was calculated for the complex between deoxyribose 5-phosphate and DEO R, and it was found that binding of one molecule of

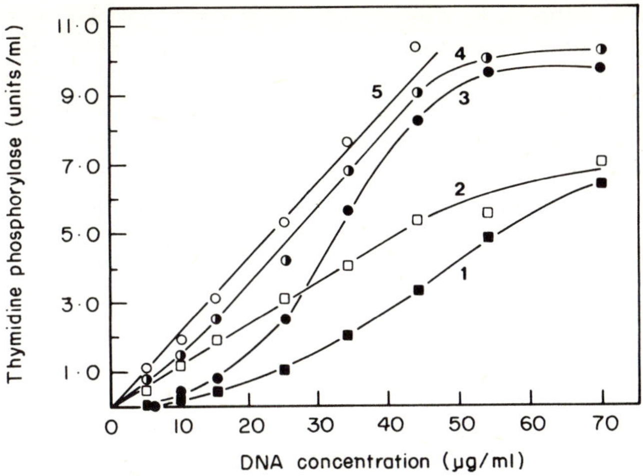

FIG. 5.4. Titration of repressors by addition of increasing amounts of *deo*DNA to the Zubay *in vitro* system. Synthesis of thymidine phosphorylase was carried out for 40 min by use of S-30 extract from a *Δ deo cytR*$^+$ *deoR*$^+$ strain and in presence of different low molecular effectors. The curves represent: ■——■ no addition (1); □——□ 250 μM deoxyribose 5-phosphate is added (2); ●——● 250 μM cAMP is added (3); ◑——◑ 250 μM cAMP + 250 μM deoxyribose 5-phosphate is added (4); ○——○ 250 μM cAMP + 250 μM deoxyribose 5-phosphate + 50 μM cytidine + 250 μM tetrahydrouridine is added (5).

10 μg/ml of *λ deo* DNA corresponds to a concentration of 4.3×10^{-10}M of *deo* DNA during the *in vitro* synthesis. Thymidine phosphorylase activity is given as μmol thymidine converted per min, per ml of *in vitro* synthesis mixture.

deoxyribose 5-phosphate per molecule of DEO R is sufficient for induction (143).

CYT R regulation. Only adenosine and cytidine are able to induce the Deo-enzymes *in vitro*, but two conditions are necessary in order to get induction with these nucleosides: (i) their deamination by cytidine deaminase and adenosine deaminase must be avoided; this may be achieved by addition of the strong competitive inhibitors tetrahydrouridine and coformycin. (ii) cAMP must be added to the *in vitro* system together with the inducer. This confirms that the cAMP–CRP complex is necessary for transcription from *cytP*.

Cytidine and adenosine cannot induce the Deo-enzymes when the S-30 extract used is prepared from a *cytR* mutant. In this case cAMP alone will induce. Addition of wild type extract to the S-30 extract from the *cytR* mutant represses the synthesis of the Deo-enzymes in the presence of cAMP, and renders them inducible by cytidine and adenosine (143). This confirms that *cytR* encodes a repressor protein. In the same way as for DEO R the following constants were found for CYT R: $K_d = 10^{-9}$ M for the *deo*DNA; $K_d = 7 \times 10^{-7}$ M for cytidine and $K_d = 6 \times 10^{-6}$ M for adenosine. Furthermore binding of one molecule of cytidine or adenosine per molecule of *CYT R* is sufficient for induction (143). CYT R has a ten-fold higher dissociation constant for the adenosine complex than for the cytidine complex. This explains the *in vivo* finding that only mutants partially blocked in adenosine degradation are induced by adenosine.

The affinity of CYT R for the *deo*-DNA is one order of magnitude smaller than that of DEO R, but on the other hand the *E. coli* cell seems to contain at least 100 molecules of CYT R i.e. four-fold the amount of DEO R (143).

Identification of promoters and transcriptional units. From the *λ deo* phages hybrid plasmids of colEl and pBR322 containing the *deo*-operon have been constructed. Use of EcoRI restriction nuclease for this subcloning results in a plasmid containing all four structural *deo*-genes and *cytP* but without a functional *deoP*. When these plasmids are used as template *in vitro* the Deo-enzymes are only regulated by CYT R and CRP. These results show that the CYT R regulated promoter *cytP* is adjacent to the *deoC* gene while the DEO R regulated promoter *deoP* is located further away (Fig. 5.3) (74).

That *cytP* and *deoP* are indeed different promoters with different properties was also shown with *λdeo* DNA as template (142). The kinetics of initiation of transcription was followed using rifampicin, which is an inhibitor of the initiation process of RNA polymerase, but not of the elongation process. The RNA polymerase forms initiation complexes immediately with *cytP*, while with *deoP* a lag period of two minutes is observed (Fig. 5.5). The kinetics of appearance of completed mRNA for the different *deo*-genes can be followed using streptolydigin, an inhibitor of the elongation process of RNA polymerase. In Fig. 5.5 the synthesis of mRNA for *deoC* and *deoA* is followed. The temporal displacement between the rifampicin and the streptolydigin curves represents the time it takes to elongate the mRNA from the respective promoters to the end of the *deoA* gene. It takes four minutes $\pm$30 seconds for both promoters. Assuming an elongation rate of 10 nucleotides per second as found for the *lacZ* gene in the same *in vitro* system, this gives an estimate of 2400 ± 300 base-pairs for the mRNA transcribed *in vitro* containing the *deoC* and the *deoA* genes. According to Table 5.9 the structural genes alone should comprise 2058 base pairs. These results as well as the induction results show that the transcription studied *in vitro* originates from the correct promoters.

In another experiment (142) the kinetics of appearance of mRNA were followed for all four *deo*-genes using streptolydigin. The expected transcription order *deoC deoA deoB deoD* was found, thus proving the existence of mRNA-1 (Fig. 5.3).

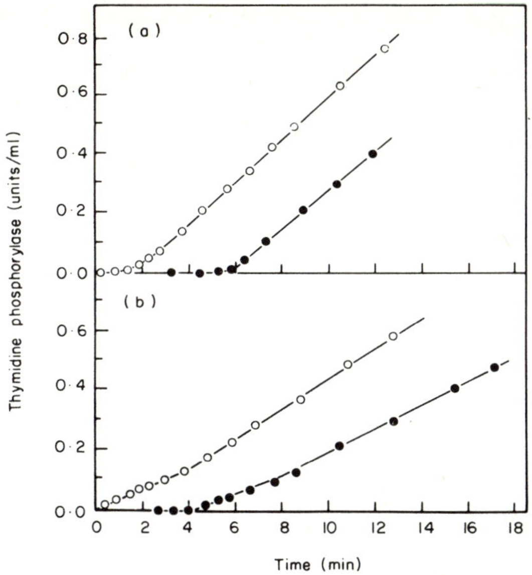

FIG. 5.5. Kinetics of initiation and elongation of *deo* mRNA-1 from *deoP* and *cytP*. The synthesis of thymidine phosphorylase was followed in the Zubay *in vitro* system using 2.1×10^{-10}M *λ deo* DNA as template. The synthesis was started at t = zero, by addition of prewarmed S-30 extract from a *Δ deo deoR*$^+$ *cytR*$^+$ *E. coli* strain. At the times indicated samples were transferred to tubes containing rifampicin (○) or streptolydigin (●) and incubation was continued for another 60 min before protein synthesis was stopped and thymidine phosphorylase activities were measured. Thymidine phosphorylase activity is given as μmol thymidine converted per min per ml of *in vitro* synthesis mixture. (a) deoxyribose 5-phosphate is added during the synthesis. (b) cAMP + cytidine + tetrahydrouridine is added during the synthesis.

Also mRNA-2 could be made *in vitro*, proving that the internal promoter P2 is active in the Zubay system. Transcription from P2 is independent of the cAMP–CRP complex (142).

Structure of the deo-*operon.* A contiguous sequence of 1800 base pairs has been established using the Maxam–Gilbert DNA sequencing method. This DNA sequence comprises *deoP*,*cytP*,*deoC* and the promoter proximal part of the *deoA* gene (140, 141). *DeoP* and *cytP* have been located by DNA–mRNA hybridization followed by digestion with nuclease S1 (140). The promoter proximal sequences of the *deoC* and *deoA* genes have been verified by amino acid sequencing of the N-termini of the purified proteins (141 and I. Svendsen and K. Hammer-Jespersen, unpublished). These structural studies show that *cytP* is located 44 base pairs before the start of the *deoC* gene while *deoP* is located 644 base pairs before *deoC*. Thus a DNA sequence of 600 base pairs separates the two promoters.

A common DNA sequence of 19 base pairs including the Pribnow box is found in both *cytP* and *deoP*. The DNA sequences are shown below, aligned with respect to the start of the mRNA's:

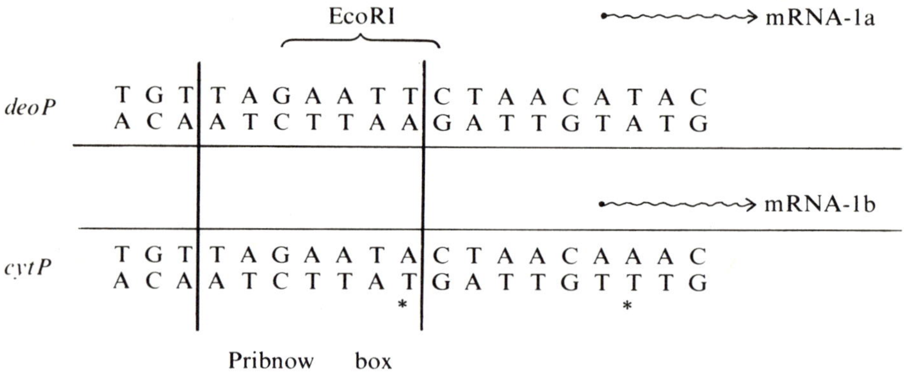

The first 16 base pairs in each sequence form a palindrome structure. The EcoRI site separating the *deo*-genes from functional *deoP* is found within the *deoP* sequence. One of the two base changes labeled with an asterisk in the homologous *cytP* region is found in this EcoRI site.

Also the *P2* region has been sequenced; it is located in an intercistronic region between the *deoA* and *deoB* genes, 33 base pairs from the start of the *deoB* gene and 19 base pairs from the TAA stop codon for the *deoA* gene (P. Valentin-Hansen, personal communication).

4. *Discussion of the* deo-*operon regulation*

The *deo*-operon exemplifies a regulation by overlapping transcriptional control. This mode of regulation imposes a greater variety of control mechanisms for gene expression. According to the model in Fig. 5.3 the *deoB* and *deoD* genes are transcribed from three different promoters each having a separate control mechanism. Apparently these gene products are needed in different amounts depending upon the growth conditions of the cell. An overlapping transcriptional control has also been found for the *CI* gene in phage λ (30, 113) and for the *gal*-operon in *E. coli* (104).

From a molecular point of view this kind of regulation gives rise to some questions. In the *deo*-operon RNA-polymerase initiating the synthesis of mRNA-1a from *deoP* has to pass two other promoter regions: *cytP* and *P2*. What happens when RNA polymerase engaged in elongation of mRNA passes a promoter site? Does it result in activation or inhibition of this promoter; and for the elongating mRNA-polymerase; does it partially recognize the new promoter site as a terminator site? Gilbert (38) calls attention to the similarity of the DNA structure of promoter and terminator sites.

It is puzzling that mRNA-1a is 600 base pairs longer than mRNA-1b. Do some of these 600 base pairs encode another protein/peptide? Could such a product have a regulatory function for the gene expression of the *deo*-operon?

Another puzzling feature is the great similarity in DNA sequence between *deoP* and *cytP*; it implies that they share a common control protein. This is also supported by the palindrome sequences found in both promoter regions. The regulatory protein could be DEO R. That would explain the ten-fold higher Deo-enzyme levels found in a *cytR deoR* mutant compared with a *cytR* mutant (Table 5.8) and the fact that the Deo-enzymes in both cases originate from mRNA-1b.

G. The physiological role of nucleoside catabolizing enzymes

1. Nucleosides as carbon sources

The presence of the nucleoside catabolizing enzymes in the cell ensures that *E. coli* is able to utilize the pentose part of all the nucleosides as carbon source (Fig. 5.2 and Table 5.5). Most of the nucleosides are also good carbon sources because the nucleoside often causes induction of the enzymes involved in its degradation. The correlation between nucleoside catabolism and induction by nucleosides is given in Table 5.10. CYT R and DEO R are the major control proteins. From the table it is seen that cytidine through the CYT R system regulates the synthesis of several more enzymes than needed for the catabolism of cytidine. Furthermore the CYT R regulation is the only control system in Table 5.10 which is dependent upon the cAMP–CRP complex (except maybe the XAP R system). One may therefore speculate whether cytidine might serve as a signal for the presence of both ribonucleosides and deoxyribonucleosides, indicating for the cell that carbon sources are available. Nucleosides will be found together when they are formed as degradation products from RNA and DNA in case of cell death.

Deoxyribonucleosides will also induce their own pathway of degradation through the DEO R regulation of the Deo-enzymes, and inosine/guanosine can separately induce their pathway of degradation. These two regulatory systems are, however, independent on the cAMP–CRP control, indicating that the corresponding pathways may serve other functions than making deoxyribose/ribose available as carbon source for the cell.

The DEO R regulation may for instance be important for the cell when it is infected with foreign DNA. It is known that such nonmodified DNA is

TABLE 5.10

Correlation between catabolism of nucleosides and their induction pattern.

Nucleoside added to growth medium	Enzymes[a] involved in catabolism of the nucleoside	Regulated structural genes	Effector	Regulatory gene
Cytidine	*cdd udp deoB*	*cdd udp deoCABD*	Cytidine	*cytR*
Adenosine	*add deoD deoB*	*cdd*[b] *udp*[b] *deoCABD*[b]	Adenosine[b]	*cytR*[b]
		add	Adenine/hypoxanthine	unknown
Deoxyribonucleosides	*deoA deoB deoC deoD (cdd add)*[c]	*deoCABD*	Deoxyribose 5-phosphate	*deoR*
Inosine	*deoD deoB*	*deoD deoB*	Inosine	unknown
Guanosine	*deoD deoB*	*deoD deoB*	Guanosine	unknown
Xanthosine	*xapA deoB*	*xapA*	Xanthosine	*xapR*

[a] Indicated by the corresponding structural genes.
[b] Only in mutants blocked in adenosine catabolism.
[c] *cdd* or *add* is only involved when the nucleoside is deoxycytidine or deoxyadenosine respectively; deoxyadenosine regulates *add*, while deoxycytidine does not regulate *cdd*.

degraded very quickly and the deoxyribonucleoside catabolism might be the last step in the pathway.

The inosine/guanosine induction might secure an efficient pathway for salvage of the purine bases through the following pathway:

$$\text{inosine} \xrightarrow{\text{hypoxanthine}} \text{ribose 1-phosphate} \longrightarrow \text{ribose 5-phosphate}$$

$$\xrightarrow[\text{ATP}]{} \text{phosphoribosyl 1-pyrophosphate} \xrightarrow{\text{hypoxanthine}} \text{IMP} + \text{PP}_i$$

The enzymes involved are purine nucleoside phosphorylase, phosphopentomutase, phosphoribosyl 1-pyrophosphate synthetase and IMP pyrophosphorylase (cf. Chapter 2).

2. Nucleosides as nitrogen source

Cytidine, adenosine and the corresponding deoxyribonucleosides may also be used as nitrogen source in *E. coli* and *S. typhimurium*. These nucleosides are substrates for cytidine deaminase and adenosine deaminase respectively (Fig. 5.2). No evidence for nitrogen control of the synthesis of these enzymes has been found.

Adenosine deaminase is inducible by hypoxanthine and adenine (Table 5.6). In order to understand this regulation it must be remembered that the cell contains two pathways of adenosine catabolism:

(a) $\text{adenosine} \xrightarrow[add]{} \text{inosine} \xrightarrow[deoD]{} \text{hypoxanthine} + \text{ribose 1-phosphate}$

(b) $\text{adenosine} \xrightarrow[deoD]{} \text{adenine} + \text{ribose 1-phosphate}$

In pathway (a) adenosine is first deaminated by adenosine deaminase (*add*). The product inosine is then cleaved by purine nucleoside phosphorylase (*deoD*). This is the main pathway of adenosine catabolism in *E. coli* (P. Nygaard, personal communication). Inosine induces purine nucleoside phosphorylase and the product of the last reaction, hypoxanthine, induces adenosine deaminase. If adenosine accumulates it may serve as effector for CYT R and induce purine nucleoside phosphorylase. In this way more adenine will be formed which induces adenosine deaminase further. By using primarily pathway (a) for adenosine/deoxyadenosine catabolism the cell is able to utilize the 6-amino group of the nucleoside as nitrogen source and hypoxanthine as purine source.

III. OTHER MICROORGANISMS

A. Pathways of nucleoside catabolism

The glycosylic bond in the nucleoside may be split in two ways: by a nucleoside phosphorylase or by a nucleoside hydrolase (Fig. 5.1). For further catabolism of the pentose moiety the nucleoside phosphorylase pathway requires the presence of phosphopentomutase while the nucleoside hydrolase pathway requires ribo- or deoxyribokinase (*see* Fig. 5.7).

There are several differences between these two pathways of nucleoside catabolism. The reactions catalyzed by nucleoside phosphorylases are reversible, while the hydrolytic cleavage of the glycosylic bond is an irreversible process. The hydrolytic pathway gives one molecule of ATP less per molecule of nucleoside consumed than the phosphorolytic pathway. Finally the substrate specificities of nucleoside phosphorylases and nucleoside hydrolases are different.

1. Distribution of nucleoside phosphorylases and nucleoside hydrolases

Nucleoside phosphorylases have been found in a variety of microorganisms (60, 108, 122). Purine nucleoside phosphorylase and uridine phosphorylase have been purified from *Bacillus subtilis*, *Erwinia carotowora* and *Corynebacterium sepedonicum* (39, 66, 123, 124, 125). From the available data also including higher organisms it may be concluded that three main types of nucleoside phosphorylases exist: (1) purine nucleoside phosphorylases, (2) thymidine phosphorylases and (3) uridine phosphorylases.

(1) Purine nucleoside phosphorylases have both purine ribonucleosides and purine deoxyribonucleosides as substrates. However, they may vary in their specificities towards the purine base (108).

(2) Thymidine phosphorylases are specific for deoxyuridine and thymidine. Deoxycytidine will not serve as substrate.

(3) Uridine phosphorylases have activity towards nucleosides containing thymine, uracil and some pyrimidine analogs, but not towards cytidine and deoxycytidine. The enzymes can act on both ribonucleosides and deoxyribonucleosides, however, the activity towards deoxyribonucleosides is usually lower. In certain microorganisms (*Bacillus*, *Haemophilus influenza*; 125, 129) the uridine and thymidine phosphorylase activities reside in one single protein. It is called pyrimidine nucleoside phosphorylase, but according

to its substrate specificity it should be classified as a uridine phosphorylase.

Some microorganisms have been shown to lack one or more nucleoside phosphorylase and a number of these organisms contain nucleoside hydrolases (Table 5.11). *Neisseria*, *Acinetobacter* and *Moraxella* are devoid of both nucleoside phosphorylases and nucleoside hydrolases while *Saccharomyces* and *Physarum* contain both these activities. It should be noted that some of the bacterial genera are heterogeneous with respect to their nucleoside catabolism (Table 5.11; 33, 34, 50, 70, 71, 72, 73, 91).

Nucleoside hydrolases appear to be abundant in fungi (Table 5.11). The properties of some nucleoside hydrolases purified from microorganisms are given in Table 5.12. According to their substrate specificities four major groups of nucleoside hydrolases can be distinguished: Ribonucleoside hydrolases, purine ribonucleoside hydrolases, purine deoxyribonucleoside hydrolases and uridine hydrolases.

Ribonucleoside hydrolase is specific for ribonucleosides but has both purine and pyrimidine nucleosides as substrates. Also cytidine is a substrate.

Purine ribonucleoside hydrolase is specific for *purine ribo*nucleosides. No activity towards pyrimidine nucleosides or purine deoxyribonucleosides is found. Enzyme g from *Aspergillus niger* is unique in that it also shows activity towards 5′-AMP (83).

Purine deoxyribonucleoside hydrolase is specific for *purine deoxy*ribonucleosides. It has recently been discovered and it is the first hydrolase found to cleave deoxyribonucleosides (79).

Uridine hydrolase is specific for uridine and 5-methyluridine. Neither cytidine nor pyrimidine deoxyribonucleosides are substrates (91).

The pattern of substrate specificity of the nucleoside hydrolases thus differs from that found for the nucleoside phosphorylases in the following respects:

1. No single nucleoside hydrolase can cleave both ribo- and deoxyribonucleosides.
2. Hydrolases acting on pyrimidine deoxyribonucleosides have so far not been found.
3. Cytidine is substrate for some hydrolases.
4. The same enzyme may show activity towards both purine and pyrimidine nucleosides.

In the following parts of this chapter some microorganisms have been selected. They include: *Bacillus* species which contain only nucleoside phosphorylases; *Lactobacillus* species, which contain trans-N-deoxyribosylases

TABLE 5.11
Presence of nucleoside catabolizing enzymes in various microorganisms.

Enzyme	Organism													
	Neisseria	*Acinetobacter*	*Moraxella*[b]	*Azotobacter*	*Rhizobium*	*Pseudomonas acidovorans*[c]	*Lactobacillus*	*Diplococcus*	*Anacystis*	*Penicillium*	*Neurospora*	*Rhodotorula*	*Saccharomyces*	*Physarum*
Purine nucleoside phosphorylase	–	nd	nd	–	nd	–	–	nd	nd	–	nd	nd	+	+
Thymidine phosphorylase	–	–	–	nd	–	–	–	–	–	nd	–	–	–	–
Uridine phosphorylase	–	nd	nd	–	nd	–	–	–	nd	nd	–	–	–	–
Trans-N-deoxyribosylase	–	–	–	nd	nd	–	+	–	nd	nd	–	nd	–	–
Purine nucleoside hydrolase[a]	–	nd	nd	+	nd	–	+	nd	nd	+	nd	nd	+	–
Pyrimidine hydrolase[a]	–	nd	nd	–	nd	–	+	–	nd	nd	+	+	+	+
Adenosine deaminase	nd	nd	nd	+	nd	–	nd	nd	nd	–	nd	nd	nd	–
Cytidine deaminase	–	nd	nd	nd	nd	–	–	+	nd	nd	+	nd	+	+
References	70 71 73	72	72	126 149	94	77	57 135 138	14 36	115	31	11 131 132	145	41, 42 50, 61 91	33 34

\+ and – means presence and absence respectively of enzyme.
nd = not determined.
[a] Indicates whether purine or pyrimidine nucleosides are degraded, not the type of hydrolase present.
[b] *Moraxella kingii* contain thymidine phosphorylase (72).
[c] *Pseudomona fluorescens* contain nucleoside hydrolase (137).

and nucleoside hydrolases; *Saccharomyces cerevisiae* which contain both a nucleoside phosphorylase and a nucleoside hydrolase and finally *Neurospora crassa* which contain a thymidine hydroxylase and a uridine hydrolase.

B. *Bacillus*

1. Pathways of nucleoside catabolism

The pathways by which nucleosides are catabolized in *B. subtilis* are given in Fig. 5.6. They differ from the pathways in *E. coli* (Fig. 5.2) in that no xanthosine cleaving activity and no adenosine deaminating activity is found in *B. subtilis*. Furthermore the phosphorolysis of uridine and thymidine is catalyzed by one single enzyme, pyrimidine nucleoside phosphorylase.

Two purine nucleoside phosphorylases have been purified from *B. subtilis*: adenosine phosphorylase and inosine phosphorylase. The adenosine phosphorylase shows activity towards deoxyadenosine (100%), adenosine (24%) and a low activity towards deoxyguanosine (4%). The molecular weight of the holoenzyme is 153 000 and 25 500 for the subunits, indicating a hexameric structure similar to the *E. coli* enzyme (66). The inosine phosphorylase has activity towards guanosine (100%), inosine (83%), deoxyguanosine (93%), and deoxyinosine (63%), but no activity towards adenine nucleosides. The enzyme has a molecular weight of 95 000 and subunits of 28 000, suggesting a trimeric structure. The inosine phosphorylase thus resembles the enzyme from human erythrocytes in both substrate specificity and molecular structure (66, 150).

From *B. cereus* a purine nucleoside phosphorylase has been purified which appears to be very similar to the inosine phosphorylase from *B. subtilis*. It has a molecular weight of 90–96 000 and shows activity towards inosine, guanosine and the corresponding deoxyribonucleosides, but no activity towards adenine nucleosides (37). Enzyme preparations purified from vegetative cells or from spores had identical properties except that the spore enzyme in absence of phosphate seemed to have a lower molecular weight (from 54 000 to 70 000) indicating a dissociation under these conditions (39). Also adenosine phosphorylase seems to be present in *B. cereus* (103).

Pyrimidine nucleoside phosphorylase from *B. stearothermophilus* has been purified and characterized (Chapter 3 and 125). It was demonstrated that both the uridine and thymidine phosphorylase activities reside in one protein with a molecular weight of 78 000. The molecular weight of pyrimidine nucleoside phosphorylase from *B. subtilis* appears to be lower than 78 000 (125).

The lack of adenosine deaminase is not general for the *Bacillus*, since this enzyme has been demonstrated in both *B. cereus* and *B. anthracis* (96, 109, 130).

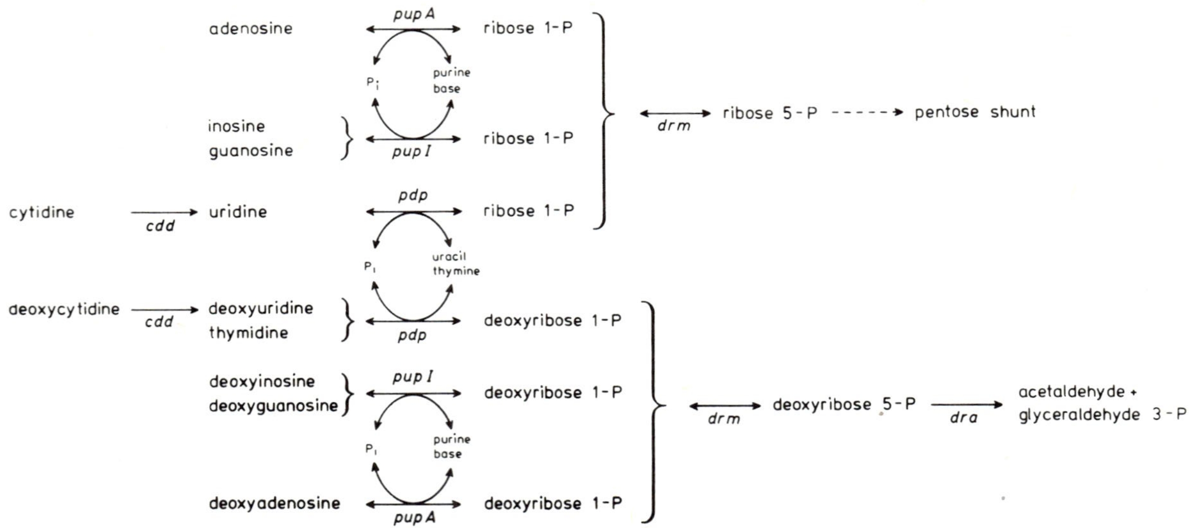

FIG. 5.6. Pathways of nucleoside catabolism in *B. subtilis*. The enzymes are listed by the code designating their respective structural genes. The six enzymes are: cytidine deaminase (*cdd*) (117); adenosine phosphorylase (*pupA*) (Neuhard, J., personal communication); inosine phosphorylase (*pupI*) (78); pyrimidine nucleoside phosphorylase (*pdp*) (120); phosphopentomutase (*drm*) (120); deoxyriboaldolase (*dra*) (120).

2. *Isolation and characterization of mutants defective in nucleoside catabolism*

Mutants defective in each of the six nucleoside catabolizing enzymes have been isolated in *B. subtilis* (Fig. 5.6) (78, 117, 120). The selection procedures employed are similar to those used in *E. coli* (compare Table 5.4). Mutants defective in inosine phosphorylase (*pupI*) have hitherto been isolated only in strains with a purine requirement, while mutations in the gene encoding adenosine phosphorylase (*pupA*) are easily obtained in prototrophic strains (J. Neuhard, personal communication, 78).

The simultaneous loss of uridine and thymidine phosphorylase activity in mutants defective in pyrimidine nucleoside phosphorylase (*pdp*) shows that only one enzyme catalyzes these reactions (120).

Mutants in cytidine deaminase (*cdd*) have lost both cytidine and deoxycytidine deaminating activities showing that the cytidine deaminase in *B. subtilis* contains both these activities (117).

Genetic mapping using phage PBS1 has shown that the *pdp* gene and the gene coding for deoxyriboaldolase (*dra*) are closely linked to each other and cotransduce with the *purA* gene. The *pupA* gene and the gene coding for phosphopentomutase (*drm*) are however not closely linked to the *pdp* and *dra* genes (J. Neuhard, personal communication, 120). Thus *B. subtilis* does not contain an operon similar to the *deo*-operon in *E. coli*.

3. *Induction of nucleoside catabolizing enzymes*

In *B. subtilis* pyrimidine nucleoside phosphorylase and deoxyriboaldolase are induced five-fold by addition of deoxyribonucleosides to the medium. The true inducing compound appears to be deoxyribose 5-phosphate (120). This conclusion is based on the following evidence: (a) Ribonucleosides do not induce. (b) In *thy drm* strains acetaldehyde but not thymidine causes induction of pyrimidine nucleoside phosphorylase, presumably because acetaldehyde promotes the formation of deoxyribose 5-phosphate through the deoxyriboaldolase reaction (Fig. 5.6). (c) In *thy dra* mutants five-fold increased levels of pyrimidine nucleoside phosphorylase were found while *thy drm* mutants contained wild type levels.

Mutants with 15-fold elevated levels of pyrimidine nucleoside phosphorylase and deoxyriboaldolase have been isolated (120), confirming that these two enzymes belong to the same regulon.

The two purine nucleoside phosphorylases, phosphopentomutase and cytidine deaminase are not induced by deoxyribonucleosides or ribonucleosides (J. Neuhard, personal communication).

In *B. cereus* the level of inosine phosphorylase increases 15-fold during sporulation. In vegetative cells the enzyme may be induced ten-fold by addition of inosine (32, 37).

The induction pattern of other nucleoside catabolizing enzymes in *B. cereus* has not been studied.

In *B. stearothermophilus* both uridine and thymidine were shown to induce pyrimidine nucleoside phosphorylase (125).

4. Physiological role of nucleoside catabolizing enzymes

Bacilli contain pathways for the utilization of most nucleosides as carbon source (Fig. 5.6). Nucleosides are poor carbon sources for *B. subtilis*, deoxyribonucleosides and uridine giving the fastest growth (J. Neuhard, personal communication). This might be explained by inefficient nucleoside transport or by the limited enzyme induction by deoxyribonucleosides alone.

As discussed in Chapter 2, 3 and 4 the nucleoside catabolizing enzymes in bacilli are involved in the salvage pathways.

Spore germination in *Bacilli* is stimulated by addition of inosine and it has been shown that the ribose part of the nucleoside is catabolized to CO_2 (145). Sporulation in *B. subtilis* may be initiated by guanine starvation (89). Thus it would be of interest to study initiation of sporulation and spore germination in purine prototrophic strains lacking inosine phosphorylase.

C. Lactic acid bacteria

1. Lactobacilli *with a growth requirement for deoxyribonucleosides*

Lactobacilli require purine and pyrimidine bases for growth and most species have an additional requirement for deoxyribonucleosides. Due to the presence of trans-N-deoxyribosylases the deoxyribonucleoside requiring strains can grow when only one deoxyribonucleoside and the nucleobases are provided in the media (*see* Chapter 4). In accordance with their growth requirement these strains (e.g. *L. delbrückii*, *L. helveticus*, *L. leichmanii*, *L. acidophilus*) are unable to catabolize deoxyribonucleosides: they do not contain any nucleoside phosphorylases or hydrolases with activity towards deoxyribonucleosides nor do they contain phosphopentomutase or deoxyriboaldolase (75, 138).

In contrast these lactobacilli do degrade ribonucleosides. Ribonucleoside hydrolases have been found in several species (35, 75, 135). The enzyme from *L. delbrückii* has been purified fifty-fold. It shows activity against both purine- and pyrimidine ribonucleosides (Table 5.12 and 135). These strains are not able to ferment exogenous ribose, suggesting that they do not contain a ribokinase and hence cannot use ribonucleosides as carbon source. The main function of the nucleoside hydrolases in these strains may therefore only serve one purpose: supply of nucleobases for salvage pathways (Chapters 2, 3, 4).

2. Lactobacilli *without a requirement for deoxyribonucleosides*

Certain *Lactobacillus* species such as *L. casei* do *not* require deoxyribonucleosides for growth. They are devoid of trans-N-deoxyribosylases and ribonucleoside hydrolases. Instead they contain nucleoside phosphorylases (75).

Catabolism of ribose and deoxyribose has been investigated in *L. plantarum*. This species contains a deoxyribokinase and a deoxyriboaldolase which are both induced by addition of deoxyribose to the growth medium. The deoxyribokinase shows activity towards ribose (40, 110). In addition a ribokinase which is specific for ribose is induced by ribose (40).

3. Streptococci

Streptococci appear to resemble *L. casei* in their mode of catabolizing nucleosides. They are devoid of trans-N-deoxyribosylases and nucleoside hydrolases, and seem to contain nucleoside phosphorylase (87, 136). Deoxyriboaldolase has been found in all group N streptococci, and both thymidine phosphorylase, phosphopentomutase and deoxyriboaldolase activities were found in *S. lactis* (87).

D. Saccharomyces cerevisiae

1. Pathways of nucleoside catabolism

Ribonucleoside catabolism. Six enzymes are involved in the catabolism of ribonucleosides in yeast: cytidine deaminase, two nucleoside hydrolases, one nucleoside phosphorylase, phosphopentomutase and ribokinase. The resulting pathways are shown in Fig. 5.7.

Cytidine deaminase (MW 57 000) has been purified. The enzyme is active against both cytidine ($K_m = 0.25$ mM) and deoxycytidine ($K_m = 0.09$ mM) (61).

Uridine hydrolase has been purified to homogeneity. It is a dimer with subunits of 17 000 daltons. The enzyme is specific for uridine and 5-methyluridine (Table 5.12, 91).

A purine ribonucleoside hydrolase (Table 5.12) and a purine nucleoside phosphorylase were partially purified from the same cells. Purine deoxyribonucleosides were not tested as substrates for any of the two enzymes (50).

The phosphopentomutase activity in yeast was demonstrated by Guarino and Sable (42). It was shown to be caused by an enzyme different from phosphoglucomutase. The activity towards deoxyribose 1-phosphate was not tested.

TABLE 5.12
Substrate specificity of nucleoside hydrolases.

		Substrate specificity (relative activity)													
Enzyme	Organism	cytidine	uridine	adenosine	inosine	guanosine	xanthosine	deoxyadenosine	deoxyinosine	deoxyguanosine	pyrimidine deoxyribonucleosides	5′-AMP	MW	Comments[a]	References
Ribonucleoside hydrolase	*Lactobacillus delbrückii*	54	7	103	100	90	0	nd	nd	nd	nd	nd	nd	I: xanthosine, uric acid riboside	135
	Leishmania donovani	41	100	0	20	0	78	0	0	0	0	0	180 000	I: adenosine and guanosine	79
	Pseudomonas fluorescense	109	100	30	5	6	7	0	nd	0	0	0	nd	pure	137

	Aspargillus[b] *niger*	0	0	25	100	7	0	0	0	0	0	29	nd	pure	83
	Azotobacter vinlandii	0	0	2	100	4	40	nd	nd	nd	nd	0	70 000	pure	149
Purine ribonucleoside hydrolase	*Leishmania donovani*	0	0	0	67	100	0	0	0	0	0	0	205 000	I: adenosine	79
	Penicillium[c]	0	0	59	100	14	42	0	0	0	0	0	nd	extra cellular location	112
	Saccharomyces cerevisiae	0	0	100	50	20	1	nd	nd	nd	nd	0	nd		50
Purine deoxy-ribonucleoside hydrolase	*Leishmania donovani*	0	0	0	0	0	0	23	100	86	0	0	33 000		79
Uridine hydrolase	*Saccharomyces cerevisiae*	0	100	0	0	0	0	0	0	0	0	0	32 500	Pure Two subunits ContainsCu^{2+}	91

nd = not determined.

[a] I = inhibitor(s).

[b] This enzyme has been named nucleosidase g (83); it has only been found in *A. niger*.

[c] This enzyme (nucleosidase a) has been found in many species of *Penicillium* as well as in several species of *Aspergillus* including *A. niger*.

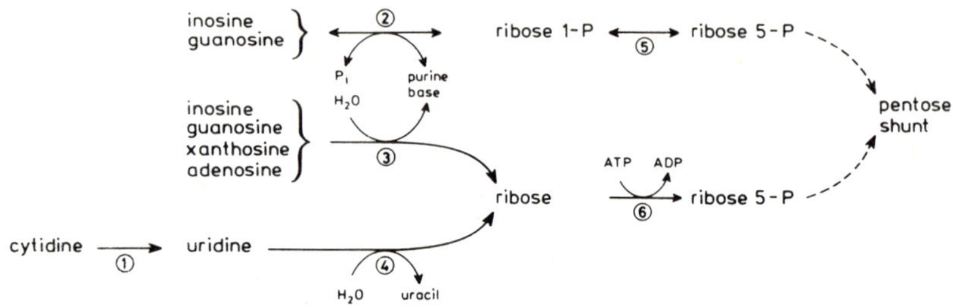

FIG. 5.7. Pathways of nucleoside catabolism in yeast. The six enzymes are: 1. cytidine deaminase (61); 2. purine nucleoside phosphorylase (50); 3. purine ribonucleoside hydrolase (50); 4. uridine hydrolase (41, 91); 5. phosphoribomutase (42); 6. ribokinase (121).

A ribokinase was purified five-fold (121). It was specific for ribose and had no activity towards deoxyribose.

As seen in Fig. 5.7 the presence of these enzymes ensures the catabolism of both purine and pyrimidine ribonucleosides by hydrolytic pathways. Inosine and guanosine, however, may be degraded either hydrolytically or phosphorolytically. The latter pathway occurs through the combined action of purine nucleoside phosphorylase and phosphopentomutase. One other group of organisms (the protozoans) also contains both purine nucleoside hydrolase and purine nucleoside phosphorylase in the same cell (79, 84). The physiological significance and possible regulation of these two pathways would be interesting to know.

Deoxyribonucleoside catabolism. Very little is known about catabolism of deoxyribonucleosides in yeast.

Deoxycytidine is a substrate for cytidine deaminase. Deoxyuridine—the product formed by this reaction—is however not a substrate for uridine hydrolase or any other known enzymes in yeast. Thymidine (deoxyuridine) kinase, uridine and thymidine phosphorylase have not been found in yeast (Table 5.11). Therefore it appears that deoxyuridine is metabolically inert in *Saccharomyces*. This is not the case in *Rhodotorula glutinis* which contains thymidine/deoxyuridine 2′-hydroxylase (148), similar to the enzyme found in *Neurospora crassa* (*see* Section E).

The abilities of purine nucleoside phosphorylase and phosphopentomutase to act on deoxyribose compounds have not been tested although these two enzymes usually show activity towards both ribose and deoxyribose compounds. It is therefore likely that yeast is able to degrade deoxyinosine and deoxyguanosine through the combined action of purine nucleoside phosphorylase, phosphopentomutase and deoxyriboaldolase. The latter enzyme has not yet been demonstrated in yeast.

2. *Regulation of uridine hydrolase*

The level of uridine hydrolase is not controlled by the presence of uridine or uracil in the growth media. But when bakers yeast is grown on ethanol instead of on glucose as carbon source the level of uridine phosphorylase is increased ten-fold (92). Furthermore, addition of glucose to ethanol grown cells results in a "catabolite inactivation" of the uridine hydrolase present. This has been shown to be caused by proteolytic degradation of uridine hydrolase by yeast proteinase A. The yeast cells also contain a protein inhibitor of this proteinase (93). Such a regulation of the uridine hydrolase activity suggests that the physiological function of the enzyme is to make ribose available as a carbon source.

Several other proteins involved in formation of glucose from other carbon sources are subject to a similar proteolytic inactivation in yeast, for instance: fructose-1,6-bisphosphatase, phosphoenolpyruvate carboxykinase and α-glucoside permease (for review see Holzer, 58).

E. *Neurospora crassa*

1. *Pathways of pyrimidine nucleoside degradation*

Neurospora crassa is devoid of enzymes capable of cleaving the glycosylic bond of pyrimidine deoxyribonucleosides (Table 5.11). Nevertheless thymidine (deoxyuridine) is degraded by this organism. A novel enzyme activity: a pyrimidine deoxyribonucleoside-2′-hydroxylase was discovered in *Neurospora*, an enzyme which catalyzes the conversion of thymidine (deoxyuridine) into 5-methyluridine (uridine) (11, 49, 131). This is the only known direct transformation of deoxyribose to ribose in living organisms. Neither deoxycytidine nor purine deoxyribonucleosides are substrates for the pyrimidine deoxyribonucleoside hydroxylase (11, 132). For further description of the enzyme *see* Chapter 4.

Uridine and 5-methyluridine are substrates for a uridine hydrolase and the ribose is further utilized (95, 131). *Neurospora crassa* contains cytidine deaminase activity (131) which ensures that both cytidine and deoxycytidine may be catabolized (Fig. 5.8).

The salvage of the pyrimidine bases are described in Chapter 3 (*see also* Fig. 3.4).

Mutants deficient in pyrimidine deoxyribonucleoside hydroxylase have been isolated. Their phenotype confirmed that both thymidine and deoxyuridine are substrates for this enzyme and that no other conversion of these nucleosides is possible in *Neurospora* (131).

2. *Pathways of purine nucleoside catabolism*

Evidence has been presented that a purine ribonucleoside hydrolase is present

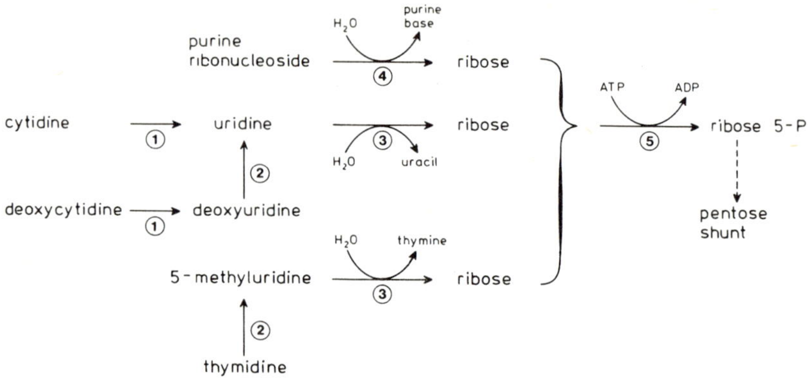

FIG. 5.8. Pathways of nucleoside catabolism in *Neurospora crassa*. The five enzymes are: 1. cytidine deaminase (131); 2. pyrimidine deoxyribonucleoside-2′-hydroxylase encoded by *uc-2* (131); 3. uridine hydrolase (132); 4. purine ribonucleoside hydrolase (95); 5. ribokinase (95).

in *Neurospora* (95). Purine ribonucleosides may therefore be utilized as carbon source through the combined action of purine ribonucleoside hydrolase and ribokinase (Fig. 5.8).

Deoxyadenosine is able to serve as adenine source in adenine requiring mutants (90, 146) but it is not known how the cleavage of the glycosylic bond is catalyzed. No 2′-hydroxylase activity towards purine deoxyribonucleosides has been found in *Neurospora* (11). Thus the fungi must contain either a purine nucleoside phosphorylase or a purine deoxyribonucleoside hydrolase.

REFERENCES

1. Ahmad, S. I., and Pritchard, R. H. (1969). A map of four genes specifying enzymes involved in catabolism of nucleosides and deoxynucleosides in *Escherichia coli. Molec. Gen. Genet.* **104**, 351–359.
2. Ahmad, S. I., and Pritchard, R. H. (1971). A regulatory mutant affecting the synthesis of enzymes involved in the catabolism of nucleosides in *Escherichia coli. Molec. Gen. Genet.* **111**, 77–83.
3. Ahmad, S. I., and Pritchard, R. H. (1973). An operator constitutive mutant affecting the synthesis of two enzymes involved in the catabolism of nucleosides in *Escherichia coli. Molec. Gen. Genet.* **124**, 321–328.
4. Aksamit, R. R., and Koshland, D. E. (1974). Identification of the ribose binding protein as the receptor for ribose chemotaxis in *Salmonella typhimurium. Biochemistry* **13**, 4473–4478.
5. Albrechtsen, H., and Ahmad, S. I. (1980). Regulation of the synthesis of nucleoside catabolic enzymes in *Escherichia coli*: Further analysis of a *deoO*^{c} mutant strain. *Molec. Gen. Genet.* **179**, 457–460.

6. Albrechtsen, H., Hammer-Jespersen, K., Munch-Petersen, A., and Fiil, N. (1976). Multiple regulation of nucleoside catabolizing enzymes: Effects of a polar *dra* mutation on the *deo* enzymes. *Molec. Gen. Genet.* **146**, 139–145.
7. Alikhanian, S. I., Iljina, T. S., Kaliaeva, E. S., Kameneva, S. V., and Sukhodolets, V. V. (1966). A genetic study of thymineless mutants of *Escherichia coli* K-12. *Genet. Res.* **8**, 83–100.
8. Anderson, A., and Cooper, R. A. (1970). Biochemical and genetical studies on ribose catabolism in *Escherichia coli* K 12. *J. Gen. Microbiol.* **62**, 335–339.
9. Baker, B. R. (1967). "Design of active-site-directed irreversible enzyme inhibitors", pp. 79–95. John Wiley and Sons, New York.
10. Baker, B. R., and Hopkins, S. E. (1970). Irreversible enzyme inhibitors. CLXVII. Thymidine phosphorylase. X. On the nature and dimensions of the hydrophobic bonding region. II. *J. Med. Chem.* **1**, 87–89.
11. Bankel, L., Lindstedt, G., and Lindstedt, S. (1972). Thymidine 2′-hydroxylation in *Neurospora crassa. J. Biol. Chem.* **247**, 6128–6134.
12. Baumanis, G. E., Smirnov, Yu. V., and Sukhodolets, V. V. (1974). Production and study of polar mutants for nucleoside catabolism linked genes in *Escherichia coli. Genetika* **10**, 81–88.
13. Beacham, I. R., Eisenstark, A., Barth, P. T., and Pritchard, R. H. (1968). Deoxynucleoside-sensitive mutants of *Salmonella typhimurium. Molec. Gen. Genet.* **102**, 112–127.
14. Bean, B., and Tomasz, A. (1971). Inhibitory effects and metabolism of 5-fluoropyrimidine derivatives in *pneumococcus. J. Bacteriol.* **106**, 412–420.
15. Beck, C. F., Ingraham, J., Neuhard, J., and Thomassen, E. (1972). Metabolism of pyrimidine and pyrimidine nucleosides by *Salmonella typhimurium. J. Bacteriol.* **110**, 219–228.
16. Blank, J., and Hoffee, P. (1972). Regulatory mutants of the *deo* regulon in *Salmonella typhimurium. Molec. Gen. Genet.* **116**, 291–298.
17. Blank, J. G., and Hoffee, P. A. (1975). Purification and properties of thymidine phosphorylase from *Salmonella typhimurium. Arch. Biochem. Biophys.* **168**, 259–265.
18. Bose, R., and Yamada, E. W. (1974). Uridine phosphorylase, molecular properties and mechanism of catalysis. *Biochemistry* **13**, 2051–2056.
19. Breitman, T. R., and Bradford, R. M. (1967). The absence of deoxyriboaldolase activity in a thymineless mutant of *Escherichia coli* strain 15. A possible explanation for the low thymine requirement of some thymineless strains. *Biochim. Biophys. Acta* **138**, 217–220.
20. Breitman, T. R., and Bradford, R. M. (1968). Inability of two thymine-requiring mutants of *Escherichia coli* lacking phosphodeoxyribomutase to be induced for thymidine phosphorylase and deoxyriboaldolase. *J. Bacteriol.* **95**, 2434–2435.
21. Budman, D. R., and Pardee, A. B. (1967). Thymidine and thymine incorporation into deoxyribonucleic acid: Inhibition and repression by uridine of thymidine phosphorylase of *Escherichia coli. J. Bacteriol.* **94**, 1546–1550.
22. Buxton, R. S. (1975). Genetic analysis of thymidine-resistant and low-thymine-requiring mutants of *Escherichia coli* K 12 induced by bacteriophage Mu-1. *J. Bacteriol.* **121**, 475–484.
23. Buxton, R. S. (1979). Fusion of the *lac* genes to the proximal promoters of the *deo* operon of *Escherichia coli* K 12. *J. General Microbiol.* **112**, 241–250.
24. Buxton, R. S., Albrechtsen, H., and Hammer-Jespersen, K. (1977). Overlapping transcriptional units in the *deo* operon of *Escherichia coli* K 12. Evidence from Phage Mu-1 insertion mutants. *J. Mol. Biol.* **114**, 287–300.

25. Buxton, R. S., Hammer-Jespersen, K., and Hansen, T. D. (1978). Insertion of bacteriophage lambda into the *deo* operon of *Escherichia coli* K 12 and isolation of plaque-forming λ *deo*$^+$ transducing bacteriophages. *J. Bacteriol.* **136**, 668–681.
26. Buxton, R. S., Hammer-Jespersen, K., and Valentin-Hansen, P. (1980). A second purine nucleoside phosphorylase in *Escherichia coli* K-12. I. Xanthosine phosphorylase regulatory mutants isolated as secondary-site revertants of a *deoD* mutant. *Molec. Gen. Genet.* **179**, 331–340.
27. Cohen, R. M., and Wolfenden, R. (1971). Cytidine deaminase from *Escherichia coli. J. Biol. Chem.* **246**, 7561–7565.
28. Cohen, R. M., and Wolfenden, R. (1971). The equilibrium of hydrolytic deamination of cytidine and N^4-methylcytidine. *J. Biol. Chem.* **246**, 7566–7568.
29. Doskočil, J., and Holý, A. (1977). Specificity of purine nucleoside phosphorylase from *Escherichia coli. Collection Czechoslov. Chem. Commun.* **42**, 370–383.
30. Echols, H., and Green, L. (1971). Establishment and maintenance of repression by bacteriophage lambda: The role of the cI, cII, and cIII proteins. *Proc. Natl. Acad. Sci. U.S.A.* **68**, 2190–2194.
31. Elzainy, T. A., and Allam, A. M. (1973). Nucleoside catabolism in *Penicillium chrysogenum. Egypt. J. Chem.* **16**, 329–335.
32. Engelbrecht, H. L. (1972). Time course of purine nucleoside phosphorylase occurrence in sporulation of *Bacillus cereus. J. Bacteriol.* **111**, 33–36.
33. Fink, K., and Nygaard, P. (1978). Pyrimidine metabolism in microplasmodia of *Physarum polycephalum. Eur. J. Biochem.* **89**, 417–424.
34. Fink, K., and Nygaard, P. (1979). Purine metabolism in microplasmodia of *Physarum polycephalum. Biochim. Biophys. Acta* **563**, 269–277.
35. Friedkin, M. (1953). Purine and pyrimidine metabolism in microorganisms. *J. Cell. Comp. Physiol.* **41** *Supplement 1*, 261–282.
36. Friedman, L. R., and Ravin, A. W. (1972). Genetic and biochemical properties of thymidine-dependent mutants of *Pneumococcus. J. Bacteriol.* **109**, 459–461.
37. Gardner, R., and Kornberg, A. (1967). Biochemical studies of bacterial sporulation and germination. V. Purine nucleoside phosphorylase of vegetative cells and spores of *Bacillus cereus. J. Biol. Chem.* 2383–2388.
38. Gilbert, W. (1976). Starting and stopping sequences for the RNA polymerase. *In* "RNA polymerase" (R. Losick and M. Chamberlin, *Eds.*), pp. 193–205. Cold Spring Harbor Laboratory, U.S.A.
39. Gilpin, R. W., and Sadoff, H. L. (1971). Physical and catalytic properties of the purine nucleoside phosphorylases from cells and spores of *Bacillus cereus T. J. Biol. Chem.* **246**, 1475–1480.
40. Ginsburg, A. (1959). A deoxyribokinase from *Lactobacillum plantarum. J. Biol. Chem.* **234**, 481–487.
41. Grenson, M. (1969). The utilization of exogenous pyrimidines and the recycling of uridine-5′-phosphate derivatives in *Saccharomyces cerevisiae*, as studied by means of mutants affected in pyrimidine uptake and metabolism. *Eur. J. Biochem.* **11**, 249–260.
42. Guarino, A. J., and Sable, H. Z. (1955). Studies on phosphomutases. II. Phosphoribomutase and phosphoglucomutase. *J. Biol. Chem.* **215**, 515–526.
43. Hammer-Jespersen, K., and Munch-Petersen, A. (1970). Phosphodeoxyribomutase from *Escherichia coli.* Purification and some properties. *Eur. J. Biochem.* **17**, 397–407.
44. Hammer-Jespersen, K., and Munch-Petersen, A. (1973). Mutants of *Escherichia coli* unable to metabolize cytidine: Isolation and characterization. *Molec. Gen. Genet.* **126**. 177–186.
45. Hammer-Jespersen, K., and Munch-Petersen, A. (1975). Multiple regulation of

nucleoside catabolizing enzymes: Regulation of the *deo* operon by the *cytR* and the *deoR* gene products. *Molec. Gen. Genet.* **137**, 327–335.
46. Hammer-Jespersen, K., and Nygaard, P. (1976). Multiple regulation of nucleoside catabolizing enzymes in *Escherichia coli*: Effects of 3 : 5′ cyclic AMP and CRP protein. *Molec. Gen. Genet.* **148**, 49–55.
47. Hammer-Jespersen, K., Buxton, R. S., and Hansen, T. D. (1980). A second purine nucleoside phosphorylase in *Escherichia coli* K-12. II. Properties of xanthosine phosphorylase and its induction by xanthosine. *Molec. Gen. Genet.* **179**, 341–348.
48. Hammer-Jespersen, K., Munch-Petersen, A., Nygaard, P., and Schwartz, M. (1971). Induction of enzymes involved in the catabolism of deoxyribonucleosides and ribonucleosides in *Escherichia coli* K 12. *Eur. J. Biochem.* **19**, 533–538.
49. Hayaishi, O., Nozaki, M., Abbott, M. T. (1975). Oxygenases: Dioxygenases. *In* "The Enzymes" (P. D. Boyer, *Ed.*), Vol. 12, 3rd Edition, pp. 119–189. Academic Press, London.
50. Heppel, L. A., and Hilmoe, R. J. (1952). Phosphorolysis and hydrolysis of purine ribosides by enzymes from yeast. *J. Biol. Chem.* **198**, 683–694.
51. Hoffee, P. A. (1968). 2-deoxyribose-5-phosphate aldolase of *Salmonella typhimurium*: Purification and properties. *Arch. Biochem. Biophys.* **126**, 785–802.
52. Hoffee, P. A. (1968). 2-deoxyribose gene-enzyme complex in *Salmonella typhimurium*. I. Isolation and enzymatic characterization of 2-deoxyribose-negative mutants. *J. Bacteriol.* **95**, 449–457.
53. Hoffee, P., and Blank, J. (1978). Thymidine phosphorylase from *Salmonella typhimurium*. *In* "Methods in Enzymology" (P. A. Hoffee and M. E. Jones, *Eds.*), Vol. 51, pp. 437–442. Academic Press, New York.
54. Hoffee, P. A., and Robertson, B. C. (1969). 2-deoxyribose gene-enzyme complex in *Salmonella typhimurium*: Regulation of phosphodeoxyribomutase. *J. Bacteriol.* **97**, 1386–1396.
55. Hoffee, P., Snyder, P., Sushak, C., and Jargiello, P. (1974). Deoxyribose-5-phosphate aldolase: Subunit structure and composition of active site lysine region. *Arch. Biochem. Biophys.* **164**, 736–742.
56. Hoffmeyer, J., and Neuhard, J. (1971). Metabolism of exogenous purine bases and nucleosides by *Salmonella typhimurium*. *J. Bacteriol.* **106**, 14–24.
57. Holguin, J., and Cardinaud, R. (1975). Trans-N-deoxyribosylase: Purification by affinity chromatography and characterization. *Eur. J. Biochem.* **54**, 505–514.
58. Holzer, H. (1976). Catabolite inactivation in yeast. *Trends in Biochem. Sciences* **1**, 178–181.
59. Hosono, H., and Kuno, S. (1973). The purification and properties of cytidine deaminase from *Escherichia coli*. *J. Biochem. (Tokyo)* **74**, 797–803.
60. Imada, A., and Igarasi, S. (1967). Ribosyl and deoxyribosyl transfer by bacterial enzyme systems. *J. Bacteriol.* **94**, 1551–1559.
61. Ipata, P. L., and Cercignani, G. (1978). Cytosine and cytidine deaminase from yeast. *In* "Methods in Enzymology" (P. A. Hoffee and M. E. Jones, *Eds.*), Vol. 51, pp. 349–401. Academic Press, New York.
62. Jargiello, P., and Hoffee, P. (1972). Orientation of the *deo* genes and the *serB* locus in *Salmonella typhimurium*. *J. Bacteriol.* **111**, 296–297.
63. Jargiello, P., Stern, M. D., and Hoffee, P. (1974). 2-deoxyribose-5-phosphate aldolase: Genetic analysis of structure. *J. Mol. Biol.* **88**, 671–691.
64. Jargiello, P., Sushak, C., and Hoffee, P. (1976). 2-deoxyribose-5-phosphate aldolase: Isolation and characterization of proteins genetically modified in the active site region. *Arch. Biochem. Biophys.* **177**, 630–641.
65. Jensen, K. F. (1976). Purine-nucleoside phosphorylase from *Salmonella typhi-*

murium and *Escherichia coli*. Initial velocity kinetics, ligand binding and reaction mechanism. *Eur. J. Biochem.* **61**, 377–386.

66. Jensen, K. F. (1978). Two purine nucleoside phosphorylases in *Bacillus subtilis*. Purification and some properties of the adenosine-specific phosphorylase. *Biochim. Biophys. Acta* **525**, 346–356.
67. Jensen, K. F., and Nygaard, P. (1975). Purine nucleoside phosphorylase from *Escherichia coli* and *Salmonella typhimurium*. Purification and some properties. *Eur. J. Biochem.* **51**, 253–265.
68. Jochimsen, B., Nygaard, P., and Vestergaard, T. (1975). Location on the chromosome of *Escherichia coli* of genes governing purine metabolism. Adenosine deaminase (*add*), guanosine kinase (*gsk*) and hypoxanthine phosphoribosyltransferase (*hpt*). *Molec. Gen. Genet.* **143**, 85–91.
69. Josephsen, J., and Hammer-Jespersen, K. (1981). Fusion of the *lac* genes to the promoter for the cytidine deaminase gene of *Escherichia coli* K-12. *Molec. Gen. Genet.* **182**, 154–158.
70. Jyssum, S. (1974). Search for thymidine phosphorylase, nucleoside deoxyribosyltransferase and thymidine kinase in genus *Neisseria*. *Acta Path. Microbiol. Scand. Section B* **82**, 53–56.
71. Jyssum, S. (1974). Purine metabolism in *Neisseria meningitidis*. 1. Utilization of exogenous adenine. *Acta Path. Microbiol. Scand. Section B* **82**, 508–520.
72. Jyssum, S., and Bøvre, K. (1974). Search for thymidine phosphorylase, nucleoside deoxyribosyltransferase and thymidine kinase in *Moraxella*, *Acinetobacter* and allied bacteria. *Acta Path. Microbiol. Scand. Section B* **82**, 57–66.
73. Jyssum, S., and Jyssum, K. (1979). Metabolism of pyrimidine bases and nucleosides in *Neisseria meningitidis*. *J. Bacteriol.* **138**, 320–323.
74. Jørgensen, P., Collins, J., and Valentin-Hansen, P. (1977). On the structure of the *deo* operon of *Escherichia coli*. *Molec. Gen. Genet.* **155**, 93–102.
75. Kalckar, H. M. (1951). Biosynthetic aspects of nucleosides and nucleic acids. *Pubbl. Staz. Zool. Napoli* **23** *Supplement*, 87–103.
76. Karlstrøm, O. (1968). Mutants of *Escherichia coli* defective in ribonucleoside and deoxyribonucleoside catabolism. *J. Bacteriol.* **95**, 1069–1077.
77. Kelln, R. A., and Warren, R. A. J. (1974). Pyrimidine metabolism in *Pseudomonas acidovorans*. *Can. J. Microbiol.* **20**, 427–433.
78. Komatsu, K., Saijo, A., Kodaira, R., and Osawa, T. (1970). Derivation of purine nucleoside-producing strains from an adenine-lacking mutant of *Bacillus subtilis*. *Amino Acid Nucleic Acid (Japan)* **22**, 54–58.
79. Koszalka, G. W., and Krenitsky, T. A. (1979). Nucleosidases from *Leishmania donovani*. Pyrimidine ribonucleosidase, purine ribonucleosidase, and a novel purine 2′-deoxyribonucleosidase. *J. Biol. Chem.* **254**, 8185–8193.
80. Krajewska, E., and Shugar, D. (1975). Pyrimidine nucleoside analogues as inducers of pyrimidine nucleoside catabolizing enzymes in *Salmonella typhimurium*. *Mol. Biol. Rep.* **2**, 295–301.
81. Krenitsky, T. A. (1976). Uridine phosphorylase from *Escherichia coli*. Kinetic properties and mechanism. *Biochim. Biophys. Acta* **429**, 352–358.
82. Krieger-Brauer, H. J., and Braun, V. (1980). Functions related to the receptor protein specified by the *tsx* gene of *Escherichia coli*. *Arch. Microbiol.* **124**, 233–242.
83. Kuwahara, M., and Fujii, T. (1978). Purification and some properties of NAD-degrading purine nucleosidase from *Aspergillus niger*. *Can. J. Biochem.* **56**, 345–348.
84. Königk, E. (1978). Purine nucleotide metabolism in promastigotes of *Leishmania*

tropica: Inhibitory effect of allopurinol and analogues of purine nucleosides. *Tropenmed. Parasit.* **29**, 435–438.

85. Leer, J. C., and Hammer-Jespersen, K. (1975). Multiple forms of phosphodeoxyribomutase from *Escherichia coli*. Physical and chemical characterization. *Biochemistry* **14**, 599–607.
86. Leer, J. C., Hammer-Jespersen, K., and Schwartz, M. (1977). Uridine phosphorylase from *Escherichia coli*. Physical and chemical characterization. *Eur. J. Biochem.* **75**, 217–224.
87. Lees, G. J., and Jago, G. R. (1977). Formation of acetaldehyde from 2-deoxy-D-ribose-5-phosphate in lactic acid bacteria. *J. Dairy Res.* **44**, 139–144.
88. Lomax, M. S., and Greenberg, G. R. (1968). Characteristics of the *deo* operon: Role in thymine utilization and sensitivity to deoxyribonucleosides. *J. Bacteriol.* **96**, 501–514.
89. Lopez, J. M., Marks, C. L., and Freeze, E. (1979). The decrease of guanine nucleotides initiates sporulation of *Bacillus subtilis*. *Biochim. Biophys. Acta* **587**, 238–252.
90. Magill, J. M., Spencer, R. R., and Magill, C. W. (1974). Relationship between [8-^{14}C] adenosine transport and growth inhibition in *Neurospora crassa* strain *ad-8*. *J. Bacteriol.* **119**, 202–206.
91. Magni, G. (1978). Uridine nucleosidase from yeast. *In* "Methods in Enzymology" (P. A. Hoffee and M. E. Jones, *Eds.*), Vol. 51, pp. 290–296. Academic Press, New York.
92. Magni, G., Santarelli, I., Natalini, P., Ruggieri, S., and Vita, A. (1977). Catabolite inactivation of bakers'-yeast uridine nucleosidase. Isolation and partial purification of a specific proteolytic inactivase. *Eur. J. Biochem.* **75**, 77–82.
93. Magni, G., Pallotta, G., Natalini, P., Ruggieri, S., Santarelli, I., and Vita, A. (1978). Inactivation of uridine nucleosidase in yeast. *J. Biol. Chem.* **253**, 2501–2503.
94. Mattes, R. (1974). Investigation of nucleoside metabolism in *Rhizobium lupini*: Obligatory thymidine phosphorylase deficiency. *Hoppe-Seylers Zeitschr. Physiol. Chem.* **355**^{10}, p. 1229.
95. Mattoo, A. K., Dandekar, A. M., Trivedi, J. P., Majmudar, G. H., and Patel, D. M. (1979). Regulation of *Neurospora crassa* nucleosidases by some environmental factors. *Zeitschrift für Allgemeine Mikrobiologie* **19**, 253–260.
96. Molnár, J., and Prágai, B. (1973). Possible role of adenosine deaminase in suboptimal growth of the *Bacillus anthracis* adenine auxotroph mutant. *Acta Microbiol. Acad. Sci. Hung.* **20**, 255–265.
97. Munch-Petersen, A. (1968). Thymineless mutants of *Escherichia coli* with deficiencies in deoxyribomutase and deoxyriboaldolase. *Biochim. Biophys. Acta* **161**, 279–282.
98. Munch-Petersen, A. (1968). On the catabolism of deoxyribonucleosides in cells and cell extracts of *Escherichia coli*. *Eur. J. Biochem.* **6**, 432–442.
99. Munch-Petersen, A. (1970). Deoxyribonucleoside catabolism and thymine incorporation in mutants of *Escherichia coli* lacking deoxyriboaldolase. *Eur. J. Biochem.* **15**, 191–202.
100. Munch-Petersen, A., and Mygind, B. (1976). Nucleoside transport systems in *Escherichia coli* K 12: specificity and regulation. *J. Cell. Physiol.* **89**, 551–559.
101. Munch-Petersen, A., and Schwartz, M. (1972). Inhibition of the catabolism of deoxyribonucleosides in *Escherichia coli* after infection by T4 phage. *Eur. J. Biochem.* **27**, 443–447.
102. Munch-Petersen, A., Nygaard, P., Hammer-Jespersen, K., and Fiil, N. (1972).

Mutants constitutive for nucleoside-catabolizing enzymes in *Escherichia coli* K 12. Isolation, characterization and mapping. *Eur. J. Biochem.* **27**, 208–215.

103. Mura, U., Sgarella, F., and Ipata, P. L. (1978). Utilization of exogenous purine compounds in *Bacillus cereus.* Translocation of the ribose moiety of inosine. *J. Biol. Chem.* **253**, 7905–7909.
104. Musso, R. E., Di Lauro, R., Adhya, S., and de Crombrugghe, B. (1977). Dual control for transcription of the galactose operon by cyclic AMP and its receptor protein at two interspersed promoters. *Cell* **12**, 847–854.
105. Neuhard, J., and Ingraham, J. (1968). Mutants of *Salmonella typhimurium* requiring cytidine for growth. *J. Bacteriol.* **95**, 2431–2433.
106. Nygaard, P. (1973). Nucleoside-catabolizing enzymes in *Salmonella typhimurium.* Induction by ribonucleosides. *Eur. J. Biochem.* **36**, 267–272.
107. Nygaard, P. (1978). Adenosine deaminase from *Escherichia coli. In* "Methods in Enzymology" (P. A. Hoffee and M. E. Jones, *Eds.*), Vol. 51, 508–512. Academic Press, New York.
108. Parks, R. E. Jr., and Agarwal, R. P. (1972). Purine nucleoside phosphorylase. *In* "The Enzymes" (P. D. Boyer, *Ed.*), Vol. 7, pp. 483–514. Academic Press, London.
109. Powell, J. F., and Hunter, J. R. (1956). Adenosine deaminase and ribosidase in spores of *Bacillus cereus. Biochem. J.* **62**, 381–387.
110. Pricer, W. E. Jr., and Horecker, B. L. (1960). Deoxyriboaldolase from *Lactobacillus plantarum. J. Biol. Chem.* **235**, 1292–1298.
111. Pritchard, R. H., and Ahmad, S. I. (1971). Fluorouracil and the isolation of mutants lacking uridine phosphorylase in *Escherichia coli*: Location of the gene. *Molec. Gen. Genet.* **111**, 84–88.
112. Reese, E. T. (1968). Extracellular purine β-ribosidases from fungi. *Can. J. Microbiol.* **14**, 377–383.
113. Reichardt, L., and Kaiser, A. D. (1971). Control of λ repressor synthesis. *Proc. Natl. Acad. Sci. U.S.A.* **68**, 2185–2189.
114. Remy, C. N., and Love, S. H. (1968). Induction of adenosine deaminase in *Escherichia coli. J. Bacteriol.* **96**, 76–85.
115. Restaino, L., and Frampton, E. W. (1975). Labeling the deoxyribonucleic acid of *Anacystis nidulans. J. Bacteriol.* **124**, 155–160.
116. Riggs, A. D., Suzuki, H., and Bourgeois, S. (1970). *lac* repressor-operator interaction. I. Equilibrium studies. *J. Mol. Biol.* **48**, 67–83.
117. Rima, B. K., and Takahashi, I. (1977). Metabolism of pyrimidine bases and nucleosides in *Bacillus subtilis. J. Bacteriol.* **129**, 574–579.
118. Robertson, B. C., and Hoffee, P. A. (1973). Purification and properties of purine nucleoside phosphorylase from *Salmonella typhimurium. J. Biol. Chem.* **248**, 2040–2043.
119. Robertson, B. C., Jargiello, P., Blank, J., and Hoffee, P. A. (1970). Genetic regulation of riboside and deoxyriboside catabolism in *Salmonella typhimurium. J. Bacteriol.* **102**, 628–635.
120. Rumyantseva, E. V., Sukhodolets, V. V., and Smirnov, Y. V. (1979). Isolation and characterization of mutants for genes of nucleoside catabolism in *Bacillus subtilis. Genetika* **15**, 594–604.
121. Sable, H. Z. (1955). Yeast pentokinase. *In* "Methods in Enzymology" (S. P. Colowick and N. O. Kaplan, *Eds.*), Vol. I, pp. 357–360. Academic Press, New York.
122. Sakai, T., Yu, T., and Omata, S. (1976). Distribution of enzymes related to cytidine degradation in Bacteria. *Agric. Biol. Chem.* **40**, 1893–1895.

123. Sakai, T., Yorifuji, T., Tochikura, T., and Ogata, K. (1967). Metabolisms of nucleosides in Bacteria. Part IV. Pyrimidine nucleoside phosphorylase in *Erwinia carotovora* and *Corynebacterium sepedonicum*. *Agric. Biol. Chem.* **31**, 525–532.
124. Sakai, T., Yorifuji, T., Tochikura, T., and Ogata, K. (1969). Metabolisms of nucleosides in bacteria. Part V. Purine nucleoside phosphorylases in *Erwinia carotovora* and *Corynebacterium sepedonicum*. *Agric. Biol. Chem.* **31**, 533–539.
125. Saunders, P. P., Wilson, B. A., and Saunders, G. F. (1969). Purification and comparative properties of a pyrimidine nucleoside phosphorylase from *Bacillus stearothermophilus*. *J. Biol. Chem.* **244**, 3691–3697.
126. Schramm, V. L., and Lazorik, F. C. (1975). The pathway of adenylate catabolism in *Azobacter vinelandii*. Evidence for adenosine monophosphate nucleosidase as the regulatory enzyme. *J. Biol. Chem.* **250**, 1801–1808.
127. Schwartz, M. (1971). Thymidine phosphorylase from *Escherichia coli*. Properties and kinetics. *Eur. J. Biochem.* **21**, 191–198.
128. Schwartz, M. (1978). Thymidine phosphorylase from *Escherichia coli*. *In* "Methods in Enzymology" (P. A. Hoffee and M. E. Jones, *Eds.*), Vol. 51, pp. 442–445. Academic Press, New York.
129. Scocca, J. J. (1978). Pyrimidine nucleoside phosphorylase from *Haemophilus influenzae*. *In* "Methods in Enzymology" (P. A. Hoffee and M. E. Jones, *Eds.*), Vol. 51, pp. 432–437. Academic Press, New York.
130. Serra, M. C., Falcone, G., Cercignani, G., and Ipata, P. L. (1971). Some regulatory properties of purine nucleoside phosphorylase of *Bacillus cereus*. *FEBS Letts.* **18**, 335–338.
131. Shaffer, P. M., Hsu, C-A., and Abbot, M. T. (1975). Metabolism of pyrimidine deoxyribonucleosides in *Neurospora crassa*. *J. Bacteriol.* **121**, 648–655.
132. Shaffer, P. M., McCroskey, R. P., and Abbott, M. T. (1972). Substrate specificity of the hydroxylase reaction in which thymidine is converted to thymine ribonucleoside. *Biochim. Biophys. Acta* **258**, 387–394.
133. Sukhodolets, V. V., Galeys, V. P., and Smirnov, Yu. V. (1973). The nature of phenotypical reversions of mutants for thymidine phosphorylase in *Escherichia coli*. *Genetika* **9**, 167–169.
134. Svenningsen, B. A. (1977). *In vitro* regulation of the *deo* operon of *Escherichia coli* at the initiation of transcription. *Carlsberg Res. Commun.* **42**, 517–524.
135. Takagi, Y., and Horecker, B. L. (1957). Purification and properties of a bacterial riboside hydrolase. *J. Biol. Chem.* **225**, 77–86.
136. Taketo, A., and Taketo, Y. (1974). Nucleoside catabolism in *Streptococcus hemolyticus*. *J. Biochem. (Tokyo)* **75**, 1341–1348.
137. Terada, M., TaTibana, M., and Hayaishi, O. (1967). Purification and properties of nucleoside hydrolase from *Pseudomonas fluorescens*. *J. Biol. Chem.* **242**, 5578–5585.
138. Uerkvitz, W. (1971). Trans-N-deoxyribosylase from *Lactobacillus helveticus*. Crystallization and properties. *Eur. J. Biochem.* **23**, 387–395.
139. Valentin-Hansen, P. (1982). Tandem CRP binding sites in the *deo* operon of *Escherichia coli* K-12. The EMBO J. **1**, 1049–1054.
140. Valentin-Hansen, P., Aiba, H., and Schümperli, D. (1982). The structure of tandem regulatory regions in the *deo* operon of *Escherichia coli* K12. The EMBO J. **1**, 317–322.
141. Valentin-Hansen, P., Boëtius, F., Hammer-Jaspersen, K., and Svendsen, I. (1982). The primary structure of *Escherichia coli* K12 2-deoxyribose 5-phosphate aldolase. Nucleotide sequence of the *deoC* gene and the amino acid sequence of the enzyme. *Eur. J. Biochem.* **125**, 561–566.

142. Valentin-Hansen, P., Hammer-Jespersen, K., and Buxton, R. S. (1979). Evidence for the existence of three promoters for the *deo* operon of *Escherichia coli* K 12 *in vitro*. *J. Mol. Biol.* **133**, 1–17.
143. Valentin-Hansen, P., Svenningsen, B. A., Munch-Petersen. A., and Hammer-Jespersen, K. (1978). Regulation of the *deo* operon in *Escherichia coli*. The double negative control of the *deo* operon by the *cytR* and *deoR* repressors in a DNA directed *in vitro* system. *Molec. Gen. Genet.* **159**, 191–202.
144. Vogels, G. D., and Van der Drift, C. (1976). Degradation of purines and pyrimidines by microorganisms. *Bacteriol. Rev.* **40**, 403–468.
145. Weinberger, S., Evenchik, Z., and Hertman, I. (1980). Utilization of purine nucleosides by germinating *Bacillus cereus* 569 spores. *J. Gen. Microbiol.* **118**, 223–228.
146. Williams, L. G., and Mitchell, H. K. (1969). Mutants affecting thymidine metabolism in *Neurospora crassa*. *J. Bacteriol.* **100**, 383–389.
147. Womach, J. E. (1977). Inhibition of thymidine phosphorylase *in vivo* provides a rapid method for switching DNA labelling. *Molec. Gen. Genet.* **158**, 11–15.
148. Wondrack, L. M., Hsu, C-A., and Abbott, M. T. (1978). Thymine 7-hydroxylase and pyrimidine deoxyribonucleoside 2′-hydroxylase activities in *Rhodotorula glutinis*. *J. Biol. Chem.* **253**, 6511–6515.
149. Yoshino, M., Tsukada, T., and Tsushima, K. (1978). Inosine nucleosidase from *Azotobacter vinelandii*. Purification and properties. *Arch. Microbiol.* **119**, 59–64.
150. Zannis, V., Doyle, D., and Martin, D. W. (1978). Purification and characterization of human erythrocyte purine nucleoside phosphorylase and its subunits. *J. Biol. Chem.* **253**, 504–510.
151. Zlotnikov, K. M., Sukhodolets, V. V., and Baumanis, G. E. (1969). Mapping of the genes controlling nucleoside catabolism in *Escherichia coli* K 12. *Genetika* **5**, 113–119.
152. Zubay, G. (1973). *In vitro* synthesis of protein in microbial systems. *Ann. Rev. Genet.* **7**, 267–287.

6

Transport of nucleic acid precursors

AGNETE MUNCH-PETERSEN
and
BENTE MYGIND

I. Introduction 259
II. Transport of nucleosides in enteric bacteria 260
A. Models for transport 261
B. Cellular location of nucleoside catabolizing enzymes . 262
C. Two nucleoside transport systems 264
D. Involvement of different parts of the cell envelope in nucleoside transport 267
E. Regulation of nucleoside transport 271
F. Energization of nucleoside transport systems 274
G. Physiological significance of nucleoside transport systems 275
III. Transport of nucleosides in other bacteria 278
A. *Bacillus subtilis* 278
VI. Transport of nucleobases in bacteria 280
A. Introduction 280
B. Cytosine transport 282
C. Uracil transport 283
D. Purine transport 285
E. Thymine transport 287
V. Transport of nucleosides and nucleobases in fungi 287
A. Transport of nucleosides in *Saccharomyces cerevisiae* . 288
B. Transport of nucleobases in *Saccharomyces cerevisiae* . 289
C. Transport of nucleosides and nucleobases in other yeasts 292
D. Transport of nucleosides in *Neurospora crassa* . . 293
E. Transport of nucleobases in *Neurospora crassa* . . 296

I. INTRODUCTION

Nucleosides and nucleobases appear in nature mainly as degradation products of nucleic acids. Usually they originate from decaying cells, but they

METABOLISM OF NUCLEOTIDES, NUCLEOSIDES
AND NUCLEOBASES IN MICROORGANISMS
ISBN: 0-12-510580-0

may also appear as excretion products from cells, which for some reason overproduce purine or pyrimidine compounds and subsequently excrete the excess of these substances to the surroundings. In both cases these nucleic acid breakdown products may be reutilized by other cells. They will serve as precursors in nucleic acid synthesis, as carbon sources and as nitrogen sources in an otherwise nitrogen-poor environment.

As a necessary basis for these reutilization processes (the socalled salvage pathways (*see* Chapters 2 and 3) a variety of cell types, procaryotic as well as eucaryotic, have evolved transport systems through which the nucleobases and the nucleosides gain entry to the cells. Some of these transport systems will accept a number of different substrates, while others are highly specific for a single substrate. A common feature, though, in the utilization of these nucleic acid derivatives is that their metabolic conversions take place immediately following their transport. In many cases it has been difficult to distinguish experimentally between the transport process proper and the subsequent enzyme-catalyzed conversion of the transported substrate. This is probably the reason why so relatively few reports have appeared, in which these transport systems in microorganisms have been studied in some detail and separate from the subsequent metabolism. Only enteric bacteria, *Bacillus subtilis*, and a few fungi (yeasts, *Neurospora crassa*) have been subject to such studies. None of the transport systems have as yet been resolved in single components, but the picture which emerges from the studies seems to include both active and passive transport functions.

II. TRANSPORT OF NUCLEOSIDES IN ENTERIC BACTERIA

The cell envelope of enteric bacteria contains two distinct membrane structures, the outer membrane and the inner membrane. The two membranes are separated by a more rigid structure, the peptidoglycan layer, and the outer membrane is partially anchored to this peptidoglycan through a lipoprotein. The space between the inner and outer membranes is termed the periplasmic space, and several soluble proteins, some of them involved in transport of nutrients, are located here.

The inner, or cytoplasmic, membrane constitutes the main permeability barrier for nutrients and other cell requirements, and this membrane also contains the energy coupling systems which are necessary for active transport. It has become increasingly evident, though, that the outer membrane may function as a permeability barrier for larger molecules ($MW > 750$) and therefore must contain sites or receptors, which in a more or less specific way allow the passage of various nutrients by diffusion.

There is evidence that the two membranes surrounding the *E. coli* cell are both involved in nucleoside transport. There is also some indication that the enzymes which catalyze nucleoside catabolism are located at or near the cytoplasmic membrane, and in the past years different hypotheses have been put forward concerning the transport mechanism and its coupling to the catabolic enzymes.

The key reaction in the catabolism of nucleosides is a phosphorolytic cleavage of the nucleoside into the pentose and base moieties. This phosphorolysis proceeds immediately when the nucleosides enter the cells, and a pertinent question in the studies of nucleoside transport has been, whether the nucleosides must penetrate a permeability barrier before they are attacked enzymatically or whether they are cleaved during the process of transport and never enter the cells in a chemically intact state.

A. Models for nucleoside transport

The first to study nucleoside transport in cells of *E. coli* were Koch and coworkers (71, 72). They investigated the kinetics of catabolism and utilization of a number of nucleosides, and in addition they studied the mutual inhibitory effect of these nucleosides, all in *E. coli* B. They concluded that the cells contained transport system(s) which could be distinguished from the catabolizing enzymes.

Hochstadt-Ozer and coworkers later advocated the hypothesis that transport of inosine, adenosine and uridine occurs through a group translocation process, mediated by the corresponding nucleoside phosphorylases (32, 75). The hypothesis was mainly based on studies of uptake of nucleosides in membrane vesicle preparations from *E. coli* and *S. typhimurium*. In these investigations the reaction products were determined after uptake of labelled nucleosides in short-time experiments. Only ribose phosphate was found inside the vesicles. The free base, formed simultaneously with the ribose phosphate in the phosphorolytic cleavage of the nucleoside, could not be detected inside the vesicles, but was quantitatively recovered in the surrounding medium. These and other results, such as response to inhibitors, kinetics of the system etc. led to the proposal that the main system for transport and utilization of nucleosides is a group translocation process. The model for such an uptake system assumes that the nucleoside phosphorylases are integrated in the cytoplasmic membrane in a spatial arrangement such that the nucleoside is cleaved during its passage through the membrane. The ribose phosphate is captured in the lumen of the vesicle, while the base is released outside the membrane from where it may be recaptured through the phosphoribosyltransferase system (31, 33). With such a transport system exogenously supplied nucleosides should not be found inside the cells in a chemically intact state. That such a mechanism cannot be the sole way of nucleoside transport is borne out by the following findings: (a) Mutant strains which lack uridine phosphorylase or purine nucleoside phosphorylase are still able to utilize uridine or inosine as pyrimidine or purine sources, respectively (pp. 49, 101). (b) Strains lacking the nucleoside phosphorylases are able to transport and concentrate nucleosides (59, 62). (c) Mutants have been isolated and characterized, which are unable to transport nucleosides although they contain a full complement of the nucleoside catabolizing enzymes (59). The mutations in the latter strains have been shown to map in regions on the *E. coli* chromosome, different from the regions where the genes for nucleoside catabolizing enzymes are located (59). From this it may be concluded that the cells contain transport systems, which operate independent of nucleoside catabolism. Since, however, the results summarized above stem from work with mutants carrying a number of different mutations, they do not exclude the possibility that in wild type cells the phosphorylases may play a role in the transport process.

B. Cellular location of the nucleoside catabolizing enzymes

In an attempt to make a decision or reconciliation between the two different hypotheses a question of obvious importance is the cellular location of the nucleoside phosphorylases in relation to the cytoplasmic membrane where the main cellular permeability barrier is located.

The first catabolic enzymes to attack the incoming nucleosides are listed in Table 6.1. Several workers have reported that these enzymes are partially released when the cells are subjected to osmotic shock (4, 60, 65). There is,

TABLE 6.1

Enzymes, catalyzing the initial catabolic attack on exogenously added nucleosides.

Nucleoside added	Initial enzyme reaction	Enzyme
Cytidine	Deamination	Cytidine deaminase
Deoxycytidine	Deamination	Cytidine deaminase
Uridine	Phosphorolysis	Uridine phosphorylase
Adenosine	Deamination	Adenosine deaminase
Deoxyadenosine	or Phosphorolysis	Purine nucleoside phosphorylase
Guanosine	Phosphorolysis	Purine nucleoside phosphorylase
Deoxyguanosine	Phosphorolysis	Purine nucleoside phosphorylase
Inosine	Phosphorolysis	Purine nucleoside phosphorylase
Deoxyinosine	Phosphorolysis	Purine nucleoside phosphorylase
Thymidine	Phosphorolysis	Thymidine phosphorylase
Deoxyuridine	Phosphorolysis	Thymidine phosphorylase

however, a general agreement between the authors that the enzymes cannot be classified as true periplasmic proteins in the sense in which this expression was formulated originally (57, 63). One important criterium for periplasmic location was release of the protein after treatment of the cells with lysozyme. None of the nucleoside catabolizing enzymes are released after lysozyme treatment (Munch-Petersen, unpublished). Moreover, the release after osmotic shock is only partial, and highly dependent on the absence of magnesium ions in the shock fluid (59).

On the basis of these results a peripheral location on the inner surface of the cell membrane appears more likely, similar to that suggested by Jacobsen *et al.* (36) for the ribosomal elongation factor Tu. Such a location would place the enzymes inside the permeability barrier of the cell. In accordance with this notion is the existence of mutants which are unable to utilize nucleosides, although they carry a full complement of the nucleoside catabolizing enzymes. It might be argued, though, that in these transport negative strains the enzymes were still able to catabolize the nucleosides, but in such a way that the products were spilled into the medium instead of being utilized by the cells. This hypothesis has been disproved by making a comparison of the enzyme activities, as measured in whole cells of transport negative and transport positive strains (Table 6.2). Clearly the ability of the four enzymes to attack their substrates is impaired in cells of the transport negative strains, while the enzyme activities, measured in sonic extracts, are unchanged by mutations in the transport systems.

TABLE 6.2

Catabolism of nucleosides in whole cells and sonic extracts of strains, positive or negative for nucleoside transport.[a]

Substrate used in assay	Activity measured		Enzyme activities: Transport negative cells	Enzyme activities: Transport positive cells
cytidine	deamination[b]	whole cells	5	38
		sonic extracts	320	297
inosine	phosphorolysis[c]	whole cells	5	11
		sonic extracts	45	45
thymidine	phosphorolysis[b]	whole cells	5	12
		sonic extracts	212	290
uridine	phosphorolysis[b]	whole cells	5	12
		sonic extracts	221	190

[a] Data taken in part from (59).

[b] Enzyme assays as described in (26). The activities are expressed as units per 4×10^8 cells, measured in cell suspensions or in extracts of the same cell suspensions.

[c] Chromatographic assay, using [^{14}C]-labelled inosine as substrate.

The results discussed here do not rule out the possibility that in wild type cells some nucleoside transport might be mediated by the nucleoside phosphorylases, but they do indicate that such a transport mechanism plays a minor role only. For this reason and since the literature concerning the group translocation hypothesis has been reviewed previously (33), only the active, non-catabolizing nucleoside transport processes will be discussed in the following.

C. Two nucleoside transport systems

The early experiments of Koch and coworkers (71, 72) led these authors to the conclusion that cells of *E. coli* transported nucleosides by a highly complex mechanism, which included both separate and overlapping processes. The actual presence of two different nucleoside transport systems was later indicated in growth experiments with the antibiotic showdomycin (2-(β-D-ribofuranosyl)maleimide) (41, 42), a compound which exerts its inhibitory effect by alkylation of unprotected sulfhydryl groups (13). The growth inhibition could be reversed by thiol compounds and also by addition of the ribonucleoside adenosine to the growth medium (41). Since showdomycin may be considered a nucleoside analog, the protective action of the nucleoside could be explained as the result of a competition between showdomycin and the nucleoside for the same transport sites (17, 42). Subsequent studies of the competitive effect of various nucleosides on the transport of labelled showdomycin in cells of *E. coli* K 12 and of *E. coli* B confirmed this hypothesis (14, 43, 80). While adenosine and pyrimidine nucleosides were strong, competitive inhibitors of showdomycin uptake, guanosine and deoxyguanosine had little effect. This implied that a separate transport system was operating for the non-competing guanine nucleosides, and the actual existence of a second transport system was demonstrated by direct measurements of nucleoside uptake in cells of showdomycin resistant strains. Here the uptake of guanosine was unimpaired, while the uptake of other nucleosides was markedly reduced (44, 80).

Besides these two transport systems, which both seem to have a high affinity for their nucleoside substrates, there is some evidence that the cells may also contain low affinity transport systems for purine nucleosides (47, Mygind, unpublished.) Since these systems have not yet been characterized, only the two high affinity systems will be discussed in the following.

1. *Isolation and characterization of nucleoside transport mutants*

The two high affinity transport systems have been designated the C system and the G system (60). The main part of the results to be described have been carried out with cytidine(deoxycytidine) and guanosine as substrates and the names C and G refer to the specificities of the transport systems towards these nucleosides, the gross distinction between the two systems being that the G system transports guanosine while the C system does not.

The previously mentioned high affinity of the C system for showdomycin and the toxicity of this compound for the cells has been the basis for selection of strains with mutations in this transport system (44, 49, 80, 83). Mutant strains that are resistant to showdomycin (80 μM) and contain only the G system are still able to transport all nucleosides and utilize them as sole carbon source. Further selection in these mutants for resistance to the toxic nucleoside analogs 5-fluorodeoxycytidine or 5-fluorodeoxyuridine (5 μg/ml) renders strains with additional mutations in the G system. These double mutants are unable to transport any nucleoside. Single mutants containing only the C system have been constructed from the double mutant by genetic procedures (59). These mutants will transport all nucleosides with the exception of guanosine, deoxyguanosine and inosine and deoxyinosine; they will, however, grow on inosine or deoxyinosine as sole carbon sources (in mM amounts), thus indicating the presence in the cells of a low affinity transport system for these nucleosides.

The specificity of the G system for guanosine and deoxyguanosine has been used for selection of *nupG* mutants. Strains which lack deoxyriboaldolase (*deoC* mutants) are sensitive to deoxyribonucleosides, because deoxyribose 5-phosphate accumulates in the cells and inhibits growth (*see* Chapter IV). In such *deoC* strains *nupG* mutants have been obtained by selection for resistance to deoxyguanosine (Mygind, unpublished).

Mutations causing deficiency in the C system have been located at 50 min on the *E. coli* chromosome (*see* Appendix) 65% cotransducible with the *ptsI* gene (7), while mutations in the G system map at 64 min, between *metC* and *serA* (*see* Appendix). The genes thus mapped have been termed *nupC* and *nupG*, respectively. The overall specificity pattern of the two systems is given in Table 6.3. In both systems the specificity pattern for deoxyribonucleosides completely matches that for the ribonucleosides.

The transport systems have been investigated in strains unable to metabolize the nucleosides. The systems show saturation kinetics (Fig. 6.1), and the apparent K_m has been determined for some of the nucleoside substrates. The values obtained all fell in the range of 0.3–0.6 μM (Munch-Petersen, unpublished).

2. *Inhibition of the nucleoside transport systems*

The specificities of the two transport systems are reflected in the mutual inhibitory actions of the various nucleosides as measured in transport assays. In these assays the transport of a labelled nucleoside in the presence of excess of the competing nucleoside was determined. In the competition experiments it is important to avoid interference from inhibition of the subsequent

TABLE 6.3
Specificity of the nucleoside transport systems for ribonucleosides and deoxyribonucleosides.[a]

Nucleoside tested	C system[b]	G system[c]	*tsx* gene product[d]
Adenosine	+	+	+
Deoxyadenosine	+	+	+
Cytidine	+	+	−
Deoxycytidine	+	+	−
Guanosine	−	+	+
Deoxyguanosine	−	+	+
Inosine	−	+	+
Deoxyinosine	−	+	+
Thymidine	+	+	+
Uridine	+	+	+
Deoxyuridine	+	+	+

[a] The results are based on tests of a number of mutants for uptake of labelled nucleosides (1.7 μM) in cell suspensions.
[b] Tested in *nupG* mutants.
[c] Tested in *nupC* mutants.
[d] Tested in strains containing both transport systems.

catabolism of the nucleosides, since it is known that the nucleosides act as competitive inhibitors of the nucleoside phosphorylases. For this reason mutant strains unable to carry out the initial enzymatic attack on the transported nucleoside have been employed whenever possible.

The C system is strongly inhibited by showdomycin. The system has a low affinity for guanosine and inosine (Table 6.3) and these two nucleosides accordingly have very little inhibitory effect on the transport of pyrimidine nucleosides and adenosine in mutants containing only this transport system. The G system accepts all nucleosides, and a mutual competition between purine and pyrimidine nucleosides is observed in strains carrying this system alone. Table 6.4 gives a survey over the inhibitory pattern found in the two classes of mutants.

The toxic pyrimidine analogs, fluorodeoxycytidine and fluorodeoxyuridine, are strong inhibitors of both systems, while tetrahydrouridine, a potent inhibitor of cytidine deaminase (10, 15) does not interfere with the transport of cytidine or any other nucleoside. The free nucleobases, added in 10–50 fold excess, have no effect on nucleoside transport (62).

The transport is completely inhibited by p-chloromercuribenzoate (0.1 mM) and partially inhibited by N-ethyl maleimide (1 mM). This was shown for cytidine transport in a cytidine deaminase less strain (62) and has later been verified for purine nucleosides in strains lacking purine nucleoside phosphorylase (Munch-Petersen, unpublished). Thus free sulfhydryl groups seem to be

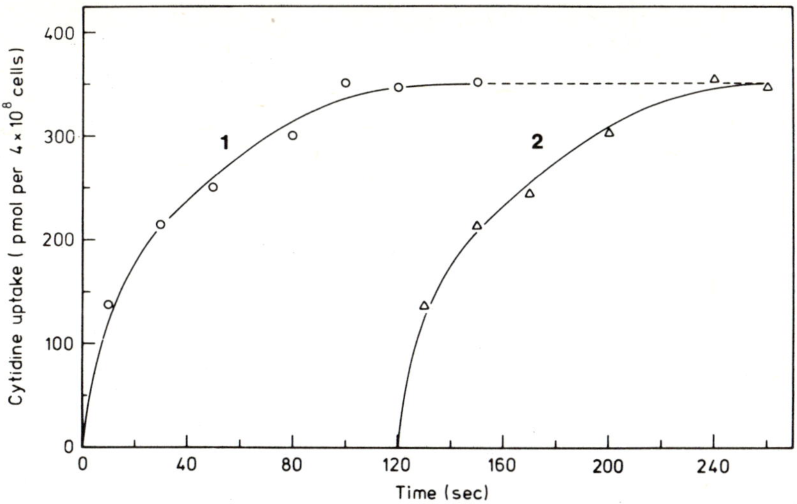

FIG. 6.1. Saturation kinetics for cytidine transport in a strain lacking cytidine deaminase and uridine (cytidine) kinase. The cells were grown in glycerol-salts medium, harvested in the exponential phase, washed and suspended in medium with glycerol as energy source. Aliquots were exposed to [^{14}C]-labelled nucleosides and transport was measured as retention of radioactivity in cell samples after rapid filtering and washing. (1) [^{14}C]-labelled cytidine (0.6 μM) added at time zero. (2) Unlabelled cytidine (0.6 μM) added at time zero. Carrier-free [^{14}C]-labelled cytidine added after 120 seconds.

required for nucleoside transport (18). It has been shown by many workers (62, 71, 72) that the transport is inhibited by energy poisons such as 2,4-dinitrophenol, sodium azide etc. indicating that nucleosides are transported actively. The energization of the transport systems will be discussed later.

D. Involvement of different parts of the cell envelope in nucleoside transport

There is some evidence that nucleoside transport involves elements in both the cytoplasmic and the outer membrane and possibly also proteins located in the periplasmic space.

1. Outer membrane

This membrane would not be expected to constitute a permeability barrier for the relatively small nucleoside molecules (average MW around 250). It has been shown, though, that at low substrate concentration, ($< 1\ \mu$M), a mutation in the *tsx* gene (*see* Appendix) specifying the T_6-receptor protein, (55) reduces the transport of certain nucleosides considerably (29, 48, 56). This

TABLE 6.4
Competition by nucleosides for transport, measured in strains containing either the C or the G system.

Labelled substrate[a]	Inhibitor[b]	Inhibition (%)[c] C system	G system
Deoxycytidine	—	—	—
	adenosine	90	90
	cytidine	90	90
	guanosine	5	90
	thymidine	56	85
	uridine	90	90
	showdomycin	90	10
Guanosine	—		
	adenosine		84
	deoxycytidine		55
	thymidine		60
	uridine		72

[a] Cells were exposed 10 sec to labelled nucleoside (1.7 μM) before filtration.
[b] Inhibitors were added in 30-fold excess of labelled substrate.
[c] Expressed as percent reduction of the uptake measured in the absence of inhibitor.

effect of the *tsx* gene product on nucleoside transport was later confirmed in experiments with strains unable to catabolize the nucleosides (60). All nucleosides, except cytidine and deoxycytidine were affected in their transport by a *tsx* mutation.

Krieger-Brauer and Braun (48) have studied the dual function of the *tsx* gene product. They found that the presence of nucleosides had no inhibitory effect on the adsorption of the T_6 phages. Furthermore there was no mutual competition between the nucleosides for the *tsx* protein dependent translocation across the outer membrane. They conclude from these results that the *tsx* gene probably specifies a pore-forming protein and not a receptor protein with binding properties.

A certain specificity for nucleosides seems to reside in the *tsx* encoded protein as compared to other pore-forming proteins. This is indicated by the co-regulation of the *tsx* protein and the active nucleoside transport systems located in the cytoplasmic membrane (48). Moreover, with adenosine as a substrate it was shown (91) that a mutation in the *ompF* gene, which specifies one of the major pore-forming proteins in the outer membrane (64), reduces the transport rate only 15–20 per cent, while a *tsx* mutation caused a rate reduction of 90 per cent. In the *tsx,ompF* double mutant the transport of the nucleoside at low concentrations was completely abolished. These results together with the observation that a mutation in the *ompC* encoded protein, another pore-forming protein, has little or no effect on nucleoside transport (2), indicate that normally the diffusion of nucleosides through the outer membrane is mediated by the *tsx* protein, but in the absence of this protein the more unspecific porins will take over the translocating function.

Clearly the composition of the outer membrane influences the transport of nucleosides. It should be emphasized, though, that neither of these outer membrane proteins are obligatory components of the overall transport system. The *tsx* encoded protein becomes necessary only at low ($<0.1\ \mu$M) substrate concentrations, when diffusion through the outer membrane is the limiting factor for transport.

2. Periplasmic proteins

The effect of osmotic shock on nucleoside transport was studied in mutants lacking the nucleoside catabolizing enzymes. A strong depression was observed in the transport of all nucleosides (60). The shock treatment is supposed to cause release of periplasmic proteins and a likely interpretation of the loss in transport capacity is that the nucleoside transport requires the participation of periplasmic binding proteins. However, attempts to demonstrate the presence of binding proteins for either cytidine or guanosine in the shock fluid from *E. coli* cells have failed so far (Mygind, unpublished). Experiments with *S. typhimurium* (89) have indicated the presence of a binding protein for uridine in this organism, but a protein with binding properties has not yet been characterized in this organism.

In this context it should be noted that nucleosides do not appear to have chemotactic properties. This would agree with the failure to demonstrate periplasmic binding proteins, since such binding proteins, involved in transport, in several cases have been shown to act in chemotaxis as well (1).

3. Cytoplasmic membrane

Nucleoside transport has been studied in membrane vesicles prepared according to Kaback (38). One of the advantages of using these vesicles is that they are free of the cytoplasmic and periplasmic cell components and thus leave the cytoplasmic membrane to be studied as a separate entity.

Komatsu and Tanaka (46) investigated transport of deoxycytidine in membrane vesicles prepared from an *E. coli* K12 wild type strain and from its showdomycin resistant derivative, i.e. a *nupC* mutant. They found that deoxycytidine was transported actively in the vesicles, stimulated by various electron donors, such as D(-)-lactate or phenazine methosulfate plus ascorbate. The transport was strongly inhibited by p-chloromercuribenzoate (0·1 mM), while other sulfhydryl reagents (N-ethylmaleimide (0·5 mM), or iodo acetate (2 mM)) showed only a slight inhibitory effect. The competitive effect of other nucleosides on the transport corresponded to the results from whole cells. In membrane vesicles from cells containing both transport systems the transport of deoxycytidine was strongly inhibited by showdomycin while the analog had very little effect on transport in vesicles from the showdomycin resistant strain.

Komatsu studied adenosine uptake in membrane vesicles from the same strains (45). From a large number of experiments in which the inhibition by other nucleosides, and by the bases adenine and caffeine, was studied, it was concluded that the vesicles contained 3 different transport systems capable of transporting adenosine. One system corresponded to the showdomycin-sensitive C system, while the second and third system were resistant to the antibiotic and thus would relate to the G system in whole cells. The third system, however, appeared to degrade adenosine and subsequently transported adenine across the membrane rather than adenosine.

Hochstadt *et al.* (31, 32) studied the transport of inosine, adenosine and uridine in membrane vesicles from *S. typhimurium* as well as from *E. coli* K12. Their observations indicated, as mentioned earlier, that the nucleosides were cleaved phosphorolytically during transport in such a way that ribose 1-phosphate entered the vesicles, while the base moiety of the nucleoside was released on the outside of the vesicle.

These reports on nucleoside transport in membrane vesicles are somewhat at variance with later results, obtained with membrane vesicles which were all prepared from strains carrying mutations in the nucleoside phosphorylases (60). Here the membrane vesicles prepared from cells containing either of the two transport systems were able to carry out some nucleoside transport, but the efficiency was low as compared to whole cells (60). Using deoxycytidine as a substrate the concentrating effect in cells was 590-fold as compared to 10-fold in the vesicles. The corresponding figures for the amino acid proline as a reference were 122-fold and 28-fold, respectively. Significant for the transport in these vesicles was also the finding that the nucleosides as substrates fell in two categories. Cytidine, deoxycytidine and uridine were transported in low, but still significant amounts, while purine nucleosides and thymidine showed little or no transport.

No clearcut explanation can be offered at present for the conflicting results concerning the transport of purine nucleosides in membrane vesicles. It is evident, though, that membrane vesicles contain residual activities of purine nucleoside phosphorylase unless they are prepared from cells lacking this enzyme (60), and this may explain the cleavage of the transported purine nucleosides, which was observed by Komatsu (45) and by Rader and Hochstadt-Ozer (75).

The insignificant transport of purine nucleosides in vesicles from strains lacking purine nucleoside phosphorylase is in accordance with the severe loss of transport capacity observed in osmotically shocked cells (60). It is conceivable that in intact cells component(s), which are specifically required for purine nucleoside transport, may be attached loosely to the cytoplasmic membrane and therefore easily are stripped off during preparation of the vesicles.

Taken together the observations reported above suggest that the nucleoside transport systems involve several proteins, some of which are attached to the cytoplasmic membrane in the intact cell, but lost or inactivated to varying degrees during shock treatment and during the preparation of the membrane vesicles.

E. Regulation of nucleoside transport

1. Regulation on the level of synthesis

As mentioned in the introduction there is an intimate coupling between transport and catabolism of the nucleosides, possibly related to the cellular location of the appropriate enzymes at or near the cytoplasmic membrane (4, 59). There is, however, also a regulatory connection between these two steps in the cellular utilization of nucleosides, since the syntheses of at least some of the components of the transport systems are controlled by the *cytR* and *deoR* genes, the same genes which are involved in regulation of the synthesis of the nucleoside catabolizing enzymes (58). Since neither of the two transport systems has as yet been resolved in single components, it has been possible only to measure the overall response of each system to the two regulatory gene products.

A regulation of the G system was first indicated in experiments by Doskočil (16), who showed that the presence of thymidine in the growth medium induced the transport of deoxycytidine and 5-azacytidine in showdomycin resistant cells of *E. coli* B. Thymidine is known to cause induction of four of the nucleoside catabolizing enzymes through the *deoR* control system, deoxyribose 5-phosphate being the actual inducer (Chapter V). The results of Doskočil therefore implied a control of the transport by the *deoR* gene product. This was verified more rigorously in mutants of *E. coli* K12 with a deletion covering the *deo* operon (Fig. 6.2). Introduction of a *deoR* mutation in such a strain caused a rise in transport rates. A *cytR* mutation had a similar effect and in double regulatory mutants the transport rates reached maximal values, much higher than the sum of the levels in the *cytR* and the *deoR* mutants. This might indicate a certain degree of cooperativity between the two repressors as suggested in the regulation of the nucleoside catabolizing enzymes (28). Alternatively, the *cytR* gene may control one component of the transport system, while the *deoR* gene controls another. Each of these components could then be the limiting factor for transport if it is synthesized in repressed amount when the other component is derepressed.

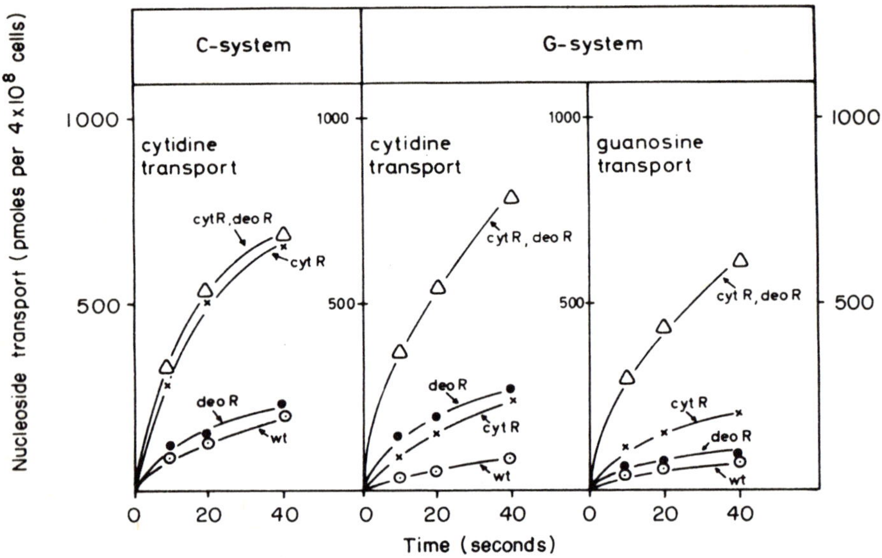

FIG. 6.2. Regulation of nucleoside transport by the *cytR* and *deoR* repressors in cells containing either the C or the G system. Strains with wild type, *cytR*, *deoR* or *cytR*,*deoR* backgrounds and in addition containing mutations in the deo enzymes were grown up in a glycerol-salts medium. The cells were harvested in the exponential phase, and transport was measured as described in legend to Fig. 6.1.

The C system seems to be controlled only by the *cytR* gene. A mutation in this gene gives rise to a significant increase in transport rates in strains containing only the C system, but subsequent introduction of a *deoR* mutation does not stimulate the transport further in such strains. A *deoR* mutation alone does not affect the transport rate in strains containing only the C system (59). The capacity of the cells to transport nucleosides is repressed through the cAMP-CRP system, when glucose is present in the growth medium (62). This repression acts in connection with the *cytR* controls system (p. 226).

The actual target proteins for the control are not known. It is likely, though, that the regulated proteins are integrated in the cytoplasmic membrane, since it has been found that vesicles prepared from strains carrying mutations in the regulatory genes showed the highest capacity for nucleoside transport (60).

It should be noted that the T_6-receptor protein, mentioned above as involved in the passage of nucleosides through the outer membrane, was found to be controlled by the *cytR* and *deoR* genes (48). The T_6-receptor, encoded by the *tsx* gene, has been shown to act as a pore in the outer membrane, specific for diffusion of all nucleosides except cytidine and deoxycytidine. Its regulation by the *cytR* and *deoR* gene products emphasizes the importance of these two control genes for the cell, since in this way they also contribute to the regulation of the composition of the cell surface.

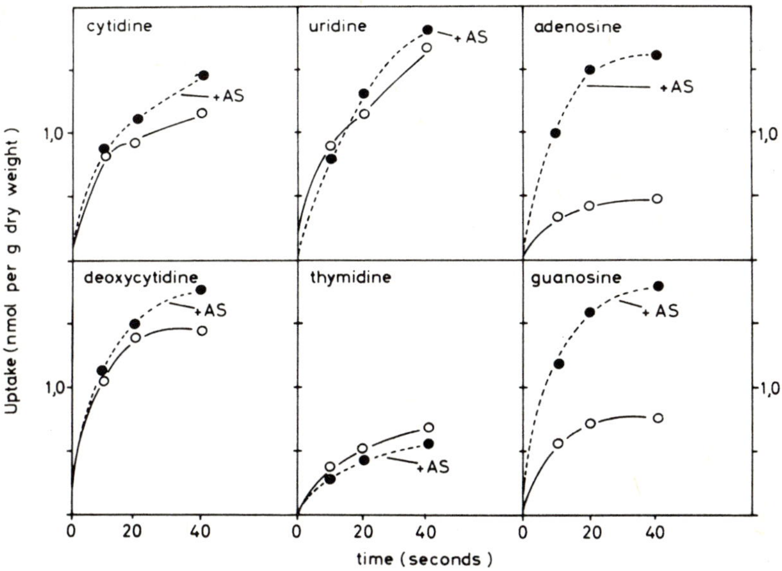

FIG. 6.3. The effect of arsenate (10 mM) on the transport of various nucleosides. Cells of a strain containing only the G system and unable to catabolize any nucleoside, were grown in glycerol-salts medium and harvested in the exponential phase. Cells were suspended in 10 mM Tris HCl buffer pH 7.3, 0.5 mM $MgCl_2$, 150 mM NaCl. Glycerol was added as energy source and the suspension was equilibrated 10 min at room temperature with or without 10 mM potassium arsenate. Transport was measured as described in Fig. 6.1.

2. Regulation on the level of activity

A specific regulation of the transport of *purine nucleosides* has been observed in mutant strains, containing only the G transport system. In these cells treatment with arsenate (10 mM) results in a substantial increase in transport rates and concentrating effect when purine nucleosides are employed as substrates (Fig. 6.3). An inverse relationship between transport rates and the cellular pools of ATP was demonstrated in experiments in which the intracellular levels of ATP were varied, either by treatment of the cells with arsenate (10 mM) prior to transport measurements or by introduction of a mutation in the Ca^{++} Mg^{++} dependent ATPase (*uncA*). In the absence of glycolytic energy sources suspensions of *uncA* cells are unable to maintain their ATP pool, but this pool may be increased simply by adding glucose or glycerol to the cell suspensions (61).

The inverse relationship between ATP-pools and transport rate suggests that ATP (or a closely related metabolite) interferes with the transport system in a regulatory way. This may be a direct allosteric effect on a protein involved in the transport. Alternatively, a component of the transport system might undergo a reversible phosphorylation. The phosphorylated state, prevailing under high ATP concentrations, would then be the inactive or inhibitory state.

Transport of pyrimidine nucleosides by the G system is not stimulated when cellular ATP pools are lowered as described above (61). This is one of the many indications that the G system is highly complex, involving components with specificities for either purine or pyrimidine nucleosides. This complexity of the G system will be discussed later.

Preliminary experiments with cells containing only the C system gave no clear cut evidence for a regulation of this system, when ATP pools were lowered (Munch-Petersen, unpublished).

F. Energization of the nucleoside transport systems

Nucleosides are transported actively against a concentration gradient and accumulated intracellularly in several hundred fold of the exogenous concentration. This has been demonstrated in mutant cells, which lack the nucleoside catabolizing enzymes (59, 62). As mentioned previously pretreatment of the cells with 10 mM arsenate stimulated the transport of purine nucleosides. Under these conditions the concentrating effect for guanosine reaches 2000 fold (61).

The energization of the G system has been studied in mutants containing only this system and in addition carrying a mutation in the $Ca^{++}Mg^{++}$ dependent ATPase (*uncA*). Energy starved cells of such a mutant retained the ability to transport purine as well as pyrimidine nucleosides (61). This together with the above mentioned enhancing effect of arsenate on purine nucleoside transport is strong evidence that ATP cannot be required as energy donor.

The contention, then, that nucleoside transport is energized by the proton motive force is supported by the finding that membrane vesicles, prepared from mutants containing the C system only or the G system only, require the presence of an electron donor system to carry out transport of nucleosides (60).

These findings with cells and membrane vesicles of *E. coli* K12 are in agreement with the results obtained by Hochstadt-Ozer *et al.* (75) with membrane vesicles from *S. typhimurium*. They found that the transport of cytidine was highly stimulated by various electron donors. The results also agree with the conclusion of Roy-Burman *et al.*, who used various metabolic inhibitors to investigate the energy coupling in the transport of uridine and deoxycytidine in cells of *E. coli* B (82).

Exit of nucleosides appears to be a passive process. The presence of 2,4-dinitrophenol (5 mM) or arsenate (10 mM) does not alter the exit rates of guanosine or cytidine (61). Addition of other nucleosides to the exit medium stimulated the exit rates, while addition of ribose had no effect. A possible interpretation of this is that the exit reaction is mediated by a carrier with some specificity, but at present no information is available about the nature of such a carrier.

G. Physiological significance of the nucleoside transport systems

In summary two high affinity systems for active transport of nucleosides operate in *E. coli*. They are designated the C system and the G system and they are characterized by different specificities and regulation. Furthermore, genes coding for essential components of the two systems have been shown to map at different locations on the *E. coli* chromosome. These essential components are probably integrated proteins in the cytoplasmic membrane since mutations in the genes abolish transport in membrane vesicles (60). Also components located outside the cytoplasmic membrane are obligatory or stimulatory for the transport.

There is no evidence of interdependence between the two transport systems, but there are indications that the G system is highly complex and includes both separate and overlapping processes. The system transports all purine and pyrimidine nucleosides, and a single mutation seems to block transport completely. Yet there is evidence that the intracellular concentration of ATP directly or indirectly regulates the transport of purine nucleosides. Thus the system must contain components with separate nucleoside specificities, while a common and essential part of the transport system is shared by all nucleosides. Such an essential component has been shown to map at 64 min.

The intimate coupling between transport and catabolism of nucleosides is potentiated by the co-regulation of these two functions by the *cytR* and the *deoR* gene products. The coregulation furthermore suggests that these systems are designed mainly for the capture and utilization of exogenous nucleosides. This is further emphasized by the *cytR* and *deoR* controlled synthesis of the *tsx* gene product, which functions as a pore in the outer membrane. The alleged attachment of the nucleoside catabolizing enzymes to the inner surface of the cytoplasmic membrane (65), thereby ensuring a rapid catabolism, is in accordance with this notion.

Nucleosides are, however, normal metabolites, arising inside the cell through breakdown of nucleic acids and nucleotides, and as such they should serve as immediate substrates for the nucleoside catabolizing enzymes. Exit reactions, proceeding in a passive way, have been demonstrated for the nucleosides (61). It is conceivable, then, that nucleosides formed intracel-

TABLE 6.5
Enzyme activities in strains, blocked in transport and/or metabolism of cytidine.

Strain number	Relevant genotype	Enzyme activities (units per mg protein) Purine nucleoside phosphorylase[a]	Thymidine phosphorylase[a]	Uridine phosphorylase[a]
SØ 430	*cdd,udk*	231	385	800
A 491	*cdd,udk,nupC*	234	332	639
A 493[b]	*cdd,udk,nupC,nupG*	118	63	232
B 119	*cdd,udk,nupG*	187	279	652
A520[b]	*cdd,udk,nupC,nupG*	121	81	169
SØ 003	wild type	118	46	90

[a] Enzyme assays were performed as described (26).
[b] Strains A 493 and A520 were two independently isolated transport negative strains.

lularly might pass through the cell membrane to be recaptured immediately by the transport systems, and that catabolism of the nucleosides occurs mainly after their re-entry into the cell.

If this was the case the cellular pools of nucleosides should in fact reflect a balance between exit and entry reactions. Certain results concerning the intracellular pools of cytidine are in favor of this hypothesis.

Cytidine functions as inducer of the nucleoside catabolizing enzymes (26). Strains, which lack the enzymes cytidine deaminase and cytidine kinase and therefore are unable to metabolize cytidine, show endogenously induced enzyme levels, probably caused by increased internal levels of the inducer (27).

The capacity for endogenous induction was investigated in derivatives of a strain lacking the enzymes cytidine deaminase and cytidine-uridine kinase (SØ 430, *cdd,udk*) in which mutations in the *C*, the *G* or in both transport systems had been introduced. Cytidine is transported by both systems, and, as seen in Table 6.5, if one of the transport systems is functioning, the cells still produce induced levels of the enzymes indicating high cellular pools of cytidine. However, when mutations in both transport systems are introduced, the enzyme levels drop to almost wild-type levels, indicating that an increased pool of cytidine, which is necessary for endogenous induction, depends strictly on the ability of the cells to transport and accumulate the nucleoside.

There are other results which in an indirect way suggest a role of the nucleoside transport systems in maintaining elevated intracellular pools of a nucleoside. It is known that the efficiency with which exogenous thymine is

TABLE 6.6

Growth pattern of *thyA* mutants, derived from strains with or without a nucleoside transport system.

	Supplement to minimal glucose-salts medium						
	Thymine				Thymidine	Thymidine monophosphate	
Parent strain	16 μM	23 μM	32 μM	40 μM	8 μM	10 μM	40 μM
SØ 128[a] *deoC*	−	+	+	+	+	−	+
AM 569[b] *deoC,nupC,nupG*	−	−	− +	+	−	−	−

[a] 24 of 24 different *thyA* derivatives.
[b] 22 of 26 different *thyA* derivatives.

utilized by thymine requiring mutants depends on the cellular pools of deoxyribose 1-phosphate. This compound in turn is formed in the cell by phosphorolysis of the nucleoside deoxyuridine (Chapter 4). Table 6.6 shows that *thyA* mutants, selected from a strain lacking the nucleoside transporting systems, require a higher thymine concentration for growth than *thyA* mutants, derived from the transport positive but otherwise isogenic strain. This difference agrees with the supposition that the transport negative cells have difficulties in maintaining the raised deoxyuridine pools, required for deoxyribose 1-phosphate formation in a situation where the cells depend on exogenous thymine for the synthesis of thymidine nucleotides.

The results reported above involve the two nucleosides cytidine and thymidine only. One might, however, speculate further that a recycling process (exit and re-entry) of the nucleosides is a common or even obligatory event in the cellular utilization of endogenously formed nucleosides through mechanisms, which were evolved for salvage of nucleosides of environmental origin. Similar findings are reported for galactose (90).

In Table 6.6 are included growth data for the two classes of *thyA* mutants under conditions where thymidine 5-phosphate is offered as an exogenous source of thymidine. The nucleotide is known to act as a slow-feeder of thymidine to a cell (86) and the results in the table demonstrate clearly that utilization of the thymidine moiety requires a functioning nucleoside transport system, thus confirming the suggestion (2) that transport of the corresponding nucleoside across the cytoplasmic membrane is an obligatory step in the cellular utilization of exogenous nucleoside monophosphates.

III. TRANSPORT OF NUCLEOSIDES IN OTHER BACTERIA

The natural habitat of enteric bacteria probably contains highly variable levels of nucleic acid components, and this may be the reason why these bacteria have evolved efficient and highly regulated mechanisms for the transport of nucleosides as well as for the subsequent metabolism. Other bacteria which in nature inhabit less well supplied environments have in a number of cases been shown to utilize exogenous nucleosides for growth or as supplement for nucleic acid synthesis (*see* Chapter 5). In general it may be concluded that such bacteria which utilize exogenous nucleosides must contain appropriate transport systems, but actual information of these systems is scarce or non-existing. Only in the case of *Bacillus subtilis* a few reports on nucleoside transport have appeared.

A. *Bacillus subtilis*

These gram positive bacteria differ from the gram negative, enteric bacteria in not having an outer membrane. Thus the transport systems in these organisms involve only proteins integrated in or spanning the cytoplasmic membrane.

Cells harvested from outgrown cultures of *B. subtilis* have been used to study the transport of purine and pyrimidine nucleosides (22, 39). Wild type cells were exposed to labelled nucleosides and the uptake of radioactivity was measured. This uptake was linear for 4–6 min and the amount of radioactivity incorporated in the initial 60 seconds was used as a measure of transport rate, although it does represent both transport *and* incorporation of the labelled compound (39).

The transport measured in this way showed saturation kinetics with apparent K_m's for cytidine and uridine of 7–8 μM, while K_m for adenosine was somewhat lower (1.8 μM). The transport system requires energy as shown by the loss of transport capacity in the presence of energy poisons, such as 2,4-dinitrophenol (10 mM) or sodium azide (50 mM). The concentrating effect could not be determined due to the metabolic processes taking place subsequent to transport, but the presence of an energy source in the medium was found necessary for transport.

The specificity of the transport system(s) was tested by measuring the mutual inhibitory effect of cytidine, uridine and adenosine on transport. A strong competition between cytidine and uridine was observed with no significant differences between apparent K_m's for transport and K_i's for mutual inhibition. Adenosine behaved differently, since it inhibited transport of the pyrimidine nucleosides strongly, while its own transport was unaffected by the two pyrimidine nucleosides. Two other purine nucleosides, guanosine and xanthosine, were without effect on the transport of the pyrimidine nucleosides.

The behaviour of adenosine indicates that cytidine and uridine share a common transport system, while adenosine has affinity for a component of or a site on this system, but requires an additional component for actual transport. The structural requirements for binding to the transport system were investigated by testing the inhibitory action of a large number of nucleosides and nucleoside analogs on cytidine uptake (22). Mainly purine nucleoside derivatives were tested. In these analogs a substitution in the 8-position of the purine ring abolished the inhibitory effect, thus emphasizing the essential role of this carbon atom, vicinal to the N-glycosylic bond, for recognition of the transport site. In contrast to this various substitutions in the 6-position did not diminish the inhibitory effect.

Surprisingly, the uridine analog showdomycin, which is an efficient inhibitor of one of the *E. coli* nucleoside transport systems (p. 264), had no effect on cytidine transport in *B. subtilis*. As for the pentose moiety of the nucleosides only unsubstituted ribose or deoxyribose nucleosides have affinity for the system.

L-nucleosides had no effect at all. This is in accordance with the findings of

Votruba *et al.* (88) that L-nucleosides do not penetrate the cell membranes of *E. coli* or *B. subtilis*.

The main conclusion from these results is that a single nucleoside transport system with a high affinity for pyrimidine nucleosides is operating in *B. subtilis*. The purine nucleoside adenosine has a high affinity for one (or more) component(s) of the transport system, but actual transport of this nucleoside requires additional factors. Thus the system is assumed to contain both separate and overlapping components.

Guanosine showed extremely low affinity for the transport system investigated. Since *B. subtilis* is not able to grow on guanosine as a carbon source, although it contains purine nucleoside phosphorylase (J. Neuhard, personal communication), it seems possible that the cells lack the capacity for transport of this purine nucleoside.

IV. TRANSPORT OF NUCLEOBASES IN BACTERIA

A. Introduction

Enteric bacteria readily incorporate exogenous nucleobases (except thymine) in their nucleic acids, and extensive incorporation takes place independent of a specific physiological defect. The incorporation is, however, strictly dependent on the presence of the appropriate phosphoribosyltransferases as well as on the intracellular supply of 5-phosphoribosyl-l-pyrophosphate (*see* Chapters 2 and 3).

This apparent coupling between uptake and phosphoribosylation of the nucleobases has lent some support to the suggestion, that nucleobases such as adenine, hypoxanthine and uracil are transported by a group translocation reaction, catalyzed by the corresponding phosphoribosyltransferase (31). The evidence presented for such a transport mechanism has not ruled out the possibility, though, that the conversion of exogenous nucleobases to intracellular nucleotides could also involve a rapid carrier-mediated translocation, followed by a phosphoribosylation of the translocated base. In this context it should be mentioned that the latter mechanism seems to operate for the overall uptake of nucleobases in cultured mammalian cells as indicated in kinetic experiments where a rapid sampling technique was employed (73).

The rapid technique has not yet been employed in bacterial investigations, but for uracil a factor involved in transport has been demonstrated, which is separable, genetically and functionally, from the uracil phosphoribosyltransferase activity.

None of the phosphoribosyltransferases will use cytosine as a substrate and utilization of this pyrimidine base requires a deamination to uracil, catalyzed

TABLE 6.7

Enzymes and transport proteins involved in the transport and initial utilization of exogenous nucleobases.

Nucleobase		Gene symbol	Product
Cytosine	Cytosine transport protein	codB	
	Cytosine deaminase	codA	uracil
Uracil	Protein involved in uracil transport	uraA	
	Uracil phosphoribosyltransferase	upp	UMP
	Uridine phosphorylase	udp	uridine
Adenine	Adenine phosphoribosyltransferase	apt	AMP
Guanine	Guanine-xanthine phosphoribosyltransferase	gpt	GMP
Xanthine			XMP
Hypoxanthine	Hypoxanthine phosphoribosyltransferase	hpt	IMP
Adenine	Purine nucleoside phosphorylase	deoD	adenosine
Guanine			guanosine
Hypoxanthine			inosine
Thymine	Thymidine phosphorylase	deoA	thymidine

by cytosine deaminase. As described below cytosine has its own transport system, specific for this base.

Table 6.7 lists the various transport proteins and enzymes, known to be involved in transport and initial utilization of the nucleobases in enteric bacteria.

B. Cytosine transport

The existence of a bacterial transport system for cytosine was first demonstrated by Lind *et al.* (50). Using a pyrimidine requiring strain of *E. coli* K12 they isolated a mutant which was resistant to low concentrations (60 μg/ml) of 6-azacytosine. The mutant had normal levels of cytosine deaminase, but was defective in its utilization of cytosine, both as a pyrimidine source and as a source of nitrogen. Phenotypically the mutant could be distinguished from mutants in cytosine deaminase, since the latter mutants are resistant to high concentrations (1 mg/ml) of 6-azacytosine and in addition are completely unable to utilize exogenous cytosine. The mutation, designated *codB*, was shown to map at 8 min on the *E. coli* chromosome close to, but different from, the *codA* gene, encoding cytosine deaminase (25, 50).

Similar mutants have been isolated as resistant to low concentrations (5 μg/ml) of 5-fluorocytosine (Andersen, L., 1979. Thesis. University of Copenhagen). Also in these mutants the utilization of exogenous cytosine was defective, while the cellular levels of cytosine deaminase were unchanged. The utilization of exogenous uracil was the same in mutant and parent strains. It was concluded that the gene in question codes for a protein involved in the transport of cytosine.

The cytosine transport system has been investigated in a *codA* strain, lacking cytosine deaminase and thus unable to metabolize cytosine (Andersen, L., 1979. Thesis. University of Copenhagen). The system shows saturation kinetics, and the transport is active with a concentrating effect of 300 fold at an exogenous concentration of 0.2 μM. Addition of other nucleobases or of nucleosides has no effect on the rate of transport, indicating that the system is specific for cytosine. This is in contrast to the cytosine transport system in yeast (9, 74) which transports purine bases as well as cytosine.

Both the synthesis of the cytosine transport system and the synthesis of cytosine deaminase are regulated. Starvation of the cells for pyrimidines or for nitrogen results in increased transport rates and a derepressed synthesis of cytosine deaminase, while addition of purines to the growth medium causes a repression of transport rate and deaminase activity. Thus there are indications that the two functions are regulated together. Mutants containing 10-fold increased levels of cytosine deaminase have been isolated, but at present it is not known whether the mutation, designated *codR*, also affects cytosine transport (Andersen, 1979. Thesis. University of Copenhagen).

C. Uracil transport

Uracil is normally incorporated into cells of *E. coli* or *S. typhimurium* through the uracil phosphoribosyltransferase reaction and mutants lacking this enzyme do not metabolize exogenous uracil (except under certain conditions as described below). This inability of *upp* mutants to utilize uracil, and the failure to demonstrate any accumulation of radioactive uracil in the cells during short time experiments led to the suggestion, that uracil phosphoribosyltransferase catalyzes a group translocation reaction, necessary for the transport of exogenous uracil across the cell membrane (33).

The following two pieces of evidence indicate that the uracil phosphoribosyltransferase cannot be the only protein involved in the transport of uracil into the cells:

1. Mutants lacking uracil phosphoribosyltransferase are resistant to 5-fluorouracil (19 μM) but may be sensitized to lower concentrations by the addition of a purine ribonucleoside (or a purine deoxyribonucleoside) to the growth medium. The toxicity is due to the intracellular formation of 5-fluorouridine through the sequential action of purine nucleoside phosphorylase and uridine phosphorylase (*see* Chapter 5).

Short time experiments in which the uptake of radioactive uracil is measured in cells lacking the uracil phosphoribosyltransferase, has confirmed this effect of purine nucleosides on uracil uptake (6, Fig. 6.5). Thus uracil may enter the cells independent of the uracil phosphoribosyltransferase.

2. Mutants of *E. coli* have been isolated, which are defective in the utilization of uracil added in concentrations below 80 μM, although they contain normal levels of uracil phosphoribosyltransferase. The mutants were isolated in a wild type background by selection for resistance to low (0.8 μM) concentrations of 5-fluorouracil (Fast, unpublished). Similar mutants have been obtained by selection for resistance to 5-fluorouracil (0.8 μM) in strains with a pyrimidine requirement, using cytosine as a pyrimidine source (Mygind, unpublished). This latter selection procedure ensures a functional uracil phosphoribosyltransferase in the mutants.

In these mutants only the uptake of uracil is affected. Thus the fluorouracil resistant, pyrimidine requiring mutants will grow normally on cytosine (18 μM) as a pyrimidine source, while the growth rate on uracil in this concentration range is strictly depending on the uracil concentration (Fig. 6.4). Furthermore the mutants fail to respond to the addition of a purine nucleoside, when uptake of uracil is measured in cell suspensions (Fig. 6.5). This failure is independent of a functional uracil phosphoribosyltransferase. In contrast the uptake of cytosine is unchanged in the mutants and markedly stimulated by purine nucleosides in both mutant and parent cells, as measured in strains lacking the uracil phosphoribosyltransferase. The utilization of the purine bases adenine and hypoxanthine is unaffected by the mutation.

From these results it is concluded that the mutants isolated as described above are deficient in a component, which is necessary and specific for the transport of uracil at low exogenous concentration (< 80 μM). The mutation, designated *uraA* has been shown to cotransduce 88 per cent with the *purG* marker at 53 min on the *E. coli* chromosome, i.e. close to the *upp* gene. The leaky phenotype of the mutants is not caused by leaky mutations, since the mutants which have been characterized harbor either the Tnlo insertion element or phage Mu in the *uraA* gene.

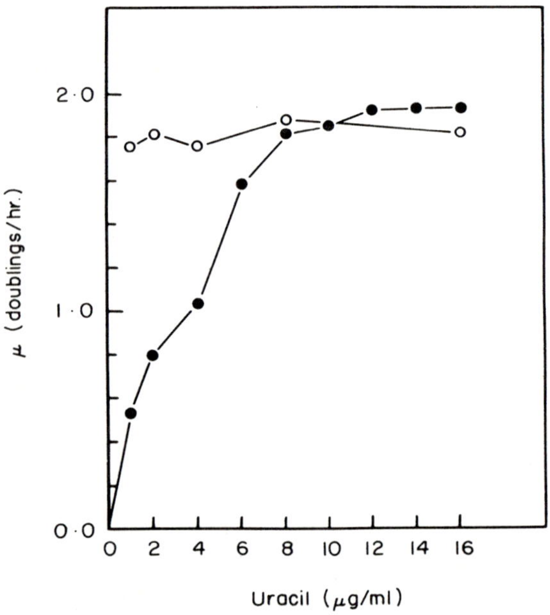

FIG. 6.4. The effect of the uracil concentration on the growth rate of a pyrimidine requiring *uraA* mutant. Cells were grown at 37°C in minimal medium with glucose (0.2%), casaminoacids (0.1%) and uracil in the concentrations indicated. ○——○ *pyrF* control strain. ●——● *pyrF*, *uraA* mutant.

The *uraA* gene product has not yet been identified, and neither its specific function in the transport of uracil nor its location in the cell envelope is known at present.

D. Purine transport

As mentioned earlier it was suggested that a group translocation reaction, catalyzed by the corresponding purine phosphoribosyltransferase might be responsible for the transport of purine bases (31, 35). Evidence for and against this hypothesis has been reviewed previously (30).

Later published reports have given no definite evidence, but they favor the existence of specific nucleobase carriers in the cell membrane. Page and Burton (65) investigated the cellular location of the purine phosphoribosyltransferases in cells of *E. coli*. They found that these enzymes were released in

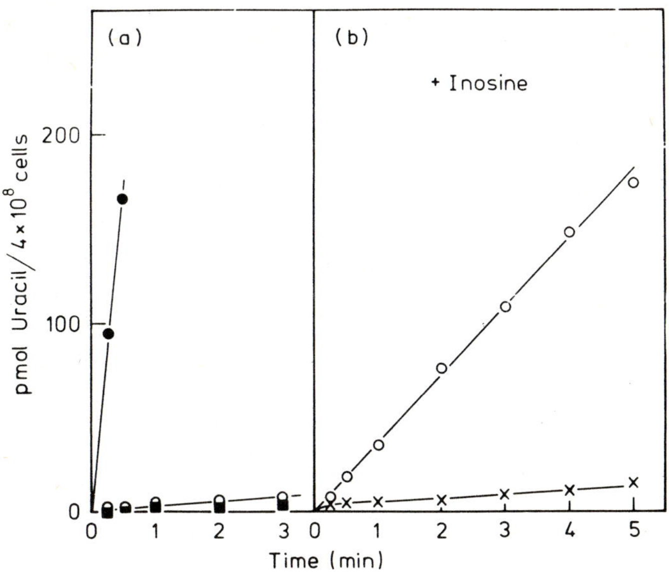

FIG. 6.5. The effect of the *uraA* mutation on uracil uptake. Cells were grown in minimal medium with glucose (0.2%) and harvested in the exponential phase of growth. Uptake measurements were performed as described in legend to Fig. 6.1, but with glucose as energy source instead of glycerol. [2-^{14}C]-uracil (0.8 μM final concentration) was added at time zero. Inosine (180 μM final concentration), when used, was added immediately before the addition of isotope.
(a) uracil uptake in a wild type strain, a *upp* and a *uraA* mutant.
(b) the effect of inosine on uracil uptake in a *upp* and a *upp,uraA* mutant.
● —— ● wild type; ○ —— ○ *upp* mutant; ■ —— ■ *uraA* mutant; × —— × *upp,uraA* mutant.

very low amounts when spheroplasts were prepared. If the spheroplasts were broken by lysis and the membrane fractions collected by gradient centrifugation, less than 10 per cent of the enzyme activities was retained by the membranes. Membrane vesicles, prepared according to Kaback (38) or Futai (23) retained 0.2–1% (depending on the number of washings) of the adenine phosphoribosyltransferase activity.

The same authors treated wild type cells or spheroplasts for up to one hour with non-penetrating probes which are known to inactivate external non-protected enzymes (e.g. trinitrobenzene sulphonate). The treatment had little or no effect on the activity of the three purine phosphoribosyltransferases, when assays were performed on sonicated cell extracts, prepared after removal of the probe and extensive washing of the cells. If the probes were added directly to the cell extracts a rapid enzyme inactivation occurred.

Thus while the phosphoribosyltransferases most likely are located inside the cell membrane, they seem to some extent to be associated with, but not integrated in the membrane. This location of the enzymes would be in accordance with a base transport involving the tandem operation of a membrane-bound carrier and the appropriate phosphoribosyltransferase, but so far there is only vague and indirect evidence for the existence of such carriers for the purine bases.

One piece of evidence comes from the work of Kocharian *et al.* (40) who isolated mutants lacking adenine phosphoribosyltransferase in an *E. coli* K12 strain with a purine requirement and an additional mutation in purine nucleoside phosphorylase. The *apt* mutants were unable to utilize exogenous adenine unless a functional purine nucleoside phosphorylase was introduced, indicating that in cells with an *apt* background adenosine is an intermediate in the utilization of adenine. The cellular location of purine nucleoside phosphorylase is believed to be inside the cytoplasmic membrane (60, 65) and the implication therefore is that adenine may enter the cells independent of the adenine phosphoribosyltransferase.

This result has been confirmed in short time experiments, where the uptake of radioactive adenine was measured in an *apt* mutant (Per Nygaard, unpublished). Adenine uptake was low but addition of nucleosides (uridine or thymidine) caused a 2–3 fold stimulation. These results resemble those obtained when uracil uptake was studied in mutants lacking uracil phosphoribosyltransferase (6) and thus suggest that a mechanism similar to the one implicated for the transport of uracil may operate for purines also.

Attempts to demonstrate an active transport and accumulation of purine bases in cells unable to metabolize these purines, i.e. lacking purine phosphoribosyltransferases and purine nucleoside phosphorylase have not given very decisive results (6).

At an external adenine concentration of 0.78 μM the uptake into an *apt* mutant was about 1% of the uptake into wild type cells and only 20% of this could be identified as free adenine. The transport seemed to be energy-dependent, but the concentrating effect was low, 30-fold or less.

The uptake of hypoxanthine in a non-metabolizing strain was even less significant. However, in both transferase-less strains a competition effect between adenine and hypoxanthine was observed, suggesting a common transport system; addition of uracil had no effect on the transport rate.

Guanine was not included in these investigations. However, Benson *et al.* (5) have reported the isolation of a purine requiring mutant of *S. typhimurium* with a specific alteration in the uptake of guanine. The mutant had retained normal levels of the purine phosphoribosyltransferases. The mutation, designated *guaP*, was located by cotransduction analysis at approximately 2 min (between *aziA* and *nadC*) on the chromosome. Preliminary evidence indicated the existence of a similar locus in *E. coli* (5).

E. Thymine transport

Thymine, when supplied exogenously, is metabolized through the thymidine phosphorylase reaction. The reaction requires a cellular supply of deoxyribose 1-phosphate, and only cells with an elevated internal pool of this compound (e.g. thymine requiring mutants) are able to convert exogenous thymine to thymidine. Wild type cells require addition of deoxyribonucleosides for incorporation of thymine into DNA. (For a detailed discussion of the reactions involved in the utilization of thymine, *see* Chapter 4.)

Thus thymine is transported into cells of both wild type and thymine requiring strains, but so far there is no published evidence for a specific transport system. Rinehart *et al.* (79) studied the ability of a prototrophic strain of *Bacillus subtilis* to take up and accumulate thymine. By incubating cell suspensions with [^{3}H]-labelled thymine and [^{14}C]-labelled dextran and subsequent determination of the $^{3}H/^{14}C$ ratio in cells and supernatant, they showed that thymine enters the cells although it is not incorporated into DNA. In other experiments where they measured uptake of thymine in washed cells, they found that the intracellular concentration was strictly proportional to the exogenous thymine concentration over the range measured (2–25 μg/ml). The uptake was unaffected by the energy poison sodium azide (10 mM). From these results the authors conclude that thymine enters the cells by passive diffusion.

V. TRANSPORT OF NUCLEOSIDES AND NUCLEOBASES IN FUNGI

The physiological organization of the cell boundaries in fungi show similarities to the bacterial cell envelope in that a cell wall, containing a structural polymer, supports and protects the cytoplasmic membrane. The chemical nature of the structural polymers in fungi show considerable diversity, but like in bacteria it is the cytoplasmic membrane in which the main permeability barrier resides, and in which the various transport systems are located.

Nucleoside and nucleobase transport in fungi has been studied mainly in yeasts and in the filamentous fungus *Neurospora crassa*. These organisms have long been known to utilize nucleosides and nucleobases, present in the growth medium. The very existence of purine and pyrimidine auxotrophs in these organisms is evidence for the presence of mechanisms for transport of the nucleic acid precursors into the cells.

In the study of these transport systems, especially in the case of *Neurospora*, there have been problems in the experimental separation of the transport process from the subsequent metabolism, similar to the difficulties encountered in the work on bacteria. It has been of importance, therefore, that both yeasts and *Neurospora* lend themselves to genetic manipulations. In yeasts various mutants in the pathways for metabolism of exogenous nucleic acid precursors are known and have been characterized. As discussed below the availability of such mutants has greatly facilitated the transport studies.

A. Transport of nucleosides in *Saccharomyces cerevisiae*

1. Uridine transport

Transport may be studied in yeast cells with the same techniques as in bacteria, i.e. as uptake of radioactive substrates. Suspensions of exponentially growing cells or of resting cells have been employed. In *Saccharomyces cerevisiae* the presence of a transport system for uridine was first demonstrated by Grenson (24) in studies of mutants impaired in the utilization of pyrimidines. Selection for resistance to 5-fluorouridine (0.4 mM) in strains lacking uridine hydrolase and thus unable to catabolize uridine gave two classes of mutants. One class consisted of strains which were deficient in uridine kinase, while mutants belonging to the second class were unable to transport uridine. The failure of these mutants to transport uridine was shown to be independent of the presence of uridine hydrolase.

Using [^{14}C]-labelled uridine the transport was studied in strains which had retained the transport system, but carried different mutations in the metabolism of uridine. The apparent rates of uridine transport were found to depend on the uridine kinase and uridine hydrolase activities, indicating that in studies of the kinetics and of the inhibition pattern for the uridine transport system it is desirable to use strains that are unable to metabolize the nucleoside.

Such an investigation was carried out by Losson *et al.* (51) on cells of a strain which lacked uridine kinase as well as uridine hydrolase. In this strain the transport system showed an apparent K_m of 0.1 mM. The same authors showed that the transport is energy requiring. In the presence of the energy poisons sodium azide (15 mM) or 2.4 dinitrophenol (1 mM) the rate of transport was inhibited 80–100%. Furthermore, the toxic agents Dio 9 (50 μg/ml) and chlorohexidine (10 μM) decreased the transport rates approximately 50%. These two compounds were also shown to inhibit the plasma membrane bound ATPase in the same cells. Although the energy providing system cannot be identified on the basis of these results, they are suggestive of a proton motive force as the energy donor.

Grenson (24) investigated transport of uridine in a pyrimidine requiring strain after starvation for uracil. The resulting increase in transport rate was taken as an indication that the system is controlled by a feedback reaction involving an intracellular pyrimidine compound.

The specificity of the uridine transport system in *S. cerevisiae* has not been investigated.

2. *Adenosine transport*

Very little is known about the transport of other nucleosides into yeast cells. It has been shown, though, that adenosine is incorporated slowly into nucleic acids in wild type cells of *S. cerevisiae* (3). The nucleoside cannot support growth of an adenine-requiring strain, but mutants may be selected which are able to utilize adenosine at a 10-fold increased rate (3). The increased capacity for adenosine transport rendered the mutant cells highly sensitive to the toxic adenosine analog 3′-deoxyadenosine (cordycepin). It is not possible from these results to decide whether the increased rate of utilization is due to a change in transport rate or to changes in the subsequent metabolism.

B. Transport of nucleobases in *S. cerevisiae*

1. *Cytosine transport*

Isolation of transport negative mutants. Utilization of cytosine by cells of *S. cerevisiae* (as by enteric bacteria) requires a deamination of cytosine to uracil, which is subsequently converted to UMP by the uracil phosphoribosyltransferase. Free uracil is exclusively utilized through this reaction since yeast cells contain no uridine phosphorylase. Uridine is catabolized in yeast by simple hydrolysis, catalyzed by uridine hydrolase (cf. Chapter 3).

Mutants deficient in cytosine transport are readily isolated in strains with a pyrimidine requirement by selection for resistance to 5-fluorocytosine (up to 0.1 mM) on media supplemented with uracil (9, 66). This selection procedure renders essentially two classes of mutants. One class is deficient in cytosine deaminase, while mutants of the second class lack a transport system for cytosine, the cytosine permease. The latter mutants contained normal amounts of cytosine deaminase.

Since uracil is necessary for growth of the cells, mutants deficient in uracil phosphoribosyltransferase do not appear in this selection.

Kinetic data. The transport of cytosine may conveniently be studied in the cytosine deaminase mutants, since in these strains the transported cytosine is accumulated without further metabolism. Chevallier *et al.* (9) studied the transport in such a mutant and found that cytosine is transported actively against a concentration gradient. Saturation was reached after approx. 10 min. At this point the concentrating effect was around 1900 at an exogenous substrate concentration of 3.7 μM. An apparent K_m of 2.4×10^{-6} M was calculated.

Specificity. The specificity of the cytosine transport system is unusual since the system, besides cytosine, will accommodate the purine bases adenine, hypoxanthine, and guanine (9). Uracil is *not* transported by this system (67). Kinetic data indicate that the element shared by the different nucleobases may be the recognition site of the system. Thus the uptake of each of the bases is competitively inhibited by the other bases. Furthermore the apparent K_m for transport of each base is not significantly different from the K_i measured when the same base is used as an inhibitor of transport of another of the substrates (9).

A polypeptide of MW approximately 80 000 and located in the plasma membrane has been identified as a likely component of the transport system (66). In these experiments cells of a permease-less mutant (carrying a frameshift mutation) and of the parent strain were independently labelled with [^{3}H]-leucine and [^{14}C]-leucine. Membrane fractions prepared from the labelled cells were mixed and subjected to electrophoresis on polyacrylamide gels. By comparison of the [^{3}H]- and [^{14}C]-labelling distribution in the gel a protein could be identified, which was present in the parent but absent in the mutant cells.

Mutants with an altered cytosine permease. The competition between the bases for the same permease has been used to select strains with a modified transport system (9). Thus mutants with altered K_m for cytosine (up to 30-fold increased) appear readily when an adenine requiring strain is used to select for resistance to growth inhibition by cytosine on a minimal medium, supplemented with adenine. The mutants were shown, by crossing and complementation tests, to be alleles of the cytosine permease gene. Similar selection procedures rendered mutants with altered V_{max} or with an altered specificity (9).

Energization. Cytosine as well as the purine bases are transported actively into the cells. Losson *et al.* (51) working with cytosine as a substrate, showed that 2.4 dinitrophenol (1 mM), sodium azide (10 mM), Dio 9 (25 μg/ml), and chlorohexidine (0.2 mM) inhibited approx. 90 per cent. Iodoacetic acid, an inhibitor of glycolysis, is without effect, while incubation with N-ethylmaleimide (0.1 mM) gradually blocks the transport (51).

The efficient inhibition by the proton conductor 2.4 dinitrophenol occurs also in yeast cells which are deficient in respiratory functions due to a mitochondrial petite mutation (ρ^{-}mutants). This suggests that the inhibitor acts directly on the plasma membrane, dissipating a proton gradient. In accordance with this Reichert *et al.* (76), have shown that the transport of purines is activated by a decrease in extracellular pH. Moreover, a detailed potentiometric investigation of the relationship between proton absorption, potassium ejection and the transport of hypoxanthine indicates a 1:1:1 stoichiometry. Thus the transport of the purine bases, and probably also of cytosine, may depend on a carrier, which acts as a proton symporter and a cation antiporter (77).

Foret *et al.* (20) studied the recognition step in the transport by measuring the initial rates under a number of different experimental conditions. These included variations in pH and the use of competitive as well as group-specific inhibitors. From their results they concluded that the association of one proton was necessary for binding of the substrate to the carrier. The assumption put forward in the paper is that a carboxyl group of the carrier, a proton and a substrate molecule associate to form a ternary complex.

Regulation. Attempts to demonstrate a regulation of the cytosine–purine transport system have failed so far. Initial rates of cytosine transport were measured in *S. cerevisiae* cells grown on minimal medium with various nitrogen sources, including cytosine. No significant differences could be detected (9).

2. Uracil transport

Uracil is transported actively into *S. cerevisiae* like other nucleobases, but by a different system, the uracil permease. This was shown by Grenson (24), who isolated a mutant, resistant to 20 μM 5-fluorouracil, which was affected in the utilization of uracil but not of cytosine. The mutation was shown genetically to be different from mutations in cytosine permease (24, 37). Uracil uptake could not be detected in the mutant under standard conditions, while the transport of cytosine and of uridine was unaffected.

The uracil permease was investigated in more detail in a strain which was unable to metabolize the transported uracil due to mutations in uracil phosphoribosyl transferase (37). An additional mutation in the uridine hydrolase blocked the cleavage of endogenously formed uridine to uracil which might otherwise be excreted and thereby interfere with the transport measurements. In cells of this strain uptake of uracil reached a steady state within 10 min, and an intracellular pool of free uracil could be determined chromatographically. The concentrating effect in the mutant strain was highly dependent on the exogenous uracil concentration. At concentrations above 0.3 mM the cells do not concentrate uracil at all. The overall concentrating effect is much lower than found for cytosine permease. At an optimal external concentration of 3.7 μM the concentrating effect does not exceed 40-fold as compared to 1900-fold for cytosine under optimal conditions.

The transport system differed further from the cytosine permease in that the relationship between transport rate and substrate concentration had a distinct biphasic character, suggesting the presence of a high affinity and a low affinity component. Both parts of the system seemed to be absent in the transport negative mutant.

When wild type cells were used in the transport assays, no steady state condition was reached and the transported uracil was quantitatively incorporated into nucleotides and nucleic acids. In this case no intracellular pool of uracil could be detected.

Specificity. The uracil permease is specific for uracil among the nucleobases. The competitive effect of other bases, purines as well as various substituted pyrimidines, including thymine, was investigated in a cytosine permease-less strain. The bases were added in 50-fold the amount of uracil. Only hypoxanthine and 6-amino uracil showed a slight inhibition, other bases tested were without effect (37).

Energization. The transport of uracil is completely inhibited by 1 mM 2.4 dinitrophenol (51), suggesting that a proton gradient is necessary for the transport to proceed. This is analogous to what was found for the cytosine permease. In contrast to the latter system the uracil transport was completely resistant to the two ATPase inhibitors chlorohexidine (10 μM) and Dio 9 (50 μg/ml). Thus the two systems seem to be energized by different mechanisms.

C. Transport of nucleosides and nucleobases in other yeasts

1. Nucleoside transport

The uptake (transport and utilization) of the nucleosides uridine and adenosine has been studied in the fission yeast *Schizosaccharomyces pombe* in glycerol-grown cells (21). Both nucleosides were utilized efficiently, but with highly different kinetic parameters. For uridine a K_m of 14 μM and V_{max} of 90 nmol/min/g protein was found, while the corresponding values for adenosine were 2 mM and 640 nmol/min/g protein, respectively. This was suggestive of two different transport systems for the two nucleosides and this notion was supported by the finding that the presence of excess adenosine did not inhibit uridine uptake. With both nucleosides the rate of uptake was stimulated two-fold by cAMP but it is not clear whether cAMP exerts its effect on transport or on metabolism.

A third nucleoside, namely guanosine, appeared not to be utilized at all by the cells, at least not under the conditions employed in the experiments just mentioned (21).

2. *Nucleobase transport*

The reports on nucleobase uptake in yeast cells other than *S. cerevisiae* indicate the presence of transport systems similar to those already described for *S. cerevisiae*. In *Candida albicans* a system with the same specificity as the cytosine permease in *S. cerevisiae* has been reported (74). In the fission yeast *Sch. pombe* a system transporting purine bases was described (11). Cytosine was not tested here.

Another aspect of the base transport was studied in *Sch. pombe*, namely the role of the phosphoribosyltransferases in the regulation of the transport rates (34). Mutants lacking guanine (hypoxanthine, xanthine) phosphoribosyltransferase shows a drastically reduced uptake of guanine as compared to the wild type cells. Both K_m and V_{max} were affected. In contrast to this adenine uptake was unimpaired. This result indicates that an active phosphoribosyltransferase enhances the base transport, but the figures given do not allow any conclusion as to whether the activation actually occurs at the level of a transport system.

In a strain lacking adenine phosphoribosyltransferase the uptake of adenine was less affected, but this again might be explained by an immediate deamination of the transported adenine to hypoxanthine which is then utilized through the guanine-hypoxanthine phosphoribosyltransferase, present in these cells.

The antibiotic formycin lowers the cellular pools of 5-phosphoribosyl-l-pyrophosphate, but cell growth is unaltered (34). The rates of purine uptake in formycin-grown cells were reduced approximately 80%. It is not possible, though, from these results to conclude whether the decrease is a direct effect of 5-phosphoribosyl-l-pyrophosphate on the transport system or an indirect effect, caused by a reduced rate of phosphoribosylation of the transported base.

D. Transport of nucleosides in *Neurospora crassa*

Germinating conidia spores have been used for the study of transport systems in *Neurospora crassa*. Mycelial preparations from cultures of *Neurospora crassa* contain hyphae of highly variable size and age, and such preparations are generally unsatisfactory for short time experiments in which uptake of labelled compounds is measured, since a large scatter of data is obtained. In contrast germinating conidia will grow exponentially in minimal medium, yielding morphologically uniform cells, and with suspensions of such cells reproducible results may be obtained. The results reported below all stem from experiments with germinating conidia. These conidia contain mechanisms for transport of nucleosides, since labelled nucleosides, when administered to the growth medium of such conidia, are taken up and utilized. The transport measurements described below are all carried out on cells capable of metabolizing the transported nucleosides, but no attempt to characterize the reaction products after uptake has been reported. It is possible, therefore, that unknown, i.e. not registered differences (genetic or regulatory) in the enzymatic background of the germinating conidia might influence the results and thus bias the conclusions, when uptake measurements from different strains are compared.

Early experiments by Schlitz and Terry (85) indicated that two nucleoside transport systems were operating in the conidia. These authors investigated the uptake of labelled thymidine and deoxyadenosine (as representatives for pyrimidine and purine nucleosides, respectively) in cultures of germinating conidia, which were sampled and tested for transport at different times during the germination period. The capacity per mg protein for transport of both nucleosides increased during germination, but competition experiments showed that the temporal pattern of thymidine uptake during the germination period was different from that of deoxyadenosine uptake in the presence or absence of thymidine. This led to the conclusion that two transport systems were generated during germination, one which will transport both purine and pyrimidine nucleosides, and a second system which will accept only purine nucleosides. The system(s) showed the characteristics of mediated, active transport: saturation kinetics, pH-dependance and inhibition by energy poisons.

These experiments were performed with a wild type strain. More detailed studies of the nucleoside uptake have been conducted by Magill *et al.* (52), using a mutant strain, *ad-8*, which lacks the adenylosuccinate synthetase and thus requires a supplement of adenine in the growth medium. When adenosine was used as supplement for growth of the mycelium the growth inhibition pattern, caused by addition of other nucleosides (in 10-fold amounts), corresponded closely to the ability of the same nucleosides to inhibit the uptake of [^{14}C]-labelled adenosine in germinating conidia. The apparent K_i was determined for uptake of a number of nucleosides into the conidia. Uridine, inosine, guanosine and their corresponding deoxyribonucleosides behaved as efficient competitive inhibitors. Thymidine had a relatively small inhibitory effect, while xanthosine had no effect at all (52).

The specificity of the nucleoside transport system was further investigated by Dunaway-Mariano and Magill (19). Using the same adenine-requiring strain, *ad-8*, they measured uptake of [^{14}C]-labelled adenosine in the presence of a large variety of nucleosides and nucleoside analogs. From the apparent K_i-values they deduced the general structural features which are required of a nucleoside to be readily accepted by the transport system. These include: (i) a purine or pyrimidine ring system, (ii) ribose or 2′deoxyribose in β-configuration, (iii) no substituents at C_8 in the purines or at C_5 or C_6 in the pyrimidines, (iv) the presence of a C_5–C_6 double bond in the pyrimidine ring, (v) the absence of a charge on the heterocyclic ring.

These specificity requirements would classify thymidine as a poor substrate for transport. This was confirmed in competition experiments with [^{14}C]-labelled adenosine. An apparent K_i of 422 μM was found, which is 60-fold higher than that found for deoxyuridine (7 μM). The K_m for thymidine transport found by Schlitz and Terry (85) was 65 μM which indicated a higher affinity for thymidine in the wild type strain than in the adenine requiring strain used by Dunaway-Mariano and Magill (19).

Another discrepancy between the two strains is the lack of a second transport system in the adenine requiring strain *ad-8*. Studies have been carried out in conidia of a different wild type strain, 74A, germinated to the stage where the system specific for purine nucleosides and the general transport system were proposed to overlap maximally in the wild type strain used by Schlitz and Terry. At this stage competition experiments were performed in which (a) [^{14}C]-adenosine uptake was inhibited by uridine, and (b) [^{14}C]-uridine uptake was inhibited by adenosine. The values for K_m (8 μM, 12 μM) and K_i (11 μM, 10 μM) for adenosine and uridine, respectively, were very close to each other. Furthermore, unlabelled uridine and unlabelled adenosine display the same efficiency to inhibit or dilute the incorporation of [^{14}C]-labelled adenosine in the conidia. It was concluded, therefore, that a single transport system, which accepts both purine and pyrimidine nucleosides is responsible for all nucleoside transport in the germinating conidia (19).

Additional evidence for this is the behaviour of a mutant strain *(ud-l,pyr-l)* which was isolated as unable to transport uridine (19). Simultaneously the strain had lost the ability to transport adenosine, suggesting that the mutation specifies loss of a transport system which is common for purine and pyrimidine nucleosides. No indication of a specific purine nucleoside transport system was observed in this strain.

In summary these results indicate that the transport of nucleosides in *Neurospora crassa* is mediated by a single transport system which will accept both purine and pyrimidine nucleosides. The indication of a second system, as proposed by Schlitz and Terry (85), may perhaps be explained by different enzymatic backgrounds in the strains employed. The conclusions are all based on the assumption that the entry of nucleosides into the cell is the rate limiting factor for uptake of nucleosides. If this is not the case, the inhibition pattern caused by other nucleosides may reflect inhibition of metabolism inside the cell or perhaps inhibition of both transport and metabolism. Such considerations should be borne in mind, when results from uptake experiments carried out in nucleoside metabolizing cells are interpreted. Little is known about the *in vivo* efficiences of the enzymes which metabolize nucleosides in *Neurospora*, especially not whether these enzymes are synthesized specifically during germination of the conidia.

E. Transport of nucleobases in *Neurospora crassa*

In germinating conidia of *N. crassa* two different transport systems for purine bases have been demonstrated. The systems differ in their specificity towards the purine bases in such a way that one system is responsible for the transport of adenine, guanine and hypoxanthine, while a second system accepts adenine specifically.

Pyrimidine bases are also taken up by *N. crassa*, but it is not known whether one or two transport systems are operating here.

1. Transport of purine bases

The kinetics and the mutual inhibitory pattern of purine base uptake in germinating conidia has been examined in some detail in a wild type strain and in two different adenine auxotrophs (53). The results were compatible with the presence of one common transport system with greater affinity for adenine than for guanine or hypoxanthine. However, another model would fit the results, namely one system common for all purine bases, along with a second system specific for adenine. The latter model was further supported by the following finding: The adenine requiring strain, *ad-8*, which lacks the adenylosuccinate synthetase, proved unable to transport hypoxanthine and in this strain addition of excess hypoxanthine did not inhibit uptake of adenine in contrast to what was found in the wild type. This was taken as an indication that in the adenine requiring mutant the system shared by hypoxanthine and adenine was absent and only an adenine specific system was operating.

The existence of two different systems for purine transport was confirmed by Pendyala and Wellman (68). They investigated uptake of labelled adenine in the absence or presence of excess (500-fold) hypoxanthine. In the presence of hypoxanthine the adenine uptake would then represent the contribution of the adenine specific carrier, while the difference in uptake with or without hypoxanthine specifies the uptake mediated by the common system. Using this test they measured uptake in conidia at different times during the germination period. The temporal pattern of uptake activities, obtained in these experiments, confirmed that two transport systems were operating in the germinating conidia. One system, common for the bases, was present initially and increased in capacity during the first 3–4 hours of germination. The second, adenine-specific system was absent initially, but developed after 2–3 hours. The specificity of the latter system was further tested in competition experiments with a number of purine and pyrimidine bases. The system was found to accept only adenine and 6-methylpurine (69).

In accordance with the findings of Magill and Magill (53) it was observed that in strain *ad-8* neither freshly harvested nor germinating conidia showed any hypoxanthine uptake. Adenine was transported to some extent in the ungerminated conidia and the adenine specific transport increased in activity during germination, but less than in the wild-type (approximately 50%).

In the ungerminated conidia of the wild type strain the adenine uptake, mediated by the common transport system showed normal Michaelis-Menten kinetics with an apparent K_m of 1.4 μM for adenine. Also 6-methylpurine was a substrate for the common transport system although the affinity is low ($K_i = 98$ μM, as compared to a K_i of 6 μM for hypoxanthine).

Regulation of purine transport. The failure of the adenine requiring strain, *ad-8*, to transport hypoxanthine and the concurrently reduced capacity for adenine transport probably reflects regulatory mechanisms in the cells. The *ad-8* strain lacks the adenylosuccinate synthetase and would be expected to accumulate IMP. Genetic studies of the *ad-8* mutation indicated that the lack of hypoxanthine transport is a consequence of this mutation and not due to a second mutation. It is probable, therefore, that accumulated IMP (or derivatives thereof) has a feed-back inhibitory effect on hypoxanthine uptake. Sabina *et al.* (84) directly demonstrated the presence of hypoxanthine and/or inosine as excretion products in the culture medium of the *ad-8* strain. These two compounds may both arise as degradation products of IMP.

The assumed regulatory effect of intracellularly accumulated IMP or its degradation products was supported by studies of a double mutant, *ad-8*, *ad-1*, which besides the mutational defect in the adenylosuccinate synthetase gene (*ad-8*) contained a lesion in a gene specifying one of the early enzymes in the purine *de novo* synthesis, amidoimidazole carboxamide formyltransferase (*ad-1*). Due to this second lesion the strain has only a limited intracellular synthesis of IMP and in this double mutant the ability to transport hypoxanthine was considerably higher than found in the *ad-8* strain, but still somewhat lower than in the wild type (84).

Relationship between purine transport and nitrogen metabolism. Uric acid and xanthine are utilized by *N. crassa* as nitrogen sources, and germinating conidia contain a transport system which by competition experiments was shown to be specific for these two bases. The transport mediated by this system was strongly repressed if the conidia were germinated in medium containing ammonium ions as nitrogen source (87). Under these circumstances transport of hypoxanthine, adenine, and guanine was unaltered.

The transport of hypoxanthine does respond, though, to changes in the nitrogen metabolism in the conidia, as seen in mutants with a lesion in the *nit-2* gene (69). The *nit-2* locus, identical to the *amr* locus (87), is a major control gene for nitrogen metabolism in *Neurospora*. A mutation in this gene results in a pleiotropic loss of several enzymes related to nitrogen metabolism and purine catabolism (78). It represses the uptake of xanthine and uric acid severely (87) and also hypoxanthine uptake is reduced while adenine uptake in both germinated and ungerminated conidia is unaltered (78). This shows that the general purine uptake system is still present. The reduced hypoxanthine transport activity in the *nit-2* mutants is believed, therefore, to be an indirect consequence of the defect in purine metabolism which is likely to create an intracellular pool pattern different from that found in wild type conidia (69).

Involvement of phosphoribosyltransferase in purine uptake. In the experiments reported above it has not been possible to distinguish between the transport process and the metabolism of the labelled bases. This poses the question, whether the phosphoribosyltransferases, the first enzymes to metabolize the purine bases, are directly involved in the transport process.

The problem was investigated in mutant strains, which were resistant to 6-methylpurine due to a deficiency in adenine phosphoribosyltransferase. This enzyme is necessary for the conversion of 6-methylpurine to the corresponding nucleotide, which is toxic to the cell. A strain lacking this enzyme (strain *mep-10*) was able to transport purines at approximately the same rate as found in wild type strains. Moreover, the transport of adenine was inhibited to the same degree (90%) by 6-methylpurine in both wild type and mutant cells, when the analog was added in excess (70). Although the possibility of an altered adenine phosphoribosyltransferase in the *mep-10* strain cannot be excluded these results suggest that, at least in the case of adenine, the transport and the subsequent phosphoribosylation are separate processes. There are indications, however, that transport and metabolism in *Neurospora* are intimately coupled processes, similar to what is found in cells of *E. coli*. In germinating conidia, which had been given a pulse of labelled purine bases, the radioactivity was found exclusively as nucleotides (70). The distribution of radioactivity in the various nucleotides in the strain lacking adenine phosphoribosyltransferase activity was different from the pattern in the wild type conidia, probably caused by lack of the enzyme and by regulatory changes due to concurrent alterations in the nucleotide pools. Further indication of a coupling between transport and utilization was the finding, that in centrifuged homogenates of the conidia the adenine phosphoribosyltransferase activity was mainly associated with the pellet, indicating that the enzyme *in vivo* might be associated with the membrane and in this way would facilitate a rapid conversion of the transported base (70).

A separation between hypoxanthine uptake and the corresponding hypoxanthine phosphoribosyltransferase activity is less clear. Investigations of conidia from wild type and from strains with mutations in purine metabolism revealed a correspondence between the levels of hypoxanthine phosphoribosyltransferase and the ability to transport hypoxanthine (69). This does not necessarily mean that the enzyme is directly involved in the transport process. Thus the *ad-8* mutant contains measurable amounts of the enzyme, while the adenine transport is virtually nil (53). An alternative explanation may be that the enzyme and the transport system are regulated by the same intermediates in the purine metabolism.

2. *Transport of pyrimidine bases*

Uptake of uracil in germinating conidia has been measured in the wild type strain and in a pyrimidine requiring strain (54). The two strains gave similar results, namely that the uptake mechanism did not exhibit simple Michaelis–Menten kinetics. A Lineweaver–Burk plot of the results indicated at least two substrate binding sites with different affinities, suggesting that either two or more different uptake systems were operating or, alternatively, one system with more than one substrate site was present.

A pyrimidine requiring strain which appears to lack also the ability to transport uracil (*pyr-1*, *uc-5*) has been isolated and investigated (54). Genetic studies of the progeny from a cross between this strain and the wild type strain, as well as the frequences with which the strain reverted to uracil uptake, indicated that the *uc-5* lesion is the result of a single mutation. Thus it appears likely that uracil transport is mediated by a single system in the germinating conidia.

Competition studies indicated that the system is shared with cytosine and thymine, while none of the purine bases tested had any inhibitory effect on the uptake of uracil (54).

The uptake of uracil was strongly repressed by the presence of ammonia ions in the germination medium, whereas growth of the original mycelial culture in ammonia-rich medium, followed by germination of the conidia in ammonium-free medium did not significantly depress the ability of the germinating conidia to transport uracil. It was reported, though, that a pyrimidine requiring strain and a wild type strain respond differently when grown in the presence of ammonia ions (54) and uracil. Under these conditions the growth rate of the pyrimidine requiring strain is significantly lowered compared to that of the wild type strain. Since uptake of uracil is necessary for the former strain but not for the latter, the difference in growth rate is probably caused by repression of the uracil uptake by the ammonium ions (54).

Involvement of uracil phosphoribosyltransferase in uracil transport. Germinating conidia of the transport negative mutant, *uc-5*, *pyr-1*, were shown to contain uracil phosphoribosyltransferase activities in the same amounts as found in the wild type conidia. Conidia from these two strains were used in a series of experiments in which the inhibition of uracil uptake was compared with the inhibition of uracil phosphoribosyltransferase. Other pyrimidine bases as well as a large number of base analogs were tested as inhibitors (12), and the uptake experiments were performed with wild type conidia germinated in ammonium-free medium. The inhibitory effect on the uracil phosphoribosyltransferase by the various base analogs was determined in crude homogenates of conidia from the *uc-5*, *pyr-1* strain. The transport negative strain was used in these specificity measurements, because the crude homogenates of conidia were found to contain certain amounts of whole cells and membrane vesicles, which might disturb the results, if wild type conidia, capable of uracil transport, were used.

More than 25 different structural analogs of uracil were tested as inhibitors in this way, and the results allowed the authors to summarize some general features of the base analogs which influenced the inhibitory effect. The characters of these general features have been used in an attempt to specify the recognition requirements of the transport system. Among the features is the presence of a substituent at the 5-position, which decreases the inhibitory effect somewhat (e.g. thymine, 5-fluorouracil). The presence of a negative charge results in a very slight inhibitory effect (e.g. orotic acid). Finally a double bond between C_5 and C_6 seems highly important for the recognition.

The specificity of the uracil phosphoribosyltransferase reaction towards the various uracil analogs was not the same as that of the uracil uptake. As an example thymine competes strongly with uracil for uptake, but is a weak inhibitor of the enzymes as measured in the *in vitro* assay. This result together with the finding that the transport negative strain, *uc-5*, *pyr-1*, contains wild type levels of the enzyme, speak against an involvement of the enzyme in the transport process, although the presence of an altered uracil phosphoribosyltransferase in the transport negative strain cannot be completely excluded.

REFERENCES

1. Adler, J., Hazelbauer, G. L., and Dahl, M. M. (1973). Chemotaxis toward sugars in *Escherichia coli. J. Bacteriol.* **115**, 824–847.
2. van Alphen, W., van Selm, N., and Lugtenberg, B. (1978). Pores in the outer membrane of *Escherichia coli* K 12. Involvement of proteins b and e in the functioning of pores for nucleotides. *Molec. Gen. Genet.* **159**, 75–82.
3. Anderson, J. M., and Roth, R. (1976). Adenosine utilization in cordycepin-sensitive mutants of *Saccharomyces cerevisiae. J. Bacteriol.* **128**, 689–691.
4. Beacham, I. R., Yagil, E., Beacham, K., and Pritchard, R. H. (1971). On the localization of enzymes of deoxynucleoside catabolism in *Escherichia coli. FEBS Letts.* **16**, 77–80.
5. Benson, C. E., Hornick, D. L., and Gots, J. S. (1980). Genetic separation of purine transport from phosphoribosyltransferase activity in *Salmonella typhimurium. J. Gen. Microbiol.* **121**, 357–364.
6. Burton, K. (1977). Transport of adenine, hypoxanthine and uracil into *Escherichia coli. Biochem. J.* **168**, 195–204.
7. Buxton, R. S., Hammer-Jespersen, K., and Valentin-Hansen, P. (1980). A second purine nucleoside phosphorylase in *Escherichia coli* K-12. *Molec. Gen. Genet.* **179**, 331–340.
8. Buxton, F. P., and Radford, A. (1982). Nitrogen metabolite repression of fluoropyrimidine resistance and pyrimidine uptake in *Neurospora crassa. Molec. Gen. Genet.* **186**, 259–262.
9. Chevallier, M. R., Jund, R., and Lacroute, F. (1975). Characterization of cytosine permeation in *Saccharomyces cerevisiae. J. Bacteriol.* **122**, 629–641.
10. Cohen, R. M., and Wolfenden, R. (1971). Cytidine deaminase from *Escherichia coli.* Purification, properties and inhibition by the potential transition state analog 3,4,5,6-tetrahydrouridine. *J. Biol. Chem.* **246**, 7561–7565.
11. Cummins, J. E., and Mitchison, J. M. (1967). Adenine uptake and pool formation in the fission yeast *Schizosaccharomyces pombe. Biochim. Biophys. Acta* **136**, 108–120.
12. Dalke, P., and Magill, J. (1979). Specificity of uracil uptake in *Neurospora crassa. J. Bacteriol.* **139**, 212–219.
13. Darnall, K. R., Townsend, L. B., and Robins, R. K. (1967). The structure of showdomycin, a novel carbon-linked nucleoside antibiotic related to uridine. *Proc. Natl. Acad. Sci. U.S.A.* **57**, 548–553.
14. von Dippe, P. J. Roy-Burman, S., and Visser, D. W. (1973). Transport of uridine in *Escherichia coli* B and a showdomycin-resistant mutant. *Biophys. Biochim. Acta* **318**, 105–112.
15. von Dippe, P. J., Leung, K-K., Roy-Burman, S., and Visser, D. W. (1975). Deoxycytidine transport in the presence of a cytidine deaminase inhibitor and the transport of uracil in *Escherichia coli* B. *J. Biol. Chem.* **250**, 3666–3671.
16. Doskočil, J. (1974). Inducible nucleoside permease in *Escherichia coli. Biochem. Biophys. Res. Commun.* **56**, 997–1003.
17. Doskočil, J., and Holý, A. (1974). Inhibition of nucleoside-binding sites by nucleoside analogues in *Escherichia coli. Nucleic Acids Res.* **1**, 491–502.
18. Doskočil, J. (1976). The role of thiol groups in nucleoside transport. *Mol. Cell. Biochem.* **10**, 137–143.
19. Dunaway-Mariano, D., Magill, J. M. (1978). Specificity of nucleoside transport in *Neurospora crassa. J. Bacteriol.* **136**, 924–928.
20. Foret, M., Schmidt, R., and Reichert, U. (1978). On the mechanism of substrate

binding to the purine-transport system of *Saccharomyces cerevisiae*. *Eur. J. Biochem.* **82**, 33–43.

21. Foury, F., and Goffeau, A. (1975). Stimulation of active uptake of nucleosides and amino acids by cyclic adenosine 3′:5′-monophosphate in the yeast *Schizosaccharomyces pombe*. *J. Biol. Chem.* **250**, 2354–2362.
22. Fucik, V., Kloudova, A., and Holý, A. (1974). Transport of nucleosides in *Bacillus subtilis*: the effect of purine nucleosides on the cytidine-uptake. *Nucleic Acids Res.* **1**, 639–644.
23. Futai, M. (1974). Orientation of membrane vesicles from *E. coli* prepared by different procedures *J. Membr. Biol.* **15**, 15–28.
24. Grenson, M. (1969). The utilization of exogenous pyrimidines and the recycling of uridine 5-phosphate derivatives in *Saccharomyces cerevisiae* as studied by means of mutants affected in pyrimidine uptake and metabolism. *Eur. J. Biochem.* **11**, 249–260.
25. de Haan, P. G., Felix, H. S., and Peters, R. (1972). Mapping of the gene for cytosine deaminase on the *Escherichia coli* chromosome. Antonie van Leeuwenhoek: *J. Microbiol. Serol.* **38**, 257–263.
26. Hammer-Jespersen, K., Munch-Petersen, A., Nygaard, P., and Schwartz, M. (1971). Induction of enzymes involved in the catabolism of deoxyribonucleosides and ribonucleosides in *Escherichia coli* K 12. *Eur. J. Biochem.* **19**, 533–538.
27. Hammer-Jespersen, K., Munch-Petersen, A. (1973). Mutants of *Escherichia coli* unable to metabolize cytidine: Isolation and characterization. *Molec. Gen. Genet.* **126**, 177–186.
28. Hammer-Jespersen, K., Munch-Petersen, A. (1975). Multiple regulation of nucleoside catabolizing enzymes: Regulation of the *deo* operon by the cytR and deoR gene products. *Molec. Gen. Genet.* **137**, 327–335.
29. Hantke, K. (1976). Phage T6—Colicin K receptor and nucleoside transport in *Escherichia coli*. *FEBS Letts.* **70**, 109–112.
30. Hays, J. B. (1978). Group translocation transport systems. *In* "Bacterial Transport" (B. P. Rosen, *Ed.*), pp. 43–102. Marcel Dekkar, Inc.
31. Hochstadt-Ozer, J., and Stadtman, E. R. (1971). The regulation of purine utilization in bacteria. II. Adenine phosphoribosyltransferase in isolated membrane preparations and its role in transport of adenine across the membrane. *J. Biol. Chem.* **246**, 5304–5311.
32. Hochstadt-Ozer, J. (1972). The regulation of purine utilization in bacteria. IV. Roles of membrane-localized and pericytoplasmic enzymes in the mechanism of purine nucleoside transport across isolated *Escherichia coli* membranes. *J. Biol. Chem.* **247**, 2419–2426.
33. Hochstadt, J. (1974). The role of the membrane in the utilization of nucleic acid precursors. *Critical Rev. Biochem.* **2**, 259–310.
34. Housset, P., and Nagy, M. (1977). Study of the role of purine phosphoribosyltransferases in the uptake of adenine and guanine by *Schizosaccharomyces pombe* cells. *Eur. J. Biochem.* **73**, 99–105.
35. Jackman, K. E., and Hochstadt, J. (1976). Regulation of purine utilization in bacteria. VI. Characterization of hypoxanthine and guanine uptake into isolated membrane vesicles from *Salmonella typhimurium*. *J. Bacteriol.* **126**, 312–326.
36. Jacobson, G. R., Takacs, B. J., and Rosenbusch, J. P. (1976). Properties of a major protein released from *Escherichia coli* by osmotic shock. *Biochemistry* **15**, 2297–2302.

37. Jund, R., Chevallier, M. R., and Lacroute, F. (1977). Uracil transport in *Saccharomyces cerevisiae*. *J. Membrane Biol.* **36**, 233–251.
38. Kaback, H. R. (1971). Bacterial membranes. *In* "Methods Enzymology" (W. B. Jakoby, *Ed.*), Vol. 22, pp. 99–120. Academic Press, New York.
39. Kloudova, A., and Fucik, V. (1974). Transport of nucleosides in *Bacillus subtilis*: characteristics of cytidine-uptake. *Nucleic Acids Res.* **1**, 629–638.
40. Kocharian, S. M., Livshitz, V. A., and Sukhodolets, V. V. (1975). Genetic study of *Escherichia coli* K-12 mutants resistant to 2,6-diaminopurine. *Genetika* **11**, 79–89.
41. Komatsu, Y., and Tanaka, K. (1970). Mechanism of action of showdomycin. II. Effect of showdomycin on the synthesis of deoxyribonucleic acid in *Escherichia coli*. *Agric. Biol. Chem.* **34**, 891–899.
42. Komatsu, Y. (1971). Mechanism of action of showdomycin. IV. Interactions between the mechanisms for transport of showdomycin and of various nucleosides in *Escherichia coli*. *Agric. Biol. Chem.* **35**, 1328–1339.
43. Komatsu, Y. (1971). Mechanism of action of showdomycin. V. Reduced ability of showdomycin-resistant mutants of *Escherichia coli* K-12 to take up showdomycin and nucleosides. *J. Antibiotics* **XXIV**, 876–883.
44. Komatsu, Y., and Tanaka, K. (1972). A showdomycin-resistant mutant of *Escherichia coli* K 12 with altered nucleoside transport character. *Biochim. Biophys. Acta* **288**, 390–403.
45. Komatsu, Y. (1973). Adenosine uptake by isolated membrane vesicles from *Escherichia coli* K 12. *Biochim. Biophys. Acta* **330**, 206–221.
46. Komatsu, Y., and Tanaka, K. (1973). Deoxycytidine uptake by isolated membrane vesicles from *Escherichia coli* K-12. *Biochim. Biophys. Acta* **311**, 496–506.
47. Komatsu, Y. (1981). A highly showdomycin-resistant mutant of *Escherichia coli* K-12 with altered nucleoside transport characteristics. *Agric. Biol. Chem.* **45**, 619–628.
48. Krieger-Brauer, H. J., and Braun, V. (1980). Functions related to the receptor protein specified by the *tsx* gene of *Escherichia coli*. *Arch. Microbiol.* **124**, 233–242.
49. Leung, K-K., and Visser, D. W. (1977). Uridine and cytidine transport in *Escherichia coli* B and transport deficient mutants. *J. Biol. Chem.* **252**, 2492–2497.
50. Lind, R. M., Sukhodolets, V. V., and Smirnov, Yu. V. (1973). Mutations affecting deamination and transport of cytosine in *Escherichia coli*. *Genetika* **9**, 116–121.
51. Losson, R., Jund, R., and Chevallier, M. R. (1978). Properties of three distinct pyrimidine transport systems in yeast. Evidence for distinct energy coupling. *Biochim. Biophys. Acta* **513**, 296–300.
52. Magill, J. M., Spencer, R. R., and Magill, C. W. (1974). Relationship between [8-^{14}C] adenine transport and growth inhibition in *Neurospora crassa*, strain ad-8. *J. Bacteriol.* **119**, 202–206.
53. Magill, J. M., and Magill, C. W. (1975). Purine base transport in *Neurospora crassa*. *J. Bacteriol* **124**, 149–154.
54. Magill, J. M., Edwards, E. S., Sabina, R. L., and Magill, C. W. (1976). Depression of uracil uptake by ammonium in *Neurospora crassa*. *J. Bacteriol.* **127**, 1265–1269.
55. Manning, P. A., and Reeves, P. (1978). Outer membrane proteins of *Escherichia coli* K-12: Isolation of a common receptor protein for bacteriophage T6 and colicin K. *Molec. Gen. Genet.* **158**, 279–286.
56. McKeown, M., Kahn, M., and Hanawalt, P. (1976). Thymidine uptake and utilization in *Escherichia coli*: a new gene controlling nucleoside transport. *J. Bacteriol.* **126**, 814–822.

57. Mitchell, P. (1961). *In* "Biological Structure and Function" (T. W. Goodwin and O. Lindberg, *Eds.*), Vol. 2, p. 581. Academic Press, New York.
58. Munch-Petersen, A., Nygaard, P., Hammer-Jespersen, K., and Fiil, N. (1972). Mutants constitutive for nucleoside-catabolizing enzymes in *Escherichia coli* K 12. Isolation, characterization and mapping. *Eur. J. Biochem.* **27**, 208–215.
59. Munch-Petersen, A., and Mygind, B. (1976). Nucleoside transport systems in *Escherichia coli* K 12: Specificity and regulation. *J. Cellular Physiol.* **89**, 551–559.
60. Munch-Petersen, A., Mygind, B., Nicolaisen, A., and Pihl, N. J. (1979). Nucleoside transport in cells and membrane vesicles from *Escherichia coli* K 12. *J. Biol. Chem.* **254**, 3730–3737.
61. Munch-Petersen, A., and Pihl, N. J. (1980). Stimulatory effect of low ATP pools on transport of purine nucleosides in cells of *Escherichia coli. Proc. Natl. Acad. Sci. U.S.A.* **77**, 2519–2523.
62. Mygind, B., and Munch-Petersen, A. (1975). Transport of pyrimidine nucleosides in cells of *Escherichia coli* K 12. *Eur. J. Biochem.* **59**, 365–372.
63. Neu, H. C., and Heppel, L. A. (1964). On the surface localization of enzymes in *E. coli. Biochem. Biophys. Res. Commun.* **17**, 215–219.
64. Nikaido, H., and Nakae, T. (1979). The outer membrane of gram-negative bacteria. *Adv. Microbial Physiol.* **20**, 163–250.
65. Page, M. G. P., and Burton, K. (1978). The location of purine phosphoribosyltransferase activities in *Escherichia coli. Biochem. J.* **174**, 717–725.
66. Parlebas, N., and Chevallier, M. R. (1976). Biochemical studies on the cytosine permease of *Saccharomyces cerevisiae. FEBS Lett.* **65**, 327–334.
67. Parlebas, N., and Chevallier, M. R. (1977). Genetic studies of the pyrimidine permeases from *Saccharomyces cerevisiae:* Lack of intragenic complementation. *Mol. Gen. Genet.* **154**, 199–202.
68. Pendyala, L., and Wellman, A. M. (1977). Developmental-stage-dependent adenine transport in *Neurospora crassa. J. Bacteriol.* **131**, 453–462.
69. Pendyala, L., and Wellman, A. M. (1978). Purine base transport in *nit-2* mutants of *Neurospora crassa. J. Bacteriol.* **133**, 401–402.
70. Pendyala, L., Smyth, J., and Wellman, A. M. (1979). Nature of 6-methylpurine inhibition and characterization of two 6-methyl-purine resistant mutants of *Neurospora crassa. J. Bacteriol.* **137**, 248–255.
71. Peterson, R. N., and Koch, A. L. (1966). The relationship of adenosine and inosine transport in *Escherichia coli. Biochim. Biophys. Acta* **126**, 129–145.
72. Peterson, R. N., Boniface, J., and Koch, A. L. (1967). Energy requirements, interactions and distinctions in the mechanisms for transport of various nucleosides in *Escherichia coli. Biochim. Biophys. Acta* **135**, 771–783.
73. Plagemann, P., and Wohlhueter, R. M. (1980). Permeation of nucleosides, nucleic acid bases, and nucleotides in animal cells. *Current Topics in Membranes and Transport*, **14**, 225.
74. Polak, A., and Grenson, M. (1973). Evidence for a common transport system for cytosine, adenine and hypoxanthine in *Saccharomyces cerevisiae and Candida albicans. Eur. J. Biochem.* **32**, 276–282.
75. Rader, R. L., and Hochstadt, J. (1976). Regulation of purine utilization in bacteria. VII. Involvement of membrane-associated nucleoside phosphorylases in the uptake and the base-mediated loss of the ribose moiety of nucleosides by *Salmonella typhimurium* membrane vesicles. *J. Bacteriol.* **128**, 290–301.
76. Reichert, U., Schmidt, R., and Forét, M. (1975). A possible mechanism of energy

coupling in purine transport of *Saccharomyces cerevisiae. FEBS Letts.* **52**, 100–102.

77. Reichert, U., and Forét, M. (1977). Energy coupling in hypoxanthine transport of yeast. Potentiometric evidence for proton symport and potassium antiport. *FEBS Letts.* **83**, 325–328.
78. Reinert, W. R., and Marzluf, G. A. (1975). Genetic and metabolic control of the purine catabolic enzymes of *Neurospora crassa. Molec. Gen. Genet.* **139**, 39–55.
79. Rinehart, K. V., and Copeland, J. C. (1973). Evidence that thymine is not a normal metabolite in wild-type *Bacillus subtilis. Biochim. Biophys. Acta* **294**, 1–7.
80. Roy-Burman, S., and Visser, D. W. (1972). Transport studies of showdomycin, nucleosides and sugars in *Escherichia coli* B and in showdomycin-resistant mutants. *Biochim. Biophys. Acta* **282**, 383–392.
81. Roy-Burman, S., and Visser, D. W. (1975). Transport of purines and deoxyadenosine in *Escherichia coli. J. Biol. Chem.* **250**, 9270–9275.
82. Roy-Burman, S., von Dippe, P., and Visser, D. W. (1978). Mechanism of energy coupling for transport of deoxycytidine, uridine, uracil, adenine and hypoxanthine in *Escherichia coli. Biochim. Biophys. Acta* **511**, 285–296.
83. Roy-Burman, S., and Visser, W. D. (1981). Uridine and uracil transport in *Escherichia coli* and transport-deficient mutants. *Biochim. Biophys. Acta* **646**, 309–319.
84. Sabina, R. L., Magill, J. M., and Magill, C. W. (1976). Regulation of hypoxanthine transport in *Neurospora crassa. J. Bacteriol.* **128**, 598–603.
85. Schlitz, J. R., and Terry, K. D. (1970). Nucleoside uptake during the germination of *Neurospora crassa* conidia. *Biochim. Biophys. Acta* **209**, 278–288.
86. Schwan, H., and Holldorf, A. W. (1975). Effective utilization of exogenous deoxythymidine-5′-monophosphate for DNA synthesis in Enterobacteria. *FEBS Letts.* **57**, 179–182.
87. Tsao, T-F., and Marzluf, G. A. (1976). Genetic and metabolic regulation of purine base transport in *Neurospora crassa. Molec. Gen. Genet.* **149**, 347–355.
88. Votruba, I., Holy, A., and Sorm, F. (1971). L-ribonucleosides do not penetrate bacterial cell walls. *FEBS Letts.* **19**, 136–138.
89. Williams, J. C., Lee, C. E., and Wild, J. R. (1980). Genetic and biochemical characterization of distinct transport systems for uracil, uridine, and cytidine in *Salmonella typhimurium. Molec. Gen. Genet.* **178**, 121–130.
90. Wu, H. C. P., Boos, W., and Kalckar, H. M. (1969). Role of the galactose transport system in the retention of intracellular galactose in *Escherichia coli. J. Mol. Biol.* **41**, 109–120.
91. Yagil, E., Beacham, I. R., Nissim, A., and Price, G. (1978). Crypticity of periplasmic enzymes. Involvement of protein b in the permeability of the outer membrane of *Escherichia coli. FEBS Letts.* **85**, 133–136.

Appendix

Linkage map of *Escherichia coli* K12 and *Salmonella typhimurium* LT2

Table A.1 lists the gene products of the *E. coli* and *S. typhimurium* genes relevant to this book. In addition it indicates the approximate location of these genes on the chromosomes using as a unit (U) one hundredth of the entire chromosomes, in accordance with the last published linkage maps of *E. coli* K12 (Bachmann and Low (1980) *Microbiol. Rev.* **44**, 1–56) and *S. typhimurium* LT2 (Sanderson and Hartman (1978) *Microbiol. Rev.* **42**, 471–519). References to work on individual genes may be found in these two articles or, if not given there, in the footnotes to Table A.1.

In Fig. A.1 the location of the genes mentioned in Table A.1 is depicted on a circular linkage map.

TABLE A.1

Genes of *E. coli* and *S. typhimurium* involved in nucleobase, nucleoside and nucleotide metabolism.

Gene symbol	Alternative symbol; phenotypic trait affected	Map position (U)	
		E. coli	*S. typhimurium*
add	Adenosine deaminase (EC 3.5.4.4)	36	nm[a]
adk	*amk* in *S. typhimurium*; AMP kinase activity (EC 2.7.4.3)	11	9
amk	see *adk*		
apt	Adenine phosphoribosyltransferase (EC 2.4.2.7)	10	nm
carA[b]	*pyrA* in *S. typhimurium*; small subunit of carbamoylphosphate synthase (EC 2.7.2.9)	1	2
carB	*pyrA* in *S. typhimurium*; large subunit of carbamoylphosphate synthase (EC 2.7.2.9)	1	2
cdd	Cytidine deaminase (EC 3.5.4.5)	46	46
cmk	CMP kinase (EC 2.7.4.14)	ni[c]	21
codA	Cytosine deaminase (EC 3.5.4.1)	8	69

METABOLISM OF NUCLEOTIDES, NUCLEOSIDES
AND NUCLEOBASES IN MICROORGANISMS
ISBN: 0-12-510580-0

Gene symbol	Alternative symbol; phenotypic trait affected	Map position (U)	
		E. coli	*S. typhimurium*
codB	Cytosine transport	8	ni
codR	Regulatory gene for cytosine utilization	8	ni
cpdB[d]	*pde* in *S. typhimurium*; 2′:3′-cyclic-nucleoside-monophosphate phosphodiesterase (EC 3.1.4.16)	95	93
cytR	Regulatory gene for the *deo*-operon and the *udp*, *nupC*, *nupG*,*tsx*, and *cdd* genes	88	88
dcd	dCTP deaminase (EC 3.5.4.13)	45	45
deoA	Thymidine phosphorylase (EC 2.4.2.4)	99	99
deoB	Phosphopentomutase (EC 2.7.5.6)	99	99
deoC	Phosphodeoxyriboaldolase (EC 4.1.2.4)	99	99
deoD	Purine nucleoside phosphorylase (EC 2.4.2.1)	99	99
deoK	Deoxyribokinase	—[e]	20
deoP	Deoxyribose transport	ni	20
deoR	Regulatory gene for the *deo*-operon, and the *nupG* and *tsx* genes	18	20
dut	dUTPase (EC 3.6.1.23)	81	ni
gmk[f]	GMP kinase	ni	20
gpt	Guanine-hypoxanthine phosphoribosyltransferase (EC 2.4.2.8)	6	7
gsk[c]	Guanosine kinase	13	13
guaA	GMP synthetase (EC 6.3.4.1)	53	54
guaB	IMP dehydrogenase (EC 1.2.1.14)	53	54
guaC[g]	GMP reductase (EC 1.6.6.8)	2	3
guaP[n]	Guanine transport	3	2
hpt	Hypoxanthine phosphoribosyltransferase	3	nm
ndk	Nucleoside diphosphokinase (EC 2.7.4.6)	ni	57
nrdA	Ribonucleoside diphosphate reductase (EC 1.17.4.1), subunit B1	48	ni
nrdB	Ribonucleoside diphosphate reductase (EC 1.17.4.1), subunit B2	48	ni
nupC	Pyrimidine nucleoside and (deoxy)adenosine transport	0	ni
nupG	Nucleoside transport	64	ni
pde[h]	*cpdB* in *E. coli*; 2′:3′-cyclic-nucleoside-monophospate phosphodiesterase (EC 3.1.4.16)	95	93
phoA	Alkaline phosphatase (EC 3.1.3.1)	9	—
phoN	Nonspecific acid phosphatase	ni	93
phoP	Nonspecific acid phosphatase	ni	25
prs	5-phosphoribosyl 1-pyrophosphate synthetase; *E. coli*: altered kinetic properties of the enzyme; *S. typhimurium*: lack of enzyme activity *in vitro*	25[i]	35[i]
purA	Adenylosuccinate synthetase (EC 6.3.4.4)	94	93
purB	Adenylosuccinate lyase (EC 4.3.2.2)	25	25
purC	Phosphoribosylaminoimidazole-succinocarboxamide synthetase (EC 6.3.2.6)	53	54

purD	Phosphoribosylglycinamide synthetase (EC 6.3.4.13)	89	90
purE	Phosphoribosylaminoimidazole carboxylase (EC 4.1.1.21)	12	12
purF	Amidophosphoribosyltransferase (EC 2.4.2.14)	49	49
purG[k]	*purI* in *S. typhimurium*; phosphoribosyl-aminoimidazole synthetase (EC 6.3.3.1)	53	54
purH	Phosphoribosylaminoimidazolecarboxamide formyltransferase (EC 2.1.2.3)	89	90
purI[k]	*purG* in *S. typhimurium*; phosphoribosylformyl-glycinamidine synthetase (EC 6.3.5.3)	55	57
purJ	IMP cyclohydrolase (EC 3.5.4.10)	ni	90
pyrA	See *carAB*		
pyrB	Aspartate carbamoyltransferase (EC 2.1.3.2)	96	98
pyrC	Dihydroorotase (EC 3.5.2.3)	23	22
pyrD	Dihydroorotase oxidase (EC 1.3.3.1)	21	20
pyrE	Orotate phosphoribosyltransferase (EC 2.4.2.10)	81	79
pyrF	OMP decarboxylase (EC 4.1.1.23)	28	34
pyrG	CTP synthetase (EC 6.3.4.2)	59	60
pyrH	UMP kinase	4	5
tdk	Thymidine kinase (EC 2.7.1.75)	27	34
thyA	Thymidylate synthase (EC 2.1.1.45)	60	62
tsx	Nucleoside uptake; receptor for phage T6 and for colicin K	9	—
udp	Uridine phosphorylase (EC 2.4.2.3)	85	84
udk	Uridine (cytidine) kinase (EC 2.7.1.48)	45	45
ung	Uracil-DNA N-glycosylase	56	ni[c]
upp	Uracil phosphoribosyltransferase (EC 2.4.2.9)	53	54
uraA[l]	Uracil transport	53	ni
ush	UDP-glucose hydrolase (5′-nucleotidase)	11	—
xapA[m]	Xanthosine phosphorylase	51	—
xapR[m]	Regulatory gene for xanthosine phosphorylase	51	—

[a] Mutants exist, but they have not yet been mapped.

[b] Gigot, D. *et al.* (1980). *J. Bacteriol.* **143**, 914–920, and Crabeel, M. *et al.* (1980). *J. Bacteriol.* **143**, 921–925.

[c] No mutants have been described.

[d] Beachham, I. R., and Garrett, S. (1980). *J. Gen. Microbiol.* **119**, 31–34.

[e] The gene product is absent.

[f] Jochimsen,B. Personal communication.

[g] Livshits, V. A. (1973). *Genetika* **9**, 134–139.

[h] Uerkvitz, W., and Beck, C. F. (1981). *J. Biol. Chem.* **256**, 382–389.

[i] Hove, B. Personal communication.

[k] In the latest *E. coli* K12 linkage map by Bachmann and Low (*Microbiol. Rev.* (1980) **44**, 1–56) the *purG* and *purI* gene products have been exchanged. The *purG* gene product should be phosphoribosylaminoimidazole synthetase, and the *purI* gene product should be phosphoribosylformylglycinamidine synthetase.

[l] Fast, R. Personal communication.

[m] Buxton, R. S. *et al.* (1980). *Molec. Gen. Genet.* **179**, 331–340.

[n] Benson, C. E. *et al.* (1980). *J. Gen. Microbiol.* **121**, 357–364.

[o] More recent data (Nygaard, P., unpublished) have shown *gsk* to be located between *apt* and *purE*.

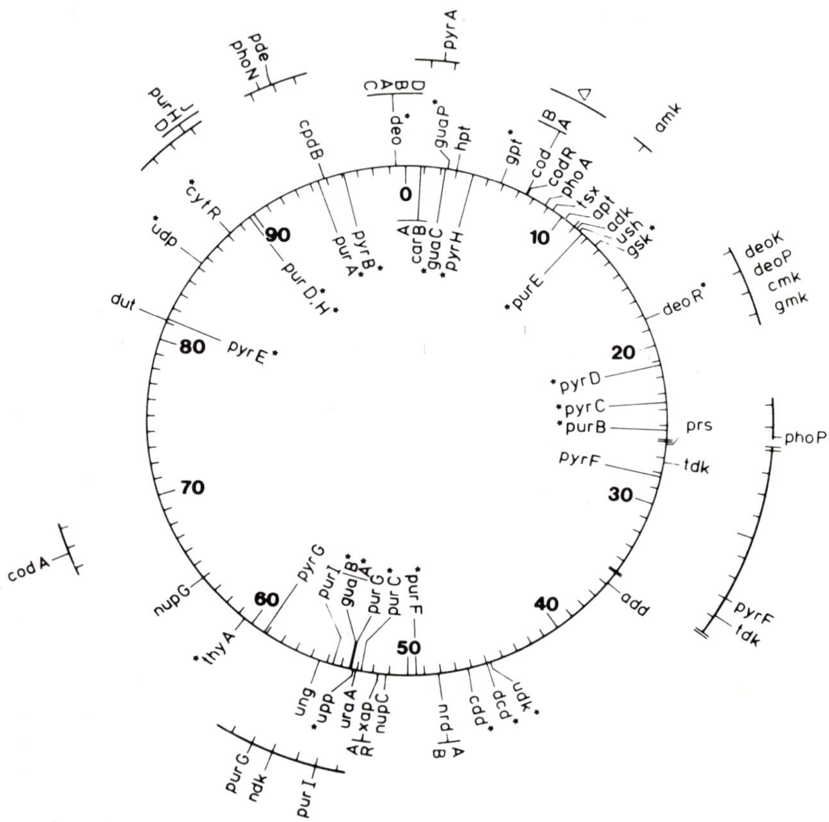

FIG. A.1. Linkage map of genes coding for proteins involved in nucleobase, nucleoside, and nucleotide metabolism in *E. coli* and *S. typhimurium*. The continuous circle represents the *E. coli* chromosome, where the symbols for genes involved in purine and pyrimidine *de novo* synthesis are presented on the inside, and the symbols for genes involved in purine and pyrimidine salvage pathways are placed on the outside of the circle. Homologous genes which are located at similar positions on the *E. coli* and *S. typhimurium* chromosomes are identified by asterisks. *S. typhimurium* genes which differ in their localization from the homologous *E. coli* genes or which have not been identified in *E. coli* are shown on the outer circle-segments. △ represents a segment of the *E. coli* chromosome which appears to be deleted from the *S. typhimurium* chromosomes. ‖——‖, represents a chromosomal segment which is inverted in *S. typhimurium* as compared to *E. coli* (Casse, F., Pascal, M., and Chippaux, M. (1973). *Molec. Gen. Genet.* **124**, 253–257).

Index

A

Acinetobacter, 189, 239, 240
Acinetobacter calcoaceticus, 112
Adenine
  effect on PRPP pool, 7–8, 14–16
  as effector, 33, 34, 44, 55, 69, 74, 219–221, 237
  excretion, 44
  toxicity, 31, 74, 77
  transport, 286, 290, 293, 296
  utilization, 41–46, 53–64
Adenine arabinoside, 63, 68
Adenine deaminase, 62, 69
  in adenine utilization, 32, 44, 54, 56, 62
  mutants defective in, 56, 57
Adenine phosphoribosyltransferase, 64–65
  in adenine utilization, 10–11, 43–45, 54–64, 281
  mutants defective in, 43, 44, 46, 56, 59, 286
Adenosine
  as carbon source, 205, 235
  as effector, 33, 35–36, 77, 211, 218–220, 223, 232, 236–237
  endogenous formation of, 69–71
  growth inhibition by, 214
  as nitrogen source, 42, 237
  transport, 264, 266, 278, 289, 294–295
  utilization, 45–46, 54–64
Adenosine deaminase, 68–69, 206, 208, 240–241
  genetic regulation, 55, 219–221, 236–237
  mutants defective in, 44, 45, 56, 212–215, 218
  in purine utilization, 32, 43–46, 54–64, 189
Adenosine kinase, 66
  mutants defective in, 58
  in purine utilization, 45–46, 55, 58, 61–64
Adenosine phosphorylase, *see also* Purine nucleoside phosphorylase, 54–55, 60, 61, 241
S-Adenosylhomocysteine, 70–71
S-Adenosylhomocysteine hydrolase, 70–71
S-Adenosylhomocysteine nucleosidase, *see* 5-Methylthioadenosine nucleosidase
S-Adenosylmethionine, 39, 56, 70–71
Adenosylmethionine decarboxylase, 71
Adenylosuccinate lyase, 33, 34, 60, 61
  mutants defective in, 33, 45
Adenylosuccinate synthetase, 33, 60, 61, 77
  mutants defective in, 33, 45, 46, 296, 297
ADP, 5–8, 17, 36
Aerobacter aerogenes, 30, 33, 34, 35, 40, 74, 75
Agrobacterium tumefaciens, 137
AICAR, *see* Phosphoribosylaminoimidazolecarboxamide
Algae, 178
Alkaline phosphatase, *see* Phosphatases
Allopurinol, 63, 66
Allosteric regulation, 6–7, 17, 38, 39, 128–129, 152–153, 158, 171, 175, 176
Amino acid starvation, *see also* ppGpp, 53, 100
Aminoimidazole ribotide, 17
Aminopropyl transferase, 71
Aminopterin, 160, 187, 189
AMP
  as effector, 12–13, 31, 33, 39, 40, 64, 75, 77
  pool, 70

AMP—*cont.*
  in purine nucleotide interconversion, 31–39, 54–64
  as purine source, 42, 52, 55
AMP deaminase, 39
  in purine nucleotide interconversion, 32, 39, 54–64
AMP glycosylase, *see* AMP nucleosidase
AMP kinase, 36–37
AMP nucleosidase, 38–39, 70, 133
Anacystis, 240
Anacystis nidulans, 189
Anthranilate, 19
Anthranilate phosphoribosyltransferase, 11, 19
Arabinofuranosylcytosine, 121
Arprinocid, 63
Arsenate, 273, 274
Ascorbate, 125, 126, 127, 269
Aspartate transcarbamylase, 12, 16, 19
Aspergillus, 78
Aspergillus nidulans, 114–115, 125, 188–189
Aspergillus niger, 239, 247
ATP
  as effector, 12, 38, 39, 40, 127, 133, 152–153
  pool 7–9, 15–16, 75–76, 273
ATP analogs, 4–6
ATPase, 273
ATP phosphoribosyltransferase, 11, 37
5-Azacytidine, 127, 128, 271
6-Azacytosine, 280
8-Azaguanine, 49
8-Aza-2-thioxanthine, 56
6-Azathymidine, 126
6-Azathymine, 126
5-Azauracil, 118, 128
6-Azauracil, 118
6-Azauridine, 127
Azide, 267, 289, 291
Azotobacter, 240
Azotobacter vinelandii, 33, 38, 69, 70

B

Bacillus anthracis, 241
Bacillus cereus, 54, 55, 67, 69, 241, 243
Bacillus megaterium, 78, 169, 187
Bacillus stearothermophilus, 123, 127, 128, 129, 241, 244
Bacillus subtilis, 64, 67, 77, 118, 120, 121, 123, 129, 131–132, 134, 187, 188
  dTTP synthesis in, 169–174, 179
  nucleoside catabolism in, 241–244
  nucleoside transport in, 278–279
  purine *de novo* synthesis in, 30–31, 33–35, 74–75
  purine salvage in, 39, 42, 54–55
  pyrimidine salvage in, 106–109
  thymine transport in, 287
Bacillus thuringiensis, 187
5-Bromodeoxyuridine, 126, 158, 166, 210
5-Bromo-dUTP, 166
5-Bromouridine, 210

C

Caldariella acidophila, 71
Candida albicans, 293
Candida utilis, 69
Carbamoylphosphate synthase, 11–12
Catalase, 126
Chlorohexidine, 291, 292
p-Chloromercuribenzoate, 266, 269
CMP pool, 135
CMP kinase, 131–132
  mutants defective in, 131, 135
  in pyrimidine utilization, 103
Cobalamin, *see* Coenzyme B_{12}
Coccidia, *see* individual organisms
Cobalt, 5, 211
Coenzyme A, 28
Coenzyme B_{12}, 28, 169, 175–176, 178
Coformycin, 68, 69, 72, 206, 231
Conidia, 58, 77
  nucleobase transport in, 296–300
  nucleoside transport in, 293–296
Cooperative binding, *see also* Allosteric regulation, 207
Copper, 124
Cordycepin, *see* 3′-Deoxyadenosine
Corynebacterium nephridii, 178
Corynebacterium sepedoniceum, 238
Crithidia fasciculata, 62, 65, 66, 69, 124
CRP, *see* 3′,5′-Cyclic AMP receptor protein

CTP
as effector, 12, 128
pool, 15–16
CTP synthetase defective mutants, 101, 104
Cyanobacterium, *see* individual organisms
3′,5′-Cyclic AMP, 28, 40, 42, 121, 123, 137, 180–181, 230–231, 272
3′,5′-Cyclic AMP receptor protein, 40, 121, 123, 180–181, 218–220, 223–226, 230–233, 235, 272
3′,5′-Cyclic GMP, 179
2′,3′-Cyclic phosphate phosphodiesterase, *see* Phosphatases
Cyclic 2′,3′-ribonucleotides, 106–107
Cysteine biosynthesis, 156
Cytidine
as carbon source, 214, 215, 217, 235, 236
as effector, 121, 123, 218–220, 223, 231, 235, 236
as nitrogen source, 103, 237
transport, 264–279
utilization, 99, 102–104, 108, 110, 113–117
Cytidine deaminase, 119, 120–121, 206, 208, 240, 242, 245, 248, 249
genetic regulation, 219–220, 223, 236, 243
localization, 262–263
mutants defective in, 99–100, 103–105, 108–109, 136, 212–215, 243, 267, 276
in pyrimidine utilization, 103, 105, 108–110, 114–115, 117
Cytidine kinase, *see* Uridine kinase
Cytokinins, 72–73
Cytosine
excretion, 136
as nitrogen source, 101, 113
transport, 280–281, 289–291
utilization, 99, 101, 108, 111, 113–114, 116–117
Cytosine deaminase, 119, 120
genetic regulation, 101, 120
mutants defective in, 99–101, 104, 113–114, 116, 136, 281–282
in pyrimidine utilization, 99, 101, 104, 110–111, 113–114, 116, 290
Cytosine permease, 281–282, 290
genetic regulation, 120
mutants defective in, 113–114, 116, 280
in pyrimidine utilization, 113–114, 116
Cytosine transport protein, *see* Cytosine permease
cyt P, 224–226, 229–235
CYT R, 121, 180–181, 219, 223–227, 231–232, 235–237, 272, 275
mutants defective in, 182, 185–186, 222–223, 226–227, 229, 231, 271–272

D

dATP
as effector, 67, 152–153, 175
pool, 168–169, 170–171, 174
dCDP as effector, 128
dCMP deaminase, 171, 176, 178–179
mutants defective in, 109, 170, 171
dCMP kinase, *see* CMP kinase
dCMP pool, 171
dCTP
as effector 128–129, 171, 175, 176
metabolism, 165, 166
pool, 162–169, 171, 174
dCTP deaminase, 158–159, 178–179
in dUMP synthesis, 165
mutants defective in, 158–159, 165, 184
Deamidation, 211
7-Deazaadenosine, 207
7-Deazainosine, 207
Decoyinine, 36, 72
Deo-enzymes, *see also* individual enzymes, 218
deo-operon, 123, 224–235, 243
structure of, 229, 233–234
transcriptional units in, 226–229, 232–233
deoP, 224–226, 230, 232–235
DEO R, 123, 180–181, 184, 219, 221, 224–227, 230, 231, 235, 236, 272, 275
mutants defective in, 182, 185, 186, 221–222, 226, 227, 230, 271–272
Deoxyadenosine
as effector, 123, 214, 218, 236, 243
as nitrogen source, 52
transport, 266, 294
utilization, 42, 52, 54–55, 189, 214, 237, 250

3′-Deoxyadenosine, 58, 72, 289
Deoxyadenosine kinase, *see also* Deoxycytidine kinase, 66–67
  in deoxyribonucleoside utilization, 55, 59, 109, 168, 189
  mutants defective in, 67, 109
5′-Deoxy-5′-adenosylcobalamin, *see* Coenzyme B_{12}
Deoxycoformycin, 61, 62, 69
Deoxycytidine
  as carbon source, 105, 205, 214–215, 235–236, 244
  as nitrogen source, 105, 237
  transport, 264–279
  utilization, 99, 105, 108–110, 115–117, 170
Deoxycytidine deaminase, *see* Cytidine deaminase
Deoxycytidine kinase, *see also* Deoxyadenosine kinase, 119, 129–130
  mutants defective in, 109
  in pyrimidine utilization, 108–109
Deoxyguanosine 42, 54, 59, 170, 265–266
  as carbon source, 204, 206, 215, 235–236, 244
Deoxyguanosine kinase, 55, 66–67, 168
Deoxyinosine, 42, 54, 265–266
  as carbon source, 204–206, 215, 235–236, 244
Deoxyriboaldolase, 180–181, 209, 212, 228, 233, 242–245, 248
  genetic regulation, 218–219, 221, 223–227, 235–236, 243, 245
  mutants defective in, 182, 185, 186, 212, 217, 229, 233, 243, 265, 277
Deoxyribokinase, 181, 217, 238, 245
Deoxyribonucleic acid, *see* DNA
Deoxyribonucleoside kinases, *see* individual enzymes
Deoxyribonucleosides, *see also* individual compounds
  as carbon sources, 204–206, 215, 235–236, 244
  as effectors, 218, 235–236, 243
  growth inhibition by, 216, 265
Deoxyribonucleoside triphosphates, *see* individual compounds
Deoxyribonucleoside triphosphate pools, 169, 170–173, 174
Deoxyribose, 164, 180, 181, 245
Deoxyribose 1-phosphate
  5-fluorouracil sensitivity and, 105, 108
  metabolism, 180–184, 187, 205
  pool, 161, 181, 182, 275
Deoxyribose 5-phosphate
  as effector, 123, 186, 218, 219, 221, 225, 230, 231, 236, 243
  growth inhibition by, 216, 265
  metabolism, 180, 205
Deoxyribose transport, 180, 181
Deoxyuridine
  as carbon source, 104, 105, 204–206, 215, 235–236, 244
  transport, 266
  utilization, 99, 104–106, 108, 111, 114–117, 166, 170
Desdanine, 157
Desulfovibrio desulfuricans, 39
dGTP
  as effector, 152–153, 175
  pool, 162, 164, 165
2,6-Diaminopurine
  metabolism, 45, 57, 64, 69
  mutants resistant to, 45, 56, 58
  toxicity, 44
2,6-Diaminopurine ribonucleoside, 68
6-Diazo-5-oxo-L-norleucine, 35
Dihydrofolate reductase, 160–161, 187
Dihydroorotase, 19
DihydroUMP, 137
2,4-Dinitrophenol, 267, 278, 289, 291, 292
Dio 9, 291, 292
Dioxygenases, *see* Hydroxylases
Diplococcus, 240
Diplococcus pneumoniae, 110–111, 119, 122, 188
DNA
  degradation, 235
  labeling of, 105, 111, 112, 185–187, 189, 211
  repair of, 167, 168
  uracil-containing, 166–168
DNA ligase, 10, 167
DNA polymerase, 153, 157–158, 166, 167
DNA replication, 153, 164, 165, 169, 186
DNA synthesis, 174
  ribonucleotide reductase synthesis and, 155, 176, 178

thymine and thymidine utilization in, 181–187
dTMP, 171, 185
dTMP kinase, 161
dTMP-requiring mutants, 188–189
dTTP
as effector, 128–129, 152–153, 158, 171, 175, 176
pool, 162–169, 171, 172, 183, 186
synthesis, 149–179
synthesis from thymine and thymidine, 179–189
dUMP
metabolism, 104, 111, 157–166, 171, 176, 178–179, 183–184
pool, 162–163, 171–173, 183, 185
dUMPase, 184
dUTP, 165, 166–168
as effector, 158
pool, 167–168
dUTPase, 159, 171, 178
mutants defective in, 159, 166–168, 184

E

Eimeria tenella, 63, 66
Endonuclease, 167
Energy charge, 7, 8, 37, 70
Enteric bacteria, *see* individual organisms
Enterobacter aerogenes, 187
Enzyme complex, 153, 169
Erwinia carotowora, 238
Erwinia herbicola, 130
Escherichia coli
dTTP synthesis from thymine and thymidine, 179–187
endogenous dTTP synthesis in, 150–159
linkage map of, 307–310
nucleobase transport in, 279–287
nucleoside catabolism in, 204–237
nucleoside transport in, 260–279
PRPP metabolism in, 1–25
purine *de novo* synthesis in, 12–13, 29–36, 74–77
purine salvage in, 41–54
pyrimidine *de novo* synthesis in, 11–12, 134–135
pyrimidine salvage in, 97–106
N-Ethylmaleimide, 266, 269, 291
1-Ethyluracil, 126
Excision repair, 159
Exonuclease III, 167

F

FAD, 28
Flavoenzymes, 156
5-Fluorocytidine, 121, 127
resistance to, 99–100, 103, 108, 111, 112
5-Fluorocytosine resistance, 99–101, 108, 111–113, 116, 280, 289
5-Fluoro-dCTP, 109
5-Fluorodeoxycytidine, 121, 266
resistance to, 99–100, 103, 105, 109, 111, 112, 213–214, 265
5-Fluorodeoxyuridine, 122, 128, 159, 210, 266
resistance to, 99–100, 105, 112, 114, 116, 213, 214, 216, 265
5-Fluoro-dUMP, 160, 172, 177, 179
5-Fluorouracil, 118
resistance to, 98–101, 105, 106, 108, 111, 113, 116, 213, 216, 217, 283, 291
5-Fluorouridine, 127
resistance to, 99–102, 108, 111–113, 288
Folate, *see also* Tetrahydrofolate, 176
Formycin, 72, 293
N-Formylmethionyl-tRNA, 188
5-Formyluracil, 114–116, 126
Fungi, *see also* individual organisms 56, 70, 113, 114, 124, 125, 178, 239, 287–288

G

GDP as effector, 33
Glucose repression, 218–220, 229, 249, 272
Glutamine phosphoribosylpyrophosphate amidotransferase, *see* PRPP amidotransferase
Glutathione, 126, 155
Glutathione reductase, 155

Glutaredoxin, 155, 156
Glycerol 3-phosphate acyltransferase, 36
GMP
  as effector, 12–13, 31, 33–36, 75, 77
  in nucleotide interconversion, 31–32, 34–36
  as purine source, 42, 52
GMP kinase, 36, 37
GMP reductase, 39–41
  mutants defective in, 40, 50
  in purine nucleotide interconversion, 17, 50–51, 56, 60–61
GMP synthetase, 34–36
  mutants defective in, 8–9, 48
Group translocation, 261, 280
GTP
  as effector, 40, 100, 118
  pool, 7–10, 15–16, 76–77
Guanine
  effect on PRPP pool, 7–9, 14–16
  as effector, 33, 34, 40, 74, 75, 77
  transport, 287, 293
  utilization, 42, 47–54, 54–64
Guanine phosphoribosyltransferase, *see* Hypoxanthine-guanine-xanthine phosphoribosyltransferase
Guanosine
  as carbon source, 204–206, 215, 235–236, 244
  as effector, 33, 34, 40, 75, 218–219, 224–225, 229, 235
  transport, 264–273, 279, 294
  utilization, 14, 17–19, 42, 49–51, 53–64
Guanosine 5′-diphosphate-3′-diphosphate, *see* ppGpp
Guanosine kinase, 66
  mutants defective in, 17, 48, 50, 220
  in purine utilization, 17, 48, 50

H

Haemophilus influenzae, 121, 123, 238
Histidine, 17, 18, 50
Histidine biosynthesis, 3, 11, 13, 31, 32, 37–39, 43, 44, 54
Hydroxylamine, 152
Hydroxylases, *see also* individual enzymes, 125
5-Hydroxymethyldeoxyuridine, 126
5-Hydroxymethyluracil, 114–116, 126
Hydroxyurea, 152, 168–169, 175, 178
Hypoxanthine
  effect on PRPP pool, 14
  as effector, 54, 69, 74, 219–221, 237
  excretion, 47
  transport, 287, 296–297
  utilization, 41–64
Hypoxanthine-guanine-xanthine phosphoribosyltransferase, 65–66
  mutants defective in, 47, 49, 56, 59, 75, 281, 293
  in purine salvage, 11, 14, 46–64

I

IDP, 36
IMP
  as effector, 11–12, 40
  pool, 297
  in purine nucleotide interconversion, 31–40, 54–64
  as purine source, 42
  synthesis *de novo*, 29–31
IMP dehydrogenase, 34, 77
  mutants defective in, 8–9, 48, 58
Inosine
  as carbon source, 204, 206, 215, 221, 222, 235–236, 244
  as effector, 33, 55, 77, 218–219, 224, 229, 235, 237, 243
  pool, 58
  transport, 265, 266
  utilization, 48, 54–64
Inosine kinase, *see* Guanosine kinase
Iodoacetamide, 35
Iodoacetate, 269, 291
Iron, 125–126, 151–152, 178
Isocytosine, 120

K

2-Ketoglutarate, 125–127

L

D-Lactate, 269
Lactic acid bacteria, *see* individual organisms

Lactobacillus, 65, 66, 68, 121, 124, 125, 129
dTTP synthesis in, 174–178
nucleoside catabolism in, 240
purine salvage in, 55–56
pyrimidine salvage in, 109–110
Lactobacillus acidophilus, 55, 66, 109, 129, 176, 179, 244
Lactobacillus casei, 55, 66, 110, 176–178, 245
Lactobacillus delbrueckii, 124, 244, 246–247
Lactobacillus helveticus, 55, 68, 125, 244
Lactobacillus lactis, 175, 245
Lactobacillus leichmannii, 41, 55, 109, 110, 175–176, 244
Lactobacillus pentosus, 55, 110
Lactobacillus plantarum, 245
Lambda, *see* Phage lambda
Leishmania, 69
Leishmania braziliensis, 61–62
Leishmania donovani, 61, 62, 65, 67, 124, 246–247
Linkage map, 307–310

M

Manganese, 211
Membranes, 106, 267–269
enzymes associated with, 156, 157, 263, 286
permeability of, 52, 105
Membrane vesicles, 269
6-Mercaptopurine, 47
Methionine, 188
6-Methylaminopurine, 44, 45
6-Methylaminopurine ribonucleoside, 68
5-Methyl-CMP, 137
5-Methylcytidine, 121, 137
N_3-Methylcytidine, 223
N_4-Methylcytidine, 223
5′-O-Methylcytidine, 223
5-Methylcytosine, 120, 126
5-Methyldeoxycytidine, 121
5,10-Methylene-tetrahydrofolate, *see* Tetrahydrofolate
6-Methylpurine, 59, 297, 298
5′-Methylthioadenosine nucleosidase, 46, 70–71
5′-Methylthioadenosine phosphorylase, 71
1-Methyluracil, 126
5′-O-Methyluridine, 223
Micrococcus radiodurans, 187
Micrococcus sodonensis, 73
Microsporum audonini, 39
Mitomycin C, 176
Moraxella, 112, 189, 239, 240
Multiple enzyme forms, 211
Mycoplasma 60, 112, 118, 119, 179
Mycoplasma mycoides, 60, 65, 112, 120
Myxococcus xanthus, 77

N

NAD, 3, 10, 14, 28
Nalidixic acid, 169
Neisseria, 39, 119, 189, 239, 240
Neisseria meningitidis, 56, 112, 118, 120, 122
Neurospora crassa, 124–126, 179, 188–189
nucleobase transport in, 296–300
nucleoside catabolism in, 240, 249–250
nucleoside transport in, 293–296
purine *de novo* synthesis in, 29, 34, 75
purine salvage in, 58–59, 73, 77
pyrimidine salvage in, 114–116
Nickel, 211
Nicotinamide coenzymes, *see* NAD
Nitrogen control, 8, 101, 120, 237
Nucleobase transport, 279–287, 289–293, 296–200
Nucleoside deoxyribosyltransferases, *see* *Trans*-N-deoxyribosylases
Nucleoside diphosphokinase, 36, 157–158, 161
Nucleoside hydrolases, 110, 111, 119, 203, 217, 238–241
substrate specificity of, 121, 124, 239, 246–247
Nucleoside phosphorylases, *see also* individual enzymes, 119, 121, 203, 206–211, 238–241
Nucleoside phosphotransferase, 117, 130–131
Nucleosides, *see also* individual nucleosides

Nucleosides—*cont.*
as carbon sources, 205, 215–217, 235–236, 244
as nucleotide sources, 14
Nucleoside transport, 265–268, 287–289, 292–296
mutants defective in, 264
regulation of, 180, 221, 223, 271
Nucleotidases, *see* Phosphatases

O

Okazaki fragments, 166
Operator, 221, 224, 225, 229
Orotate phosphoribosyltransferase, 11, 16
Outer membrane proteins, 106, 267–269

P

Penicillium, 240
Periplasmic enzymes, 38, 52, 105–106, 135, 263, 269
Petite mutation, 291
Phage lambda, 212, 221, 223
Phage Mu, 212, 227, 229
Phage T4, 211
Phage T6-receptor protein, 267, 268, 272
Phenazine methosulfate, 269
Phosphatases, 38, 52–53, 55, 70, 105–107
Phosphate, 3–4, 19, 38, 39, 211
Phosphopentomutase, 180–181, 209, 211–212, 243–245
genetic regulation, 218–219, 221, 223, 224–234, 243
mutants defective in, 182, 185–187, 212, 214–217, 229, 243
Phosphoribosylaminoimidazole, *see* Aminoimidazole ribotide
Phosphoribosylaminoimidazolecarboxamide, 30–32, 37, 43
5-Phosphoribosyl-1-pyrophosphate, *see* PRPP
Phosphoribosyltransferases, *see also* individual enzymes
in nucleobase uptake, 280–287, 293, 298–300
Phosphothioredoxin, 156
Physarum, 239, 240
Physarum polycephalum, 59, 73, 116–117, 120, 129
Plasmodium, 179
Plasmodium berghei, 62–63
Plasmodium chabaudi, 62–63, 66
Plasmodium lophurae, 62–63
Pneumococci, 110, 129
Polynucleotide phosphorylase, 135
Pool, *see* under individual compounds
ppGpp, 28
effect on purine nucleotide interconversion, 33, 34, 38
effect on purine salvage, 53, 54, 65
effect on pyrimidine salvage, 100, 118
Promoter, *see cytP* and *deoP*
Protein synthesis, 188
Proteolytic inactivation, 249
Proteus, 130
Proteus mirabilis, 187
Protozoa, *see also* individual organisms, 60, 65, 179, 248
purine salvage in, 61–64
pyrimidine salvage in, 117
PRPP, 2
biosynthesis, 3–10
as effector, 11–12, 12–13
pool, 9, 14–16, 31, 75–77, 293
utilization, 3, 10–14
PRPP amidotransferase, 12–13, 30–31, 75–77
PRPP synthetase
catalytic properties of, 4–5
cloning of structural gene of, 18
control of synthesis of, 8–10
enzyme structure of, 1–4
mutants defective in, 16–21, 50
regulatory properties of, 5–8
Pseudomonas, 111, 119, 121, 124, 130
Pseudomonas acidovorans, 111, 120, 122, 188, 240
Pseudomonas aureofaciens, 120
Pseudomonas fluorescens, 124, 240, 246
Pseudomonas oleovorans, 132
PseudoUMP, 137
PseudoUMP synthetase, 137
Pseudouridine kinase, 137
Psicofuranine, 36, 72
Purine auxotrophs, 59–64
Purine degradation, 29

Purine deoxyribonucleoside hydrolase, 62, 239, 247, 250
Purine deoxyribonucleoside kinases, *see* individual enzymes
Purine deoxyribonucleosides, *see* individual compounds
Purine deoxyribonucleotide synthesis, 40–41
Purine nucleoside analogs, 71–73
Purine nucleoside kinases, *see* individual enzymes
Purine nucleoside phosphorylase, 67, 72, 77, 206–208, 238, 241
  genetic regulation of, 218–219, 221, 223, 224–234, 243
  location of, 263, 286
  mutants defective in, 44–50, 105, 185–186, 212, 214–217, 243
  in nucleoside catabolism, 180–181, 204–207, 235–248
  in purine utilization, 42–64
Purine nucleosides, *see also* individual compounds
  in pyrimidine salvage, 100, 105, 106, 108
  storage pool of, 58, 77
  as purine sources in *prs* mutants, 17–19
Purine nucleotide synthesis *de novo*, 12–13, 29–31, 74–77
Purine phosphoribosyltransferases, *see* individual enzymes
Purine ribonucleoside hydrolase, 67, 239, 245, 247–249
Purine starvation, 7–9, 33, 40, 75
Purine toxicity, 74
Putrescine, 71
Pyridine nucleotide coenzymes, *see* NAD
Pyrimidine deoxyribonucleoside 2′-hydroxylase, 125–126, 241, 248, 249
  mutants defective in, 116
  in pyrimidine utilization, 114–115
Pyrimidine nucleoside phosphorylase, *see also* Uridine phosphorylase and Thymidine phosphorylase, 121, 123, 239, 241, 243–244
  mutants defective in, 108, 243, 249
  in pyrimidine utilization, 106, 108–109
Pyrimidine nucleosides, *see also* individual compounds, 116–117
Pyrimidine nucleotides, *see also* individual compounds
  as pyrimidine sources, 105–107
Pyrimidine nucleotide synthesis *de novo*, 11–12, 96, 134–135
  organisms lacking, 112, 117
Pyrimidine ribonucleoside hydrolase, *see* Nucleoside hydrolases and Uridine hydrolase
Pyrophosphate, 35
Pyruvate kinase, 158

R

Recombination, 159, 167, 168
Relaxed mutants, 53, 100
Rhizobium, 240
Rhizobium lupini, 188
Rhodotorula, 240
Rhodotorula glutinis, 114–115, 125, 127, 189, 248
Ribokinase, 216, 238, 244, 245, 248, 250
Ribonucleic acid, *see* RNA
Ribonucleoside diphosphate reductase, 151–155, 165, 178
  genetic regulation of, 154–155, 165, 166
  hydrogen donor of, 155–157
  mutants defective in, 153, 168, 183
Ribonucleoside hydrolase, 239, 244, 246
Ribonucleoside monophosphate glycosylase, *see also* AMP nucleosidase, 104, 119, 132–133, 136
Ribonucleoside triphosphate reductase, 175–176, 178
Ribonucleotide reductase, 40–41, 169–171, 174, 178, 186
Ribose 1-phosphate, 98, 108, 183, 204–206, 237
  as effector, 123
Ribose 1,5-diphosphate, 211
Ribose 5-phosphate, 124, 204–206, 237, 248
Ribosylthymine, *see* Thymine ribonucleoside
RibosylTMP, 137
RNA
  labelling of, 73, 112, 189
  as pyrimidine source, 106

RNA—*cont.*
turnover, 28, 69–70, 73, 132, 135–137, 235
RNAse, 106, 135
RNA polymerase, 10, 12, 20, 51, 232–235

S

Saccharomyces cerevisiae, 65, 124, 239
dTMP utilization by, 188–189
nucleobase transport in, 289–292
nucleoside catabolism in, 240, 245–249
nucleoside transport in, 288–289
purine salvage in, 57–58
pyrimidine salvage in, 113–114, 119
Salmonella typhimurium
dTTP synthesis from thymine and thymidine in, 179–187
endogenous dTTP synthesis in, 150–159
linkage map of, 307–310
nucleobase transport in, 282, 287
nucleoside catabolism in, 204–237
nucleoside transport in, 269
PRPP metabolism in, 1–25
purine *de novo* synthesis in, 12–13, 29–36, 74–77
purine salvage in, 41–54
pyrimidine salvage in, 97–106
pyrimidine *de novo* synthesis in, 11–12, 134
Schizosaccharomyces pombe, 33, 35, 36, 65, 69
nucleobase transport in, 293
nucleoside transport in, 292
purine salvage in, 56
Serine, 188
Serratia, 130
Serratia marcescens, 120, 187
Showdomycin, 264, 265, 279
Spermidine, 71
Spermine, 71
Spores, 77, 174, 241, 244
Staphylococcus, 130
Staphylococcus aureus, 179
Streptococci, 245
Streptococcus faecalis, 187, 188
Sulfanilamide, 189

T

Tetrahydrofolate, 28, 160–161, 176–178, 187–189
3,4,5,6-Tetrahydrouridine, 121, 206, 231
Tetrahymena, 117, 119, 120
Tetrahymena pyriformis, 118, 127–128, 130–132, 137, 179
purine salvage in, 63–64
Thermobacterium, 68
Thiamine, 28, 31, 77
6-Thioguanine, 65
Thioredoxin, 153, 155–157, 175
Thioredoxin reductase, 155–156, 175
2-Thiouracil, 118
Thymidine
as carbon source, 204–206, 215, 235–236, 244
as effector, 123, 218, 235, 236, 243, 244
growth inhibition by, 216
transport, 184, 188, 266, 276, 294, 295
as uracil source, 114–117, 189
utilization for DNA synthesis, 111, 112, 184–189
Thymidine kinase, 119, 128–129, 181, 188–189, 248
mutants defective in, 99, 100, 104–105, 108, 213, 214, 216
in pyrimidine utilization, 104–105, 108, 110–113, 117, 180, 184, 188
Thymidine phosphorylase, *see also* Pyrimidine nucleoside phosphorylase, 122–123, 180, 207, 209–211, 221
distribution of, 114, 119, 121, 123, 188–189, 238, 240, 248–249
genetic regulation of, 180, 186, 218–219, 221, 223–236
mutants defective in, 99–100, 104–105, 182, 183, 185–186, 212, 214–217, 223, 229, 233
in pyrimidine utilization, 102, 104–105, 112
in thymine and thymidine utilization, 181–187
Thymidine-requiring mutants, 111, 188, 223
Thymidylate kinase, *see* dTMP kinase
Thymidylate synthase, 153, 160–161, 172, 176–179
mutants defective in, *see* Thymine-,

Thymidine-, and dTMP-requiring mutants
organisms lacking, 112, 179
Thymine
conversion to dTMP, 116, 179–181
as pyrimidine source, 114–116, 189
transport, 276, 287, 299
utilization for DNA synthesis, 111, 117, 181–189
Thymine arabinoside, 128
Thymine 7-hydroxylase, 126–127
genetic regulation of, 115–116
mutants defective in, 116
in pyrimidine utilization, 114–115
Thymineless death, 164, 170
Thymine-requiring mutants
isolation of, 160–161, 172, 187–188
nucleotide pools in, 161–165, 172
thymine and thymidine utilization by 182–187, 216, 221, 222, 276–277
thymine starvation of, 154, 156, 162–165, 183
Thymine ribonucleoside, 114–116, 124–126, 128, 137
Toxoplasma gondii, 63
Trans-N-deoxyribosylase, 55, 68, 109–110, 119, 121, 124–125, 175, 240, 241, 244, 245
genetic regulation of, 125
Transition state analog, 121, 206
Trimethoprim, 160, 172, 187–188
Trinitrobenzene sulfonate, 286
Trypanosoma brucei gambiensis, 61
Trypanosoma cruzi, 61
Tryptophan, 3, 11, 13, 17–18

U

UMP
as effector, 11–12, 118
as pyrimidine source, 102–103
UMP kinase, 131
Uracil
in DNA, 159, 167–168
excretion, 136
transport, 281–284, 291–292, 299–300
utilization, 98–101, 106, 109–117
Uracil arabinoside, 123
Uracil 5-carboxylic acid, 114–115, 126–127
Uracil 5-carboxylic acid decarboxylase, 127
genetic regulation of, 115–116
in pyrimidine utilization, 114–115
Uracil-DNA N-glycosylase, 167
Uracil permease, 113–114, 116, 291–292
mutants defective in, 113–114, 136
Uracil phosphoribosyltransferase, 100, 118–120
mutants defective in, 98–106, 108, 110, 111, 113, 114, 116, 136, 213, 216, 217, 282
in pyrimidine utilization, 98–117
in uracil transport, 282
Uridine
as carbon source, 205, 215, 222, 223, 235, 236, 244
as effector, 102, 121, 123, 211, 219–223, 244
excretion, 136
transport, 266, 269, 288, 292, 295
utilization, 99–117
Uridine hydrolase, *see also* Nucleoside hydrolase 119, 121, 124, 239–240, 245–247
catabolite inactivation of, 249
mutants defective in, 113–114, 136, 288
in pyrimidine utilization, 113–114, 116, 289
Uridine kinase, 119, 127–128, 217
mutants defective in, 98–104, 108, 113–114, 127, 136, 213
in pyrimidine utilization, 98–104, 108, 110, 113–114, 116
Uridine permease, 113–114
Uridine phosphorylase, *see also* Pyrimidine nucleoside phosphorylase, 122, 181, 184, 207, 209, 210, 223, 238
distribution of, 119, 121, 238–240
genetic regulation of, 122, 181, 182, 219–220, 223, 235, 236, 276
location of, 263
mutants defective in, 98–105, 183, 185, 212, 213, 215, 217
in pyrimidine utilization, 98–105, 281
in thymidine catabolism, 182–184
UTP
as effector, 9–10, 12, 128
pool, 14–16

V

Valinc, 51
Vitamin B_{12}, *see also* Coenzyme B_{12}, 109

X

Xanthine
excretion, 58
utilization, 34, 42, 51–52, 56–57, 61–63
Xanthine phosphoribosyltransferase, *see* Hypoxanthine-guanine-xanthine phosphoribosyltransferase
Xanthosine
as effector, 52, 220–221, 224, 235–236
utilization, 52, 57, 215, 217, 235–236
Xanthosine phosphorylase, 52, 207–208, 210
genetic regulation of, 219–220, 224, 235–236
mutants defective in, 212, 213, 215, 217
XAP R, 222, 224, 236
XMP amidotransferase, *see* GMP synthetase
XMP aminase, *see* GMP synthetase

Y

Yeast, *see also* individual organisms 29, 30, 34, 37, 39, 66, 70, 74, 120, 245, 288, 292

Z

Zink, 65, 159, 171